Springer Proceedings in Mathematics & Statistics

Volume 194

Springer Proceedings in Mathematics & Statistics

This book series features volumes composed of selected contributions from workshops and conferences in all areas of current research in mathematics and statistics, including operation research and optimization. In addition to an overall evaluation of the interest, scientific quality, and timeliness of each proposal at the hands of the publisher, individual contributions are all refereed to the high quality standards of leading journals in the field. Thus, this series provides the research community with well-edited, authoritative reports on developments in the most exciting areas of mathematical and statistical research today.

More information about this series at http://www.springer.com/series/10533

Raffaele Argiento · Ettore Lanzarone
Isadora Antoniano Villalobos
Alessandra Mattei
Editors

Bayesian Statistics in Action

BAYSM 2016, Florence, Italy, June 19–21

Springer

Editors
Raffaele Argiento
ESOMAS Department
University of Torino
Turin
Italy

and

Collegio Carlo Alberto
Moncalieri (TO)
Italy

Ettore Lanzarone
CNR–IMATI
Milan
Italy

Isadora Antoniano Villalobos
Department of Decision Sciences
Bocconi University
Milan
Italy

Alessandra Mattei
Department of Statistica, Informatica, Applicazioni "G. Parenti"
University of Florence
Florence
Italy

ISSN 2194-1009 ISSN 2194-1017 (electronic)
Springer Proceedings in Mathematics & Statistics
ISBN 978-3-319-54083-2 ISBN 978-3-319-54084-9 (eBook)
DOI 10.1007/978-3-319-54084-9

Library of Congress Control Number: 2017933863

Mathematics Subject Classification (2010): 62F15, 97K80

Printed on acid-free paper

This Springer imprint is published by Springer Nature
The registered company is Springer International Publishing AG
The registered company address is: Gewerbestrasse 11, 6330 Cham, Switzerland

Preface

This volume includes a collection of peer-reviewed contributions from among those presented at the third Bayesian Young Statisticians Meeting (BAYSM 2016). The conference was organized by the Istituto di Matematica Applicata e Tecnologie Informatiche "E. Magenes" (IMATI) of the National Research Council of Italy (CNR), in collaboration with the Dipartimento di Statistica, Informatica, Applicazioni "G. Parenti", University of Florence, and the Department of Decision Sciences, Bocconi University. The conference was held in Florence from 19 to 21 June 2016.

The Bayesian approach to statistics is becoming increasingly popular within the scientific community. Over the years, various problems involving complex data (e.g., *big data*) and posing challenging methodological issues have stimulated the scientific community to develop and apply novel Bayesian statistical methods. BAYSM represents an opportunity for MS and Ph.D. students, post-docs, and young researchers interested in Bayesian statistics to connect with their scientific community at the beginning of their careers. The goal of BAYSM is to stimulate collaborations and encourage discussions both with colleagues at the same level of experience and with more senior researchers, in order to promote research in a wide spectrum of fields where Bayesian methods can be employed.

The scientific program of BAYSM 2016 included contributions that develop and apply Bayesian methods in a variety of fields, ranging from the traditional (e.g., biostatistics and reliability) to the most innovative (e.g., big data and networks).

BAYSM 2016 included four plenary sessions and a keynote session as well as contributed talk sessions and a poster session. The plenary sessions offered four brilliant lectures by Fabrizia Mealli (University of Florence, Italy), Peter Müller (The University of Texas at Austin, TX, USA), Steve Scott (Google Inc., CA, USA), and Marina Vannucci (Rice University, TX, USA). The exciting keynote speech on the foundations of Bayesian statistics was delivered by Alessandra Guglielmi (Politecnico di Milano, Italy). Each standard session was chaired by a senior discussant, who provided helpful feedback to the young speakers on their current activity, encouraging them to pursue their research. Networking among the young participants was supported by vivid and inspirational discussions.

Three prizes were assigned: best methodological talk, best talk on applications, and best poster.

We acknowledge all participants who made BAYSM 2016 an outstanding scientific event and an enjoyable experience. We thank all of the speakers, young and senior. We express our gratitude to the discussants (Daniela Cocchi, Emanuela Dreassi, Brunero Liseo, Antonio Pievatolo, and Fabrizio Ruggeri) for their valuable work. Special acknowledgments are also owed to Ilaria Bianchini (Politecnico di Milano, Italy) and Michela Tizzani (Promoest S.r.l., Italy) for their support in organizing the conference. Finally, we give credit to our sponsors for their outstanding support: Google Inc., Comune di Firenze, International Society for Bayesian Analysis (ISBA), and Società Italiana di Statistica (SIS).

Hosting this meeting was an exciting and rewarding experience for the organizers. We expect that future BAYSM conferences will continue to have the same success as the first three editions, inspiring the new generations of Bayesian statisticians. Stay tuned at www.baysm.org!

This volume includes contributions from younger Bayesian statisticians in which theoretical aspects and application issues, often considered separately, are blended together to successfully tackle complex problems. It is structured in two parts. The first, *Theory* and *Methods*, is mainly devoted to mathematical statistics, model building, and methodological works. The second, *Applications and Case Studies*, deals with applications of complex methods to real-world problems and data.

Turin, Italy — Raffaele Argiento
Milan, Italy — Ettore Lanzarone
Milan, Italy — Isadora Antoniano Villalobos
Florence, Italy — Alessandra Mattei

Contents

Part I Theory and Methods

Sequential Monte Carlo Methods in Random Intercept Models for Longitudinal Data ... 3
Danilo Alvares, Carmen Armero, Anabel Forte and Nicolas Chopin

On the Truncation Error of a Superposed Gamma Process ... 11
Julyan Arbel and Igor Prünster

On the Study of Two Models for Integer-Valued High-Frequency Data ... 21
Andrea Cremaschi and Jim E. Griffin

Identification and Estimation of Principal Causal Effects in Randomized Experiments with Treatment Switching ... 31
Emanuele Gramuglia

A Bayesian Joint Dispersion Model with Flexible Links ... 39
Rui Martins

Local Posterior Concentration Rate for Multilevel Sparse Sequences ... 51
Eduard Belitser and Nurzhan Nurushev

Likelihood Tempering in Dynamic Model Averaging ... 67
Jan Reichl and Kamil Dedecius

Localization in High-Dimensional Monte Carlo Filtering ... 79
Sylvain Robert and Hans R. Künsch

Linear Inverse Problem with Range Prior on Correlations and Its Variational Bayes Inference ... 91
Ondřej Tichý and Václav Šmídl

Part II Applications and Case Studies

Bayesian Hierarchical Model for Assessment of Climate Model Biases 105
Maeregu Woldeyes Arisido, Carlo Gaetan, Davide Zanchettin and Angelo Rubino

An Application of Bayesian Seemingly Unrelated Regression Models with Flexible Tails 115
Charles Au and S.T. Boris Choy

Bayesian Inference of Stochastic Pursuit Models from Basketball Tracking Data 127
Harish S. Bhat, R.W.M.A. Madushani and Shagun Rawat

Identification of Patient-Specific Parameters in a Kinetic Model of Fluid and Mass Transfer During Dialysis 139
Camilla Bianchi, Ettore Lanzarone, Giustina Casagrande and Maria Laura Costantino

A Bayesian Nonparametric Approach to Ecological Risk Assessment 151
Guillaume Kon Kam King, Julyan Arbel and Igor Prünster

Approximate Bayesian Computation Methods in the Identification of Atmospheric Contamination Sources for DAPPLE Experiment 161
Piotr Kopka, Anna Wawrzynczak and Mieczyslaw Borysiewicz

Bayesian Survival Analysis to Model Plant Resistance and Tolerance to Virus Diseases 173
Elena Lázaro, Carmen Armero and Luis Rubio

Randomization Inference and Bayesian Inference in Regression Discontinuity Designs: An Application to Italian University Grants 183
Federica Licari

Bayesian Methods for Microsimulation Models 193
Consuelo R. Nava, Cinzia Carota and Ugo Colombino

A Bayesian Model for Describing and Predicting the Stochastic Demand of Emergency Calls 203
Vittorio Nicoletta, Ettore Lanzarone, Alessandra Guglielmi, Valérie Bélanger and Angel Ruiz

Flexible Parallel Split-Merge MCMC for the HDP 213
Debora Parisi and Stefania Perego

Bayesian Inference for Continuous Time Animal Movement Based on Steps and Turns 223
Alison Parton, Paul G. Blackwell and Anna Skarin

Explaining the Lethality of Boko Haram's Terrorist Attacks in Nigeria, 2009–2014: A Hierarchical Bayesian Approach 231
André Python, Janine Illian, Charlotte Jones-Todd and Marta Blangiardo

Optimizing Movement of Cooperating Pedestrians by Exploiting Floor-Field Model and Markov Decision Process 241
Vladimíra Sečkárová and Pavel Hrabák

Part I
Theory and Methods

Sequential Monte Carlo Methods in Random Intercept Models for Longitudinal Data

Danilo Alvares, Carmen Armero, Anabel Forte and Nicolas Chopin

Abstract Longitudinal modelling is common in the field of Biostatistical research. In some studies, it becomes mandatory to update posterior distributions based on new data in order to perform inferential process on-line. In such situations, the use of posterior distribution as the prior distribution in the new application of the Bayes' theorem is sensible. However, the analytic form of the posterior distribution is not always available and we only have an approximated sample of it, thus making the process "not-so-easy". Equivalent inferences could be obtained through a Bayesian inferential process based on the set that integrates the old and new data. Nevertheless, this is not always a real alternative, because it may be computationally very costly in terms of both time and resources. This work uses the dynamic characteristics of sequential Monte Carlo methods for "static" setups in the framework of longitudinal modelling scenarios. We used this methodology in real data through a random intercept model.

Keywords Bayesian analysis · IBIS algorithm · Marginal likelihood · Particle filter

1 Introduction

Observations of subjects that are measured repeatedly over time is a broad definition for longitudinal data. The modelling for this correlated data structure had its starting point through work of [10] and since then the statistical methods to this data type

D. Alvares (✉) · C. Armero · A. Forte
Universitat de València - Calle Dr. Moliner 50, 46100 Burjassot, Spain
e-mail: daldasil@alumni.uv.es

C. Armero
e-mail: carmen.armero@uv.es

A. Forte
e-mail: anabel.forte@uv.es

N. Chopin
CREST-ENSAE and HEC Paris, 3, Avenue Pierre Larousse, 92245 Malakoff, France
e-mail: nicolas.chopin@ensae.fr

R. Argiento et al. (eds.), *Bayesian Statistics in Action*, Springer Proceedings in Mathematics & Statistics 194, DOI 10.1007/978-3-319-54084-9_1

have been constantly growing. Furthermore, in many applications as well as in this study, new data are routinely incorporated in the analysis, making it necessary to update the inferential process for decision-making.

In this context, the Bayesian approach is natural and intuitive on the concept of updating the previous information from the data. However, in the majority of cases the posterior distribution has not analytic form and we only have an approximated sample of it. The most common mechanism is to remake the inference but this procedure may be computationally expensive.

An alternative approach based on simulation of posterior distributions is the Sequential Monte Carlo (SMC) methods [5]. In addition, for the our particular case of "static" models, we have the iterated batch importance sampling (IBIS) algorithm [3]. Given the need to save computational time and resources, it is fundamental to use a dynamic inferential methodology, in which we focus on SMC methods through the IBIS algorithm. As an illustration, we explore the longitudinal framework of data on the growth of Sitka spruce trees through a random intercept model.

2 Bayesian Random Intercept Model

A particular case of linear mixed-effects models, and also our interest to study, is the random intercept model, which allows each subject to deviate from the overall mean response by a subject-specific term that applies equally over time [7]:

$$\mathbf{y}_i = x_i\boldsymbol{\beta} + \mathbf{1}_{n_i} b_{0i} + \boldsymbol{\varepsilon}_i, \tag{1}$$

where $\mathbf{y}_i$ is the vector of observations associated to ith subject, $i = 1, \ldots, I$, x_i known design matrix for the fixed-effects regression coefficients vector $\boldsymbol{\beta}$, and $\mathbf{1}_{n_i}$ is the unit vector of length n_i. We assume prior independence for $\boldsymbol{\beta}$ with $\beta_k \sim \mathrm{N}(0, \nu_k^2)$ for $k = 1, \ldots, K$. We consider that the vector of random effects $\mathbf{b}_0$ are independently normally distributed with standard deviation σ_0 and set an uniform prior distribution $\mathrm{U}(0, u_0)$ for this deviation. The error terms, ε_{ij}'s, where $j = 1, \ldots, n_i$ (number of observations for the ith subject), are independent and identically normally distributed with standard deviation $\sigma \sim \mathrm{U}(0, u)$ and these errors are assumed independent of the random-effect b_{0i}, i.e., $\mathrm{Cov}(\boldsymbol{\varepsilon}_i, \mathbf{1}_{n_i} b_{0i}) = 0$ for $i = 1, \ldots, I$. The values ν_k^2 for $k = 0, \ldots, K$, u_0, and u are predefined constants and their values characterize the level of information in the prior distributions involved.

3 Sequential Monte Carlo Methods

Sequential Monte Carlo (SMC), also known as particle filters, methods are a set of simulation-based methods which provide a convenient and attractive approach to computing posterior distributions [5]. SMC strategies have become quite popular due the limitations of the Markov chain Monte Carlo (MCMC) techniques in dynamic settings, since the MCMC approach involves accessing the entire sequence of data for each iteration and becomes computationally unusable for complex models and/or massive data sets [2].

Generally, SMC methods are designed to update dynamic parameters, i.e., parameters that depend on the time somehow. Therefore, the application of the (default) particle filter algorithm [1] in the model (1) needs to be adapted for the configuration of "static" parameters. Based on resample-move particle filter [8] we have an update procedure for "static" models, the Iterated Batch Importance Sampling (IBIS) algorithm [3]. Even though this approach does not consider models with random effects, the latent variable context provides a smart way to deal with them.

To show the operating of the IBIS algorithm in random effects models, we used the model (1) to develop a toy example, since, in this case, the exact form of the posterior distribution is known. Hereafter, we present the different stages of the IBIS algorithm for our particular example

Step 1. Simulate $\boldsymbol{\theta}^{(s)}$, for $s = 1, \ldots, S$, from the joint posterior distribution for the common parameters and hyper-parameters in model (1), $\pi(\boldsymbol{\theta} \mid \mathbf{y}_{(1:n)})$, based on initial data $\mathbf{y}_{(1:n)} = (\mathbf{y}_{(1,1:n_1)}, \ldots, \mathbf{y}_{(I,1:n_I)})$ where $\mathbf{y}_{(i,1:n_i)} = (y_{(i,1)}, \ldots, y_{(i,n_i)})^\top$.

Step 2. REWEIGHTING: given q_i new data for the ith subject $\mathbf{y}_{(i,n_i+1:n_i+q_i)}$ compute the importance weights

$$w^{(s)} \leftarrow p\left(\mathbf{y}_{(i,n_i+1:n_i+q_i)} \mid \boldsymbol{\theta}^{(s)}, \mathbf{y}_{(i,1:n_i)}\right)$$

that comes from the marginal likelihood of the model (1).

Step 3. RESAMPLING: resample

$$\left(\boldsymbol{\theta}^{(s)}, w^{(s)}\right)_{s=1,\ldots,S} \leftarrow \left(\tilde{\boldsymbol{\theta}}^{(s)}, 1\right)_{s=1,\ldots,S}$$

according to the multinomial selection scheme [9].

Step 4. MOVE: draw

$$\boldsymbol{\theta}_m^{(s)} \sim K\left(\tilde{\boldsymbol{\theta}}^{(s)}, \cdot\right)$$

where K is a transition kernel with stationary distribution $\pi(\boldsymbol{\theta} \mid \mathbf{y}_{(1:n)}, \mathbf{y}_{(i,n_i+1:n_i+q_i)})$ of the marginal model that comes from (1).

Step 5. LOOP: $n_i \leftarrow n_i + q_i$, $[\boldsymbol{\theta}^{(s)} \leftarrow \boldsymbol{\theta}_m^{(s)}]_{s=1,\ldots,S}$, sample $(b_{01}, \ldots, b_{0I})^{(s)}$ for $s = 1, \ldots, S$ from their full conditional posterior distributions and if we have new data then return to **Step 2.**

The strategy is to integrate out the random intercepts to return to the standard IBIS algorithm. Thus we are able to update the population parameters and then the random effects from their full conditional posterior distributions (**Step 5**).

In order to illustrate the algebraic manipulation required, we consider the marginal likelihood for model (1) with p covariates when integrating out the random intercept

$$
\begin{aligned}
L(\mathbf{y}_{(1:n)} \mid \boldsymbol{\theta}) &= \int L\left(y_{(1:n)} \mid \boldsymbol{\theta}, \mathbf{b}_0\right) \pi\left(\mathbf{b}_0 \mid \sigma_0^2\right) \mathrm{d}\mathbf{b}_0 = \int L\left(y_{(1:n)} \mid \boldsymbol{\theta}, \mathbf{b}_0\right) \prod_{i=1}^{I} \pi\left(b_{0i} \mid \sigma_0^2\right) \mathrm{d}\mathbf{b}_0 \\
&= \prod_{i=1}^{I} \left(\frac{1}{2\pi\sigma^2}\right)^{n_i/2} \left(\frac{\sigma^2}{\sigma^2 + n_i\sigma_0^2}\right)^{1/2} \exp\left\{\frac{1}{2\sigma^2}\left[\frac{\sigma_0^2}{\sigma^2 + n_i\sigma_0^2}\left(\sum_{j=1}^{n_i}\psi_{ij}\right)^2 - \sum_{j=1}^{n_i}\psi_{ij}^2\right]\right\}
\end{aligned} \tag{2}
$$

where $\psi_{ij} = y_{(i,j)} - \beta_0 - \beta_1 X_{1ij} - \beta_2 X_{2ij} - \ldots - \beta_p X_{pij}$. Equation (2) is partially (only for the subject that contain the new data) used to compute the importance weights (**Step 2**). Its complete form is required in independent Metropolis–Hastings kernel (**Step 4**), in which the proposed particle is generated independently from a Gaussian instrumental distribution as suggested in [3].

4 Application

To exemplify the sequential approach in our longitudinal model, we used data from a study of the effect of ozone pollution on the growth of Sitka spruce trees [4]. In particular, we considered 25 trees growing under normal conditions, and selected calendar time in days, since 1 January 1988. Our response variable is the logarithm of the product of tree height and diameter square (from now on log-size). Additionally, each tree was measured 5 times (days: 152, 174, 201, 227, 258).

Figure 1 shows the spaghetti plot of these data. Two features of the data are quickly identified. Firstly, there is a clear quasi-linear relationship between the growth of the trees and the time. The second pattern is the high degree of parallelism between the growth of each tree. Consequently, model (1) seems a sensible model for our Sitka data.

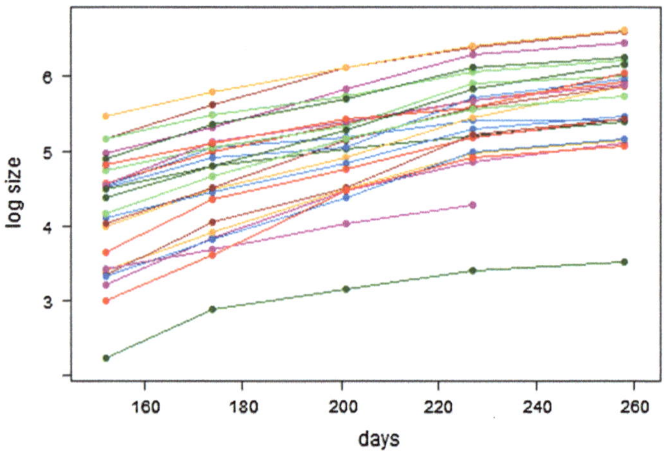

Fig. 1 Longitudinal measures of log-size for Sitka spruce trees grown in normal environments

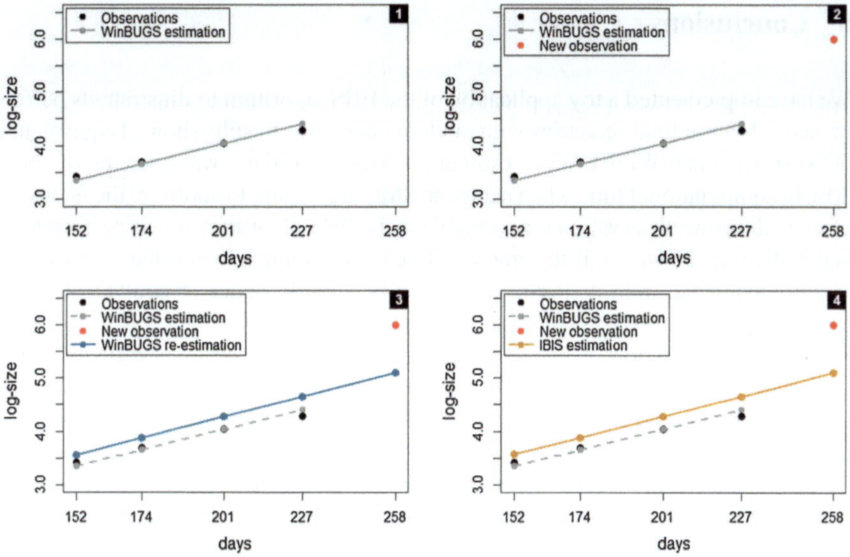

Fig. 2 Update of the posterior mean for the log-size of tree 23, after the inclusion of a fifth new observation, through WinBUGS and IBIS algorithm

We estimated the model and explored the behaviour of the dynamic update. Firstly, we excluded the last measurement (day 258) of one of the trees (tree 23). Then, we update the estimation through the incorporation of this 'new' data, but replacing its actual value (log-size 4.541) by an 'outlier' (log-size 6) to identify the accuracy of the IBIS algorithm in the presence of extreme values.

The performance of the IBIS algorithm was compared with the approximated posterior distribution by MCMC methods [6] through the software WinBUGS [12]. In the IBIS setting, we used 10 000 particles and 4 iterations for the independent Metropolis–Hastings kernel. In WinBUGS, we ran one MCMC chain for 200 000 iterations plus 20 000 dedicated to the burn-in period. The sample was thinned by only storing one of every 5 iterations in order to reduce autocorrelation in the saved sample. Both configurations are minimal to achieve convergence and accuracy. This process is sequentially presented in Fig. 2. Figure 2(1) shows the posterior mean of the log-size based on the data without information from day 258 through WinBUGS. Figure 2(2) incorporates the measurements corresponding to day 258. Finally, Fig. 2(3) and (4) show the subsequent posterior means computed from WinBUGS (integrates initial and new data into a single set and estimates the parameters again) and IBIS, respectively.

Notice that the Fig. 2(3) and (4) are practically identical (except a minimal difference after the third decimal of the estimated regression coefficients). This resemblance meets all our expectations of the IBIS algorithm for the marginal model combined with the computation of the random effects from their full conditional posterior distributions. Another essential factor was the reduction in the computational time, with an average saving of about 20%.

5 Conclusions

We have implemented a toy application of the IBIS algorithm to illustrate its performance in longitudinal scenarios with random intercepts. Results showed equivalence between IBIS and WinBUGS re-estimation. Moreover, there was a saving of about 20% in computational time when using the SMC algorithm. In addition, the inclusion of more than one observation is acceptable to the IBIS algorithm and its performance is not affected. However, if the amount of new data is much larger than the quantity initially available, then its performance can be slightly better or equivalent to the MCMC approach.

The implementation of the IBIS algorithm is simple. The great difficulty is to integrate out the random effects because in most cases the integral is intractable (no closed form) and hence some method of numerical integration or simulation is required [11].

Anyway, the success for integration procedure gives evidence that SMC methods could be extended to more complex and flexible longitudinal models, such as the joint models of longitudinal and survival data [13], maintaining a good performance in processing time.

Acknowledgements Alvares is partially supported by the research grants Coordination for the Improvement of Higher Level Personnel (BEX: 0047/13-9), Brazil. Armero and Forte are partially funded by MTM2016-77501-P from the Spanish Ministry of Economy and Competitiveness and ACOMP/2015/202 from the Generalitat Valenciana. Chopin is partially funded by Labex ECODEC ANR-11-LABEX-0047 grant from the Agence Nationale de la Recherche.

References

1. Arulampalam, M.S., Maskell, S., Gordon, N., Clapp, T.: A tutorial on particles filters for online nonlinear/non-Gaussian Bayesian tracking. Proc. IEEE Trans. Signal Process. **50**(2), 174–188 (2002)
2. Ba-Ngu, V., Ba-Tuong, V., Sumeetpal, S.: Sequential Monte Carlo methods for static parameter estimation in random set models. In: Proceedings of the Intelligent Sensors, Sensor Networks and Information Processing Conference, pp. 313–318 (2004)
3. Chopin, N.: A sequential particle filter for static models. Biometrika **89**(3), 539–551 (2002)
4. Diggle, P.J., Heagerty, P., Liang, K.-Y., Zeger, S.L.: Analysis of Longitudinal Data, 2nd edn. Oxford University Press, New York (2002)
5. Doucet, A., Freitas, N., Gordon, N.: Sequential Monte Carlo Methods in Practice. Statistics for Engineering and Information Science. Springer, New York (2001)
6. Gelman, A., Carlin, J., Stern, H., Dunson, D., Vehtari, A., Rubin, D.: Bayesian Data Analysis. Texts in Statistical Science, 3rd edn. Chapman and Hall/CRC, London (2013)
7. Gibbons, R.D., Hedeker, D., DuToit, S.: Advances in analysis of longitudinal data. Annu. Rev. Clin. Psychol. **6**, 79–107 (2010)
8. Gilks, W.R., Berzuini, C.: Following a moving target: Monte Carlo inference for dynamic Bayesian models. J. R. Stat. Soc. Ser. B **63**(1), 127–146 (2001)
9. Gordon, N.J., Salmond, D.J., Smith, A.F.M.: Novel approach to nonlinear/non-Gaussian Bayesian state estimation. IEE Proc. Radar Signal Process. **140**(2), 107–113 (1993)

10. Laird, N.M., Ware, J.H.: Random-effects models for longitudinal data. Biometrics **38**(4), 963–974 (1982)
11. Lee, Y., Nelder, J.A.: Likelihood for random-effect models. Stat. Oper. Res. Trans. **29**(2), 141–164 (2005)
12. Lunn, D.J., Thomas, A., Best, N., Spiegelhalter, D.: Winbugs - a Bayesian modelling framework: concepts, structure, and extensibility. Stat. Comput. **10**, 325–337 (2000)
13. Rizopoulos, D.: Joint Models for Longitudinal and Time-to-Event Data: With Applications in R. Chapman and Hall/CRC, Boca Raton (2012)

On the Truncation Error of a Superposed Gamma Process

Julyan Arbel and Igor Prünster

Abstract Completely random measures (CRMs) form a key ingredient of a wealth of stochastic models, in particular in Bayesian Nonparametrics for defining prior distributions. CRMs can be represented as infinite series of weighted random point masses. A constructive representation due to Ferguson and Klass provides the jumps of the series in decreasing order. This feature is of primary interest when it comes to sampling since it minimizes the truncation error for a fixed truncation level of the series. In this paper we focus on a general class of CRMs, namely the *superposed gamma process*, which suitably transformed has already been successfully implemented in Bayesian Nonparametrics, and quantify the quality of the approximation in two ways. First, we derive a bound in probability for the truncation error. Second, following [1], we study a moment-matching criterion which consists in evaluating a measure of discrepancy between actual moments of the CRM and moments based on the simulation output. To this end, we show that the moments of this class of processes can be obtained analytically.

Keywords Bayesian Nonparametrics · Completely random measures · Ferguson and Klass algorithm · Moment-matching · Normalized random measures · Posterior sampling · Superposed gamma process

1 Introduction

Completely random measures (CRMs), also known as independent increment processes, have blossomed in the last decades in modern stochastic modeling and inference as a basic building block of countless popular models. A prominent usage

J. Arbel (✉)
Inria Grenoble - Rhône-Alpes, Grenoble, France
e-mail: julyan.arbel@inria.fr

J. Arbel
Université Grenoble Alpes, Grenoble, France

I. Prünster
Department of Decision Sciences, BIDSA and IGIER, Bocconi University, Milan, Italy
e-mail: igor@unibocconi.it

R. Argiento et al. (eds.), *Bayesian Statistics in Action*, Springer Proceedings in Mathematics & Statistics 194, DOI 10.1007/978-3-319-54084-9_2

of CRMs is within Bayesian nonparametric statistics (see [15, 18]). For instance, the popular Dirichlet process [9] can be obtained as normalization or exponentiation of suitable CRMs (see [10]). Survival analysis, [2, 17], random sparse networks, [6], biology [20, 28], are only a few of the various modern applications tackled with CRMs.

Implementation of CRM-based models usually requires to simulate the CRMs trajectories. As infinite dimensional objects, they need to be truncated, leading to an approximation error. The representation due to Ferguson and Klass [11] (see also [30]) is arguably one of the most useful ones in that it displays the weights in decreasing order. This implies that the approximation error is minimized over the whole sample space for a given truncation level. This appealing feature was exploited in many works, including [3, 4, 7, 8, 13, 22–26] to cite just a few in Bayesian Nonparametrics. The quality of the approximation, hardly addressed by those previous works, is the focus of this paper.

Many classical methods in statistics and econometrics use moments, such as, for instance, the method of simulated moments [21] and the general method of moments [12]. In this paper, we follow another research line and show how moments of the CRMs can be used in order to assess the quality of the approximation due to the truncation [1]. It is based on the observation that moments of CRMs are simple to compute, hence one can quantify the quality of the approximation by evaluating a measure of discrepancy between the actual moments of the CRM at issue and the moments computed based on the sampled realizations of the CRM. The truncation level is then selected so that the measure of discrepancy does not exceed a given threshold, say 5%. In Arbel and Prünster [1] the methodology is illustrated on two classes of CRMs, namely the *generalized gamma process* and the *stable-beta process*. In the present paper we focus on another broad class called the *superposed gamma process* (see [19, 27]). More specifically, after a brief presentation of CRMs and of the Ferguson and Klass algorithm in Sect. 2, we derive a bound in probability on the truncation error in Sect. 3.1 and then show the applicability of the moment-matching criterion by deriving analytically the moments of the superposed gamma process in Sect. 3.2.

2 Sampling Completely Random Measures

2.1 Completely Random Measures

A CRM $\tilde{\mu}$ on $\mathbb{X}$ is a random measure which spreads out mass independently in the space. More precisely, the random variables $\tilde{\mu}(A_1), \ldots, \tilde{\mu}(A_n)$ are mutually independent for any disjoint sets $A_1, \ldots, A_n$.

Kingman [16] showed that the only way to spread out mass in a *completely random* fashion (without deterministic components) is by randomly scattering point masses in the space. In other words, CRMs select (almost surely) discrete measures and hence can be represented as

$$\tilde{\mu} = \sum_{i \geq 1} J_i \delta_{Z_i} \tag{1}$$

where the jumps J_i and locations Z_i are random. Both jumps and locations are controlled by the so-called Lévy intensity which characterizes the CRM. It is a measure on $\mathbb{R}^+ \times \mathbb{X}$ which can be written as $\nu(\mathrm{d}v, \mathrm{d}x) = \rho(\mathrm{d}v)\alpha(\mathrm{d}x)$ for so-called homogeneous CRM, which are considered here and correspond to the case of jumps independent of the locations. The function ρ controls the intensity of the jumps. The measure α, if the CRM is (almost surely) finite, which is assumed throughout, splits up in $\alpha = aP_0$ where $a > 0$ is called the total mass parameter and the probability distribution P_0 tunes the locations.

Ever-popular CRMs include the *generalized gamma process* introduced by Brix [5] and the *stable-beta process*, or *three-parameter beta process*, defined by Teh and Gorur [29] as an extension of the beta process [14]. Here we consider another large class of completely random measures called *superposed gamma process*, introduced by Regazzini et al. [27]. It is identified by the jump intensity

$$\rho(\mathrm{d}v) = \frac{1 - \mathrm{e}^{-\eta v}}{1 - \mathrm{e}^{-v}} \frac{\mathrm{e}^{-v}}{v} \mathrm{d}v, \quad \eta > 0. \tag{2}$$

As noted by Lijoi et al. [19], one usually restricts attention to the case of positive integer η. Under this assumption, the superposed gamma processtakes the form of a genuine *superposition* of independent gamma processes with increasing integer-valued scale parameter, with jump intensity $\rho(\mathrm{d}v) = \frac{1}{v}\left(\mathrm{e}^{-v} + \mathrm{e}^{-2v} + \ldots + \mathrm{e}^{-\eta v}\right) \mathrm{d}v/v$. The specification of integer values for η has also the advantage to lead to analytical computation of the moments. Note that the special case $\eta = 1$ reduces to the gamma process, which gives rise to the Dirichlet process by normalization. Alternatively, the normalization of the superposed gamma processfor unspecified η provides the so-called generalized Dirichlet process [19].

2.2 Ferguson and Klass Algorithm

Ferguson and Klass [11] devise a constructive representation of a CRM which produces the jumps in decreasing order. This corresponds to the (almost surely unique) ordering of the sum elements in (1) where $J_1 > J_2 > \cdots$. Indeed, the jumps are obtained as $\xi_i = N(J_i)$, where $N(v) = \nu([v, \infty), \mathbb{X})$ is a decreasing function, and $\xi_1, \xi_2, \ldots$ are jump times of a standard Poisson process (PP) of unit rate: $\xi_1, \xi_2 - \xi_1, \ldots \overset{\text{i.i.d.}}{\sim} \mathrm{Exp}(1)$. Figure 1 illustrates the function $N(\,\cdot\,)$ which takes the following form in the superposed gamma process case

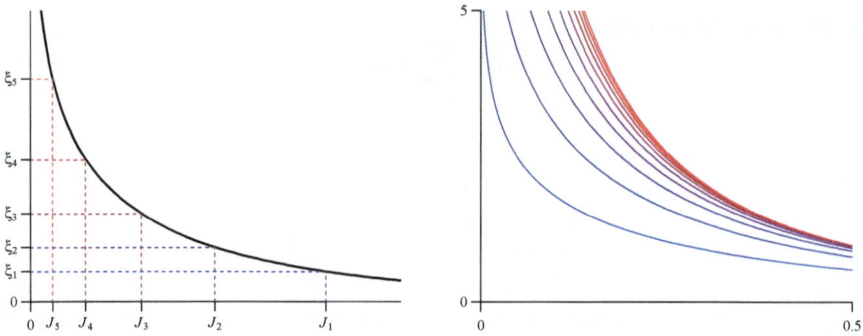

Fig. 1 *Left* illustration of Ferguson and Klass representation through the inversion of the jumps times $\xi_1, \ldots, \xi_5$ for a homogeneous Poisson process on $\mathbb{R}_+$ to the jumps $J_1, \ldots, J_5$ of a CRM. *Right* tail of the Lévy measure $N(\cdot)$ of the superposed gamma processwith $\eta \in \{1, \ldots, 10\}$, $\eta = 1$ for the *lowest curve*, $\eta = 10$ for the *highest curve*

$$N(v) = aE_\eta(v), \text{ where } E_\eta(v) = \sum_{k=1}^{\eta} E_1(kv) \text{ and } E_1(v) = \int_v^\infty u^{-1}\mathrm{e}^{-u}\mathrm{d}u, \tag{3}$$

and where the function E_1 denotes the *exponential integral function.*

Since it is impossible to sample an infinite number of jumps, approximate simulation of $\tilde{\mu}$ is in order. This becomes a question of determining the number M of jumps to sample leading to the truncation $\tilde{\mu}_M$ and truncation error T_M as follow

$$\tilde{\mu}_M = \sum_{i=1}^{M} J_i \delta_{Z_i}, \quad T_M = \sum_{i=M+1}^{\infty} J_i. \tag{4}$$

The Ferguson and Klass representation has the key advantage of generating the jumps in decreasing order implicitly minimizing such an approximation error. However, a precise evaluation of T_M, for example in expectation, is a daunting task due to the non independence of the jumps in the Ferguson and Klass representation. The algorithm is summarized in Algorithm 1.

Then, the natural path to determining the truncation level M would be the evaluation of the Ferguson and Klass tail sum

$$\sum_{i=M+1}^{\infty} N^{-1}(\xi_i). \tag{5}$$

Brix ([5], Theorem A.1) provided an upper bound in probablity for (5) in the generalized gamma case. In Proposition 1 of Sect. 3 we derive also an upper bound for the tail sum of the superposed gamma process. However, both bounds are far from sharp and therefore of little practical use as highlighted in Sect. 3. This motivates the idea of looking for a different route and our proposal consists in the moment-matching technique detailed in the next section.

From Fig. 1 it is apparent that increasing η leads to larger jumps which in turn leads to the need of a higher truncation level in order to match a given precision level. This is not surprising given the CRM at hand can be thought of as a superposition of η gamma CMRs. Such an intuition is made precise in the next section.

Algorithm 1 Ferguson & Klass algorithm

1: sample $\xi_i \sim \text{PP}$ for $i = 1, \ldots, M$
2: define $J_i = N^{-1}(\xi_i)$ for $i = 1, \ldots, M$
3: sample $Z_i \sim P_0$ for $i = 1, \ldots, M$
4: approximate $\tilde{\mu}$ by $\sum_{i=1}^{M} J_i \delta_{Z_i}$

3 Truncation Error of the Superposed Gamma Process

3.1 Bound in Probability

We provide an evaluation in probability of the truncation error T_M in (4).

Proposition 1 *Let $(\xi_j)_{j\geq 1}$ be the jump times for a homogeneous Poisson process on $\mathbb{R}^+$ with unit intensity. Then for any $\epsilon \in (0, 1)$, the tail sum of the superposed gamma process* (4) *satisfies*

$$\mathbb{P}\Big(T_M \leq t_M^\epsilon\Big) \geq 1 - \epsilon, \ \textit{for}\ t_M^\epsilon = \frac{C}{(\eta!)^{1/\eta}} \mathrm{e}^{1-\frac{M}{C}}, \ \textit{where}\ C = \frac{2ea\eta}{\epsilon}.$$

Proof The proof follows along the same lines as the proof of Theorem A.1. by Brix [5] for the generalized gamma process and Proposition 4 by Arbel and Prünster [1] for the stable-beta process. Let q_j denote the $\epsilon 2^{M-j}$ quantile, for $j = M+1, M+2, \ldots$, of a gamma distribution with mean and variance equal to j. Then

$$\mathbb{P}\Bigg(\sum_{j=M+1}^{\infty} N^{-1}(\xi_j) \leq \sum_{j=M+1}^{\infty} N^{-1}(q_j)\Bigg) \geq 1 - \epsilon.$$

Denote $\tilde{t}_M^\epsilon = \sum_{j=M+1}^{\infty} N^{-1}(q_j) = \sum_{j=M+1}^{\infty} E_\eta^{-1}(q_j/a)$, and let us upper bound E_η^{-1}. By using $E_1(u) \leq 1 - \log(u)$, one gets

$$E_\eta(u) = \sum_{l=1}^{\eta} E_1(lu) \leq \eta - \sum_{l=1}^{\eta} \log(lu) = \eta - \log\big(\eta! u^\eta\big),$$

which can be inverted to obtain

$$E_{\eta}^{-1}(x) \leq \frac{1}{(\eta!)^{1/\eta}} e^{1-\frac{x}{\eta}}.$$

Additionally, since the quantiles satisfy $q_j \geq \frac{\epsilon}{2e} j$, we can conclude that

$$\tilde{t}_M^{\epsilon} \leq \frac{1}{(\eta!)^{1/\eta}} \sum_{j=M+1}^{\infty} e^{1-\frac{q_j}{a\eta}} \leq \frac{1}{(\eta!)^{1/\eta}} \sum_{j=M+1}^{\infty} e^{1-\frac{\epsilon j}{2ea\eta}} \leq \frac{2ea\eta}{\epsilon(\eta!)^{1/\eta}} e^{1-\frac{\epsilon M}{2ea\eta}}. \quad \square$$

Remark It is interesting to note that the bound t_M^{ϵ} for the superposed gamma process is equal to its counterpart for the beta process with concentration parameter c set to η, all else things being equal (total mass parameter a and threshold ϵ). See Proposition 4 in [1]. This finding provides a nice connection between both processes otherwise seemingly unrelated.

The bound t_M^{ϵ} obtained in Proposition 1 is exponentially decreasing with M, which is reminiscent of the results obtained by Brix [5] and Arbel and Prünster [1], respectively, for the generalized gamma process and the stable-beta process with no stable component. As already pointed out by these authors, the bound t_M^{ϵ} is very conservative due to a crude lower bound on the quantiles q_j (notation of the proof). The left panel of Fig. 2 displays this bound t_M^{ϵ}, while the right panel illustrates the truncation level M (in log-scale) required in order to guarantee with 95% probability an upper bound on T_M of $t_{max} \in \{1, 10, 100\}$, for varying values of η. Inspection of the plots demonstrates the rapid increase with η of the number of jumps needed in order to assess a given bound in probability.

As suggested by a Referee, a possible strategy for improving the result in Proposition 1 is to rely on concentration of measure techniques. This will be the object of future investigations. A numerical strategy to improve the approximation consists in

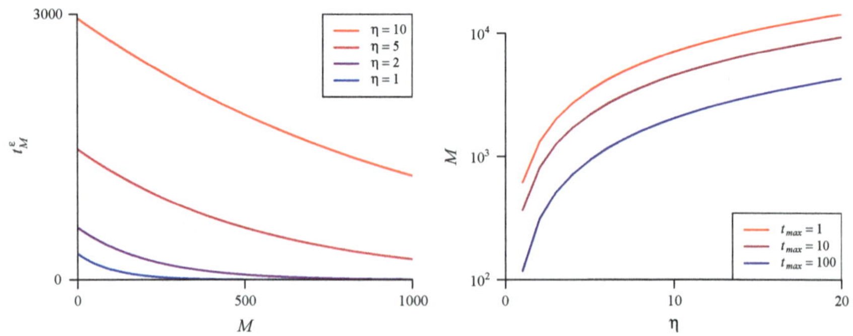

Fig. 2 *Left* variation of $M \mapsto t_M^{\epsilon}$ for $\eta \in \{1, 2, 5, 10\}$. *Right* variation of the threshold function $\eta \mapsto M$ needed to match an error bound of $t_{max} \in \{1, 10, 100\}$ with $\eta \in \{1, \ldots, 20\}$, log scale on y-axis

directly calculating the quantiles q_j (instead of resorting to the lower bound), thus loosing the closed form expression of the bound.

3.2 Moment-Matching Criterion

We first concisely recall the moment-matching methodology introduced by Arbel and Prünster [1] and then tailor it to the superposed gamma process. We assess the quality of approximation of the Ferguson & Klass algorithm by comparing the actual distribution of the random total mass $\tilde{\mu}(\mathbb{X}) = \sum_{i=1}^{\infty} J_i$ with its empirical distribution (obtained by the sampled trajectories). Motivated by the fact that the first moments carry much information about a distribution, the comparison is made by comparing theoretical and empirical moments of $\tilde{\mu}(\mathbb{X})$. As measure of discrepancy, we use the mean squared error between theoretical and empirical moments. We refer to [1] for illustrations of this moment-matching criterion on the generalized gamma process and the stable-beta process.

In order to apply this methodology also to the superposed gamma process, we need to derive its theoretical moments. The n-th (raw) moment of the random total mass is defined as

$$m_n = \mathbb{E}\big[\tilde{\mu}^n(\mathbb{X})\big].$$

For general homogeneous CRMs, it takes on the form (see, e.g., Proposition 1 in [1])

$$m_n = \sum_{(*)} \binom{n}{k_1 \cdots k_n} \prod_{i=1}^{n} \big(\kappa_i / i!\big)^{k_i}, \tag{6}$$

where the sum $(*)$ is over all n-tuples of nonnegative integers $(k_1, \ldots, k_n)$ satisfying the constraint $k_1 + 2k_2 + \cdots + nk_n = n$ and where κ_i is the ith cumulant defined by

$$\kappa_i = a \int_0^{\infty} v^i \rho(\mathrm{d}v).$$

In the case of the superposed gamma process, simple algebra leads to the following expression for the cumulants

$$\kappa_i = a(i-1)!\zeta_\eta(i) \tag{7}$$

which are in terms of the *incomplete Euler–Riemann zeta function* $\zeta_\eta(i) = \sum_{l=1}^{\eta} \frac{1}{l^i}$. Hence the moment-matching methodology introduced by Arbel and Prünster [1] can be readily applied by resorting (6) and (7).

Acknowledgements The authors are grateful to two anonymous referees for valuable comments and suggestions. Julyan Arbel was a postdoc at Bocconi University and Collegio Carlo Alberto, Italy, when this article was submitted. This work is supported by the European Research Council (ERC) through StG "N-BNP" 306406.

References

1. Arbel, J., Prünster, I.: A moment-matching Ferguson and Klass algorithm. Stat. Comput. **27**(1), 3–17 (2017). doi:10.1007/s11222-016-9676-8, arxiv:1606.02566
2. Arbel, J., Lijoi, A., Nipoti, B.: Full Bayesian inference with hazard mixture models. Comput. Stat. Data Anal. **93**, 359–372 (2016)
3. Argiento, R., Bianchini, I., Guglielmi, A.: A blocked Gibbs sampler for NGG-mixture models via a priori truncation. Stat. Comput. **26**(3), 641–661 (2016)
4. Barrios, E., Lijoi, A., Nieto-Barajas, L.E., Prünster, I.: Modeling with normalized random measure mixture models. Stat. Sci. **28**(3), 313–334 (2013)
5. Brix, A.: Generalized gamma measures and shot-noise Cox processes. Adv. Appl. Prob. **31**, 929–953 (1999)
6. Caron, F., Fox, E.B.: Sparse graphs using exchangeable random measures (2015). arXiv:14011137
7. De Blasi, P., Favaro, S., Muliere, P.: A class of neutral to the right priors induced by superposition of beta processes. J. Stat. Plan. Inference **140**(6), 1563–1575 (2010)
8. Epifani, I., Lijoi, A., Prünster, I.: Exponential functionals and means of neutral-to-the-right priors. Biometrika **90**(4), 791–808 (2003)
9. Ferguson, T.S.: A Bayesian analysis of some nonparametric problems. Ann. Stat. **1**(2), 209–230 (1973)
10. Ferguson, T.S.: Prior distributions on spaces of probability measures. Ann. Stat. **2**(4), 615–629 (1974)
11. Ferguson, T.S., Klass, M.J.: A representation of independent increment processes without Gaussian components. Ann. Math. Stat. **43**(5), 1634–1643 (1972)
12. Gallant, A.R., Tauchen, G.: Which moments to match? Econometr. Theor. **12**, 657–681 (1996)
13. Griffin, J.E., Walker, S.G.: Posterior simulation of normalized random measure mixtures. J. Comput. Graph. Stat. **20**(1), 241–259 (2011)
14. Hjort, N.L.: Nonparametric bayes estimators based on beta processes in models for life history data. Ann. Stat. **18**(3), 1259–1294 (1990)
15. Jordan, M.I.: Hierarchical models, nested models and completely random measures. Frontiers of Statistical Decision Making and Bayesian Analysis: In Honor of James O Berger, pp. 207–218. Springer, New York (2010)
16. Kingman, J.: Completely random measures. Pac. J. Math. **21**(1), 59–78 (1967)
17. Lijoi, A., Nipoti, B.: A class of hazard rate mixtures for combining survival data from different experiments. J. Am. Stat. Assoc. **109**(506), 802–814 (2014)
18. Lijoi, A., Prünster, I.: Models beyond the Dirichlet process. In: Hjort, N.L., Holmes, C.C., Müller, P., Walker, S.G. (eds.) Bayesian Nonparametrics, pp. 80–136. Cambridge University Press, Cambridge (2010)
19. Lijoi, A., Mena, R.H., Prünster, I.: Bayesian nonparametric analysis for a generalized Dirichlet process prior. Stat. Inference Stoch. Process. **8**(3), 283–309 (2005)
20. Lijoi, A., Mena, R.H., Prünster, I.: Bayesian nonparametric estimation of the probability of discovering new species. Biometrika **94**(4), 769–786 (2007)
21. McFadden, D.: A method of simulated moments for estimation of discrete response models without numerical integration. Econometrica **57**(5), 995–1026 (1989)
22. Nieto-Barajas, L.E.: Bayesian semiparametric analysis of short- and long-term hazard ratios with covariates. Comput. Stat. Data Anal. **71**, 477–490 (2014)

23. Nieto-Barajas, L.E., Prünster, I.: A sensitivity analysis for Bayesian nonparametric density estimators. Stat. Sin. **19**(2), 685 (2009)
24. Nieto-Barajas, L.E., Walker, S.G.: Markov beta and gamma processes for modelling hazard rates. Scand. J. Stat. **29**(3), 413–424 (2002)
25. Nieto-Barajas, L.E., Walker, S.G.: Bayesian nonparametric survival analysis via Lévy driven Markov processes. Stat. Sin. **14**(4), 1127–1146 (2004)
26. Nieto-Barajas, L.E., Prünster, I., Walker, S.G.: Normalized random measures driven by increasing additive processes. Ann. Stat. **32**(6), 2343–2360 (2004)
27. Regazzini, E., Lijoi, A., Prünster, I.: Distributional results for means of normalized random measures with independent increments. Ann. Stat. **31**(2), 560–585 (2003)
28. Ren, B., Bacallado, S., Favaro, S., Holmes, S., Trippa, L.: Bayesian nonparametric ordination for the analysis of microbial communities. J. Am. Stat. Assoc. (2017)
29. Teh, Y.W., Gorur, D.: Indian buffet processes with power-law behavior. In: Advances in Neural Information Processing Systems, pp. 1838–1846 (2009)
30. Walker, S.G., Damien, P.: Representations of Lévy processes without Gaussian components. Biometrika **87**(2), 477–483 (2000)

On the Study of Two Models for Integer-Valued High-Frequency Data

Andrea Cremaschi and Jim E. Griffin

Abstract Financial prices are usually modelled as continuous, often involving geometric Brownian motion with drift, leverage and possibly jump components. An alternative modelling approach allows financial observations to take integer values that are multiples of a fixed quantity, the ticksize - the monetary value associated with a single change during the price evolution. In the case of high-frequency data, the sample exhibits diverse trading operations in a few seconds. In this context, the observables are assumed to be conditionally independent and identically distributed from either of two flexible likelihoods: the Skellam distribution - defined as the difference between two independent Poisson distributions - or a mixture of Geometric distributions. Posterior inference is obtained via adaptive Gibbs sampling algorithms. Comparisons of the models applied to high-frequency financial data is provided.

Keywords Time series · High-frequency data · Integer-valued random variables · Bayesian Econometrics · Adaptive MCMC

1 Introduction

The last decades of financial activities have seen rapid technological change in the execution of financial transactions. This has allowed the recent introduction of high-frequency trading (HFT), an automated way of performing transactions based on the use of algorithms interacting with electronic price books, called limit order books (see [5] for an overview of market dynamics and existing statistical literature). The electronic system generating the financial data is characterised by a very high transaction speed, conveying a large volume of data, making high-frequency trading an increasingly promising field for statistical analysis. Indeed, being able to observe the

A. Cremaschi (✉) · J.E. Griffin
Department of Biostatistics,
Universitetet i Oslo, Sognsvannsveien 9, Domus Medica, 0372 Oslo, Norway
e-mail: andrea.cremaschi@biotek.uio.no

J.E. Griffin
e-mail: J.E.Griffin-28@kent.ac.uk

R. Argiento et al. (eds.), *Bayesian Statistics in Action*, Springer Proceedings in Mathematics & Statistics 194, DOI 10.1007/978-3-319-54084-9_3

price processes so often in time could provide more information about their dynamics. From a structural point of view, bid and ask prices are characterised by being integer multiples of a fixed quantity, the ticksize, the minimum monetary value tradable in the market. This discrete interpretation of the price processes represents one of the peculiarity of high-frequency data modelling (we refer to [7] for a statistical analysis based on ticksize). In particular, the focus of this work will be on the New York Stock Exchange (NYSE), where the ticksize is of one dollar cent (0.01). Electronic markets such as the NYSE are characterised by their high-depths, meaning that the different bid/ask prices present in the limit order book take values in a wide range (i.e., the *depth* of a market is the number of different prices available for transaction), allowing for market and limit orders to be fulfilled very quickly and hence increase the liquidity while reducing the bid-ask spread (usually equal to a single tick). This feature of the market affects the dynamic of the price processes, influencing the relationships between returns at different time points. This paper is structured as follows: the next Section will introduce the object of the study, the integer-valued returns, and two conditional distributions used to model them. Section 3 will introduce the algorithms used to perform posterior inference, while an application to real stock market data is presented in Sect. 4. Section 5 concludes.

2 Distributions for Tick Data

The nature of the prices makes them interpretable as positive integers, indicating the multiple of the ticksize at which a specific transaction takes place. Let p_t be the price process at time $t > 0$, and P_t the discretised version of it, simply computed as $P_t = \frac{p_t}{ticksize}$. It is assumed that $T > 0$ transactions are observed in a finite time interval, hence the observed prices are $P_1, \ldots, P_T$, taking values over the set of positive integers $\mathbb{N}$. Despite the notation, time dependency will not be assumed in the analysis, in order to present a preliminary study of the alternative discrete transformation of the data, together with an outline of the pros and cons related with the different modelling choices. It is of interest to study the behaviour of the price process, with particular attention to its fluctuations in terms of variability. In order to do so, in Financial Econometrics the concept of a return is introduced: a transformation of the raw price data that is able to provide information about the variability of the data generating process underlying the observations. When the data are assumed to be continuous, the standard definition of return is the log-ratio of two consecutively observed prices. Clearly, this is not useful in the context of this work, due to the discretisation adopted at the basis of the study. Hence, following [3], the ticksize returns are defined as the difference of two consecutive discrete valued prices, such that for each $t > 0$, $Y_t := P_t - P_{t-1} \in \mathbb{Z}$. While, in the continuous case, the returns can be easily modelled by using a Normal distribution, this alternative definition of returns requires a suitable probability distribution defined on $\mathbb{Z}$ in order to build a valid statistical model. In this work, two such distributions are specified, namely the Skellam distribution and the Folded Geometric distribution, and their inference properties are compared. The first distribution is defined as the difference of two independent Poisson distributed random variables, which has been used, among others, in the analysis of intra-day high-frequency trading data in the work of [4, 10]. A feature of their model is the presence of a zero-inflation parameter to accommodate a mode at zero. This choice is motivated by the high-depth of the electronic market. In this work, a similar issue is tackled by the introduction of the Folded Geometric distribution, defined as a mixture of two Geometric distributions on the positive and negative integers, with two additional parameters to represent centrality and modality. In the rest of this section, the two distributions are introduced, as well as the two models arising from these choices.

2.1 Skellam Distribution

With reference to [3], the integer returns are modelled by using the difference between two positive quantities representing the positive and the negative jumps governing the price evolution, such that $Y_t = L_t^+ - L_t^-$, with $Y_t \in \mathbb{Z}$, for $t = 1, \ldots, T$. Notice that here the jump processes L_t^+ and L_t^- are used to support the modelling choice that will follow, and have a different meaning from the integer-valued prices P_t and P_{t-1} defined above. The two independent distributions can be interpreted as responsible for the evolution of the returns, by making it move up (L_t^+) or down (L_t^-). When $\{L_t^+\}_{t=1}^T$ and $\{L_t^-\}_{t=1}^T$ are two independent Poisson distributions of intensities ϕ^+ and ϕ^-, then Y_t is Skellam distributed, with probability mass function as follows:

$$\mathbb{P}(Y_t = k) = e^{-t(\phi^+ + \phi^-)} \left(\frac{\phi^+}{\phi^-}\right)^{k/2} I_{|k|}(2t\sqrt{\phi^+ \phi^-}),$$

$$I_k(x) = \left(\frac{1}{2}x\right)^k \sum_{n=0}^{\infty} \frac{(\frac{1}{4}x^2)^n}{n!(n+k)!},$$

where $I_k(x)$ is the modified Bessel function of the first kind of positive arguments x and k (see [1]). Figure 1a presents the p.m.f.'s of the Skellam distribution for different combinations of the intensity parameters. In this work, we consider an alternative parameterisation of the Skellam distribution in terms of the variance and skewness of the distribution, such as:

$$\begin{aligned} \phi^+ &= \tfrac{1+a}{2} e^h \\ \phi^- &= \tfrac{1-a}{2} e^h \end{aligned} \quad \rightarrow \quad \begin{aligned} a &= \tfrac{\phi^+ - \phi^-}{\phi^+ + \phi^-} = \tfrac{\mathrm{E}(L)}{\mathrm{Var}(L)} \\ h &= \log(\phi^+ + \phi^-) = \log(\mathrm{Var}(L)) \end{aligned}$$

Notice that the newly introduced parameters can be interpreted using the moments of the Skellam distribution. In particular, the real-valued parameter h represents the log-volatility of the distribution, while $a \in (-1, 1)$ can be seen as a scaled skewness parameter, since when $Y \sim Sk(\phi^+, \phi^-)$, then $\mathrm{Skew}(Y) = \frac{\phi^+ - \phi^-}{(\phi^+ + \phi^-)^{3/2}} = e^{-h/2} a$.

A computational issue with the Skellam model is the dependence on the modified Bessel function of the first kind, $I_k(x)$. The computation of this hypergeometric series greatly affects the accuracy of the computed probabilities hence influencing the inference. This is usually the case for large values of (k, x), such as when a rapid change in price is observed, i.e. $k = |y_t|$, or when the intensity parameters are such that $x = \sqrt{2\psi^+\psi^-}$ takes large values. It is worth noting that, in the latter case, large values of x correspond to large values of the variance of the returns, but could as well be associated to small values of the conditional mean. In order to avoid this problematic aspect, a latent variable is introduced in the model to represent the negative jumps in the evolution of the returns, $\{L_t^-\}_{t=1}^T$. The resulting likelihood for the returns is therefore a shifted Poisson, shifted by $-L_t^-$ units. In a Gibbs sampler targeting the posterior distribution, this yields a much easier expressions for the full conditionals of the parameters, and hence eases the computation. The final Skellam model analysed in this work is the following:

$$
\begin{aligned}
&Y_t|L_t^-, h, a \overset{\text{ind}}{\sim} ShPoi\left(-L_t^-, \frac{1+a}{2}e^h\right), && h \sim N(\mu, \psi),\\
&L_1^-, \ldots, L_T^-|h, a \overset{\text{iid}}{\sim} Poi\left(\frac{1-a}{2}e^h\right), && \mu \sim N(0, 1),\\
&\frac{a+1}{2} \sim Beta(0.5, 0.5), && \psi \sim inv-gamma(3, 2),
\end{aligned}
$$

where $X \sim ShPois(s, \eta)$ is distributed according to a shifted Poisson with shifting parameter s and intensity η, if $(X - s) \sim Poi(\eta)$. Moreover, $N(m, s^2)$ indicates the normal distribution with mean m and variance s^2, and $inv - gamma(a, b)$ indicates the inverse-Gamma distribution with mean $\frac{b}{a-1}$ and mode $\frac{b-1}{a+b-2}$. Notice how the parameter a is modelled as a linear transformation of the Beta distribution $Beta(a, b)$ with mean $\frac{a}{a+b}$ and mode $\frac{a-1}{a+b-2}$, as done in [9].

2.2 *Folded Geometric Distribution*

As mentioned before, it is useful to avoid the computation of the Bessel function $_k(x)$. An alternative modelling approach defines a different distribution to model the returns. As mentioned above, the market considered in this work is a one-tick high-depth market (such as the New York Stock Exchange market), meaning that the bid-ask spread is usually equal to one tick, producing transaction returns that fluctuate very little, and that present a lot of zeros. In order to capture this behaviour, a probability distribution F is introduced satisfying the following requirements:

(a) it has support on $\mathbb{Z}$,
(b) it allows the presence of a mode at zero,
(c) it does not include convolution terms (such as the Bessel function $I_k(x)$),
(d) it is flexible enough to represent the evolution of ticksize normalised returns.

To start, define the discrete random variables X^+ and X^- with support on $\mathbb{N} \setminus 0$ and $\mathbb{Z} \setminus \mathbb{N}$, distributed according to F^+ and F^-, respectively. Hence, assume that the two distributions F^+ and F^- admit probability measures indicated with $\mathbb{P}^+$ and $\mathbb{P}^-$. Let X be a discrete random variable defined on $\mathbb{Z}$ with the following p.m.f.:

$$\mathbb{P}(X = k) = \frac{1}{c} \begin{cases} \mathbb{P}^-(X^- = k + l) & k < l \\ a & k = l \\ \mathbb{P}^+(X^+ = k - l) & k > l \end{cases}$$

where a is proportional to the probability of taking the value l, representing the centre of the distribution. The term c is the normalising constant of the distribution, and is equal to $c = 2 + a - \mathbb{P}^+(X^+ = l) - \mathbb{P}^-(X^- = l)$. The mixture of these three random variables covers the whole sample space $\mathbb{Z}$, satisfying (a), and can be constructed such that there is a mode at $l = 0$, satisfying (b). Condition (c) and (d) are also satisfied, since this is a mixture density without any convolution, and the two halves can be chosen arbitrarily, providing suitable flexibility for different applications. The resulting distribution is hereby called a *Folded* distribution. In this work, a mixture of two Geometric distributions with success probabilities denoted as p^+ and p^- is considered, together with a mode at $l = 0$, and it will be called *Folded Geometric* distribution, indicated as $FG(p^+, p^-, l, a)$. The first three moments of the random variable X, when $l = 0$, are:

$$\begin{aligned} \mathrm{E}(X) &= \frac{\mathrm{E}(X^+) - \mathrm{E}(X^-)}{c}, \\ \mathrm{E}(X^2) &= \frac{\mathrm{E}(X^{+2}) + \mathrm{E}(X^{-2})}{c}, \\ \mathrm{E}(X^3) &= \frac{\mathrm{E}(X^{+3}) - \mathrm{E}(X^{-3})}{c}. \end{aligned}$$

Consider using this p.m.f. to describe the distribution of the returns. For $t = 1, \ldots, T$:

$$\mathbb{P}(Y_t = k) = \frac{1}{c} \begin{cases} p^-(1 - p^-)^{l-k} & k < l \\ a & k = l \\ p^+(1 - p^+)^{k-l} & k > l \end{cases}$$

notice that $a \geq \frac{1}{4}$ guarantees the unimodality of the distribution at $l = 0$, since $p(1 - p) \leq \frac{1}{4}$ for the Geometric distribution, and that the normalising constant is $c = 2 + a - p^- - p^+$. Finally, notice how the choice of the two success probabilities is completely arbitrary, and no restriction is imposed, apart from the obvious $p^+, p^- \in (0, 1)$. Figure 1b and c show the different shapes of the Folded Geometric distribution, when the Symmetric ($p^+ = p^-$) or Asymmetric ($p^+ \neq p^-$) setting is chosen. Finally, the Folded Geometric model can be outlined as:

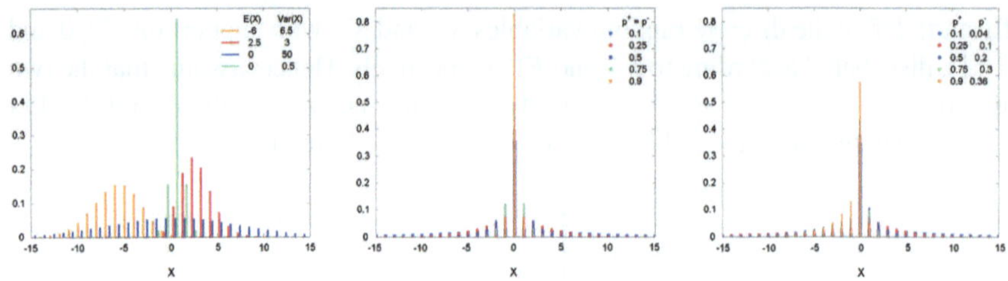

(a) *Skellam* (b) $FG(p,1-p,0,1)$ (c) $FG(p,p/2.5,0,3.5)$

Fig. 1 Different p.m.f.'s used to describe the conditional distribution of the integer-valued returns

$$\begin{aligned}
&y_1,\ldots,y_T|h^+,h^-,a \overset{\text{iid}}{\sim} FG\left(\tfrac{e^{h^+}}{1+e^{h^+}},\tfrac{e^{h^-}}{1+e^{h^-}},l=0,a\right),\\
&\left(a-\tfrac{1}{4}\right)\sim Exp(\tfrac{3}{4}),\\
&h^+\sim N(\mu^+,\psi^+), && h^-\sim N(\mu^-,\psi^-),\\
&\mu^+\sim N(0,1), && \mu^-\sim N(0,1),\\
&\psi^+\sim inv-gamma(3,2), && \psi^-\sim inv-gamma(3,2).
\end{aligned}$$

Notice how the centrality parameter a is set to be positive, and greater than the value $\frac{1}{4}$, in order to guarantee the unimodality of the conditional distribution of the observations. Prior specification is analogous to the one used for the Skellam model.

3 Algorithms

Posterior computations for the specified models are obtained via the implementation of adaptive Gibbs sampler algorithms, outlined in this section. The adaptive part of the algorithm scheme is an implementation of the adaptive random walk Metropolis algorithm described in [2]. In their work, the authors present an innovative sampling scheme for the random walk Metropolis-Hastings, where the proposal kernel is allowed to depend on the history of the process, i.e. to depend from the previously sampled values of the parameters. An extensive review of various adaptive algorithms can be found in [8]. To give an illustration of such a procedure, the **AMH** algorithm is reported in the frame below, outlining the sequence of steps necessary to update a non-conjugate parameter q of a given model.

Adaptive random walk Metropolis-Hastings algorithm

Choose a burn-in value g_0, and initialise θ at iteration $g=1$;
run $g_0>0$ iterations with a fixed value of the proposal variance s_θ^2;
for $g>g_0$, perform the following log-scale update: $\log(s_\theta^2(g+1))=\log(s_\theta^2(g))+(g^{-0.55}(\alpha_\theta-\overline{\tau}))$.

In the algorithm, α_θ is the acceptance rate of the Metropolis-Hastings step, and $\overline{\tau}$ is a reference optimal acceptance rate fixed equal to the value 0.234, following the work of [11, 12]. The parameters for which it is possible to adopt the **AMH** algorithm are the non-conjugate ones, that is a and h in this work (for all the models).

4 Application

In this Section, we present an application to a subset of the Disney transaction data, originally sampled every minute during the years 2004–2015 from the New York Stock Exchange market, and here restricted to the months of September and thinned each 10 min. The reduction and the thinning are adopted in order to reduce the computational burden. Figure 2a shows the whole dataset for the Disney stock (2004–2015), while Fig. 2b shows the year 2008 only, from which the month of September (Fig. 2) is extracted for the analysis.

Gibbs sampler algorithms are run for 525.000 iterations, of which 500.000 constitute the burn-in period, and 5000 are subsequently saved every fifth iteration.The

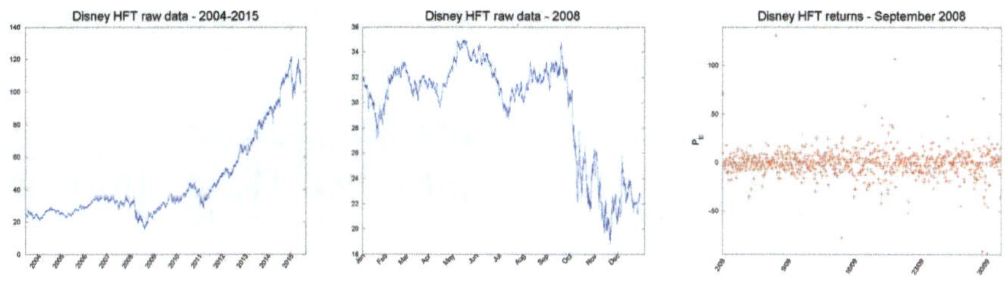

(a) from 2004 to 2015 (b) 2008 (c)

Fig. 2 Disney HFT raw data

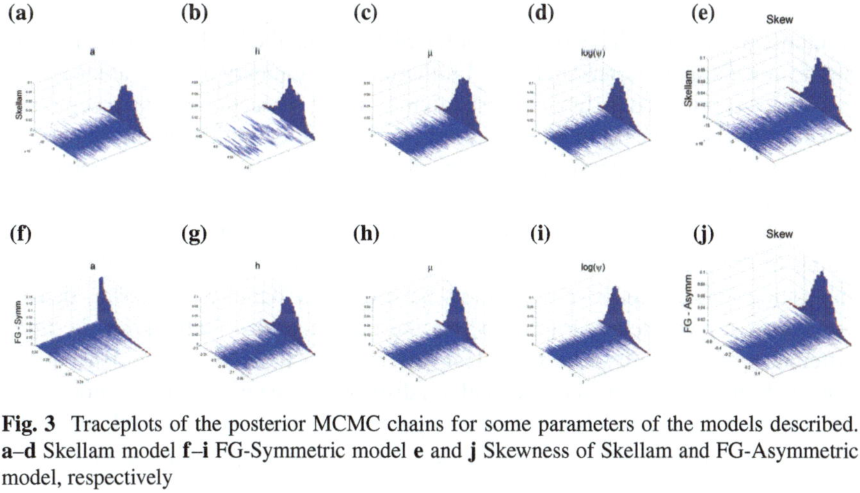

Fig. 3 Traceplots of the posterior MCMC chains for some parameters of the models described. **a–d** Skellam model **f–i** FG-Symmetric model **e** and **j** Skewness of Skellam and FG-Asymmetric model, respectively

Table 1 Acceptance rates and predictive indices log-**BF** and **LPML**

Acceptance rates		log-**BF** \ **LPML**	Skellam	FG - Symm	FG - Asymm
a	0.22602	Skellam		−170.323	−157.1342
h	0.23423				
a	0.23407	FG - Symm	$2.9389 \cdot 10^3$		13.1887
h	0.2352				
a	0.23359	FG - Asymm	$2.9213 \cdot 10^3$	−17.6052	
h^+	0.23356				
h^-	0.2341				

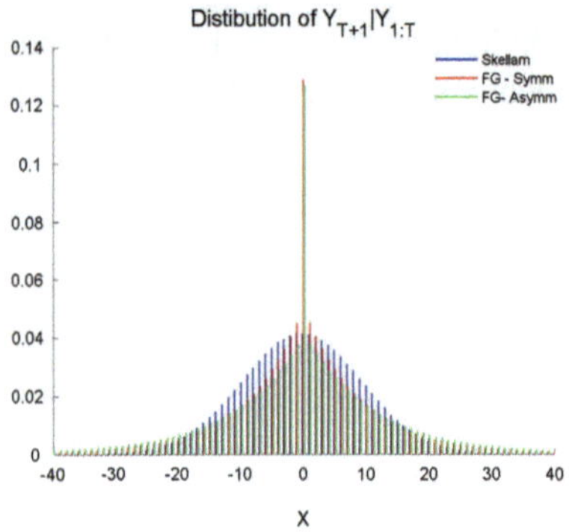

(a) Predictive distribution under different models

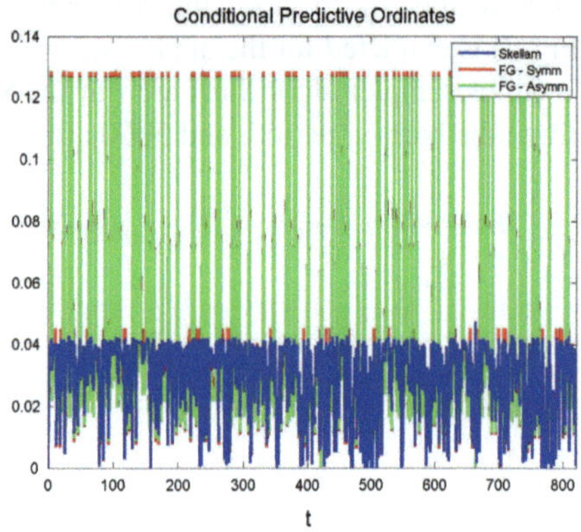

(b) CPO under different models

Fig. 4 Predictive comparisons

resulting traceplots with histograms for the Skellam and the Folded Geometric Symmetric models are presented in Fig. 3. Furthermore, a summary of the average acceptance rates is provided in Table 1 (first column). From inspection of the traceplots, it can be argued that the posterior chains have converged for all the parameters, except for the parameter h in the Skellam model, clearly affected by the introduction of the latent variables $L_1^-, \ldots, L_T^-$ (the Asymmetric case is not reported here for unavailability of space, but leads to the same conclusions as the Symmetric case), while the values of the acceptance rates are all very close to the gold standard value 0.234, suggesting that the adaptive algorithms have reached stability. Moreover, density estimation is presented in Fig. 4a, where the predictive distributions for the three different models are displayed. A clear difference in the tails of the distributions can be observed. In particular, the Skellam distribution seems to allocate more mass away from zero, possibly because of the aggregation effect induced by thinning the observations, and the ability of the Skellam distribution to approximate the Normal distribution when the intensity parameters are equal (see [3] for an analogous result on the Skellam process). Not surprisingly, the Folded Geometric models are instead able to capture the zero-inflated aspect of the distribution of the returns much better than the Skellam model does. Further analysis considered the choice of the most suitable model among the suggested ones, evaluating the *Log Pseudo-Marginal Likelihood* (LPML), as defined by [6] in terms of the *Conditional Predictive Ordinate*:

$$(CPO_t^j)^{-1} = \frac{1}{G}\sum_{g=1}^{G}\frac{1}{f^j(y_t|\theta^j(g))}$$

$$LPML_j = \sum_{t=1}^{T}\log(CPO_t^j),$$

where $G = 5000$ is the number of iterations saved, $f^j(y_i|\theta^j(g))$ is the likelihood function for the j-th model, and $\theta^j(g)$ is the g-th MCMC sample of the parameter vector for the j-th model. The higher the values of the LPML, the better the fitting of the data to the j-th model. As it can be seen from Fig. 4b, the Skellam model has higher CPO values for some of the observations, while the two Folded Geometric models look more stable, and in agreement with each other. A measure indicating whether a model is suitable to describe the data at hand is indeed the log-ratio of LPMLs. These values are reported in the right hand side of Table 1, together with the estimates of the log-Bayes Factors for each pair of models. From such values, it appears that the Folded Geometric models are to be preferred to the Skellam model, probably as a consequence of introducing the latent variables $L_1^-, \ldots, L_T^-$. Between the two Folded Geometric models, it seems like the Symmetric one is performing better than the Asymmetric one, suggesting little evidence of asymmetry in the data, as it is also shown by the traceplots of the skewness parameters for the Skellam and Folded Geometric Asymmetric models, in Fig. 3e and j.

5 Discussion

In this work, two different models for the statistical analysis of discretised high-frequency data are presented. The conditional distribution of the observables in the two scenarios is set to be either the Skellam distribution or the Folded Geometric distribution, the latter being in turn distinguishable between its Symmetric and Asymmetric case. An adaptive Gibbs sampling algorithm is described that is able to provide good mixing properties for the posterior chains of the parameters of the different models. Model comparison revealed some discrepancies between the performances of the different models. In particular, the predictive distribution of $Y_{T+1}|\boldsymbol{Y}_{1:T}$ seems to be quite different for the Skellam and the Folded Geometric models, probably due to the heavier-tailed Geometric mixture distribution, that is capable of capturing more extreme behaviours and outliers in the returns, while the predictive distribution for the Skellam model is closer to a Normal distribution centered at zero. As expected, the Folded Geometric distribution is able to capture the zero-inflated aspect of the returns, differently from the Skellam one. Furthermore, some predictive quantities such as log-LPML and log-Bayes Factor are compared, supporting the idea that the Folded Geometric model might be a better choice for the analysis of high-frequency data when no time dependency is included in the parameter space, this point being crucial in interpreting the results obtained so far. The results provided by the Skellam model can be explained by recalling that the Skellam distribution can be seen as a discretised version of the Normal distribution. On the contrary, in the Folded Geometric case, there is space for detection of extremal behaviours. To conclude, the analysis presented in this work has shown how, under suitable algorithmic conditions and standard prior elicitation choices, the assumption of independent and identically distributed data is accommodated in different ways by different model choices. In order to deepen the study of this matter, it is our intention to study the property of models where time dependence is included at parameter level, as well as in the likelihood term, via the introduction of a stochastic volatility process. In this case, the two models might provide more consistent results, not so distant from one another.

References

1. Abramowitz, M., Stegun, I.A.: Handbook of Mathematical Functions with Formulas, Graphs, and Mathematical Tables. Dover, New York (1972)
2. Atchadé, Y.F., Rosenthal, J.S.: On adaptive Markov chain Monte Carlo algorithms. Bernoulli **11**(5), 815–828 (2005)
3. Barndorff-Nielsen, O.E., Pollard, D.G., Shephard, N.: Integer-valued Lévy processes and low latency financial econometrics. Quant. Finance **12**(4), 587–605 (2012)
4. Barra, I., Koopman, S.J.: Bayesian dynamic modeling of high-frequency integer price changes. Tinbergen Inst., 16-028 (2016)
5. Cont, R.: Statistical modeling of high-frequency financial data. IEEE Signal Process. Mag. **28**(5), 16–25 (2011)
6. Geisser, S., Eddy, W.F.: A predictive approach to model selection. J. Am. Stat. Assoc. **74**(365), 153–160 (1979)
7. Griffin, J.E., Oomen, R.C.A.: Sampling returns for realized variance calculations: tick time or transaction time? Econ. Rev. **27**(1–3), 230–253 (2008)
8. Griffin, J.E., Stephens, D.A.: Advances in Markov chain Monte Carlo. In: Bayesian Theory and Applications, pp. 104–144 (2013)
9. Kim, S., Shephard, N., Chib, S.: Stochastic volatility: likelihood inference and comparison with ARCH models. Rev. Econ. Stud. **65**(3), 361–393 (1998)
10. Koopman, S.J., Lit, R., Lucas, A.: The dynamic Skellam model with applications (2014)
11. Roberts, G.O., Rosenthal, J.S.: Optimal scaling for various Metropolis-Hastings algorithms. Stat. Sci. **16**(4), 351–367 (2001)
12. Roberts, G.O., Gelman, A., Gilks, W.R.: Weak convergence and optimal scaling of random walk Metropolis algorithms. Ann. Appl. Probab. **7**(1), 110–120 (1997)

Identification and Estimation of Principal Causal Effects in Randomized Experiments with Treatment Switching

Emanuele Gramuglia

Abstract In randomized clinical trials designed to evaluate the effect of a treatment on patients with advanced disease stages, treatment switching is often allowed for ethical reasons. Because the switching is a prognosis-related choice, identification and estimation of the effect of the actual receipt of the treatment becomes problematic. Existing methods in the literature try to reconstruct the ideal situation that would be observed if the switchers had not switched. Rather than focusing on reconstructing the a-priori counterfactual outcome for the switchers, had they not switched, we propose to identify and estimate effects for (latent) subgroups of units according to their switching behaviour. The reference framework of the proposed method is the potential outcome approach. In order to estimate causal effects for sub- groups of units not affected by treatment, we rely on the principal stratification approach (Frangakis and Rubin in Biometrics 58(1): 21–29 2002) [1]. To illustrate the proposed method and evaluate the maintained assumptions, we analyse a dataset from a randomized clinical trial on patients with asymptomatic HIV infection assigned to immediate (the active treatment) or deferred (the control treatment) Zidovudine (ZDV). The results, obtained through a full-Bayesian estimation approach, are promising and emphasize the high heterogeneity of the effects for different latent subgroups defined according to the switching behaviour.

Keywords Bayesian causal inference · Rubin causal model · Treatment switching

1 Introduction

Randomized clinical trials are the most widely used designs to assess the effect of a new treatment versus placebo or standard treatment. In studies focusing on

E. Gramuglia (✉)
University of Oslo, Oslo, Norway
e-mail: emanueg@math.uio.no

R. Argiento et al. (eds.), *Bayesian Statistics in Action*, Springer Proceedings in Mathematics & Statistics 194, DOI 10.1007/978-3-319-54084-9_4

survival outcomes for patients suffering from Acquired Immune Deficiency Syndrome (AIDS)-related illnesses and particularly painful cancers in advanced stages, patients are often allowed to switch from their allocated arm for ethical reasons. This situation can be in general viewed as a non-adherence to the protocol. Often switching is a prognosis-related choice; for instance in some patients, during the follow-up period, the assigned treatment causes unwanted side effects preventing the continuation of the study or, for weaker units, a sudden worsening of the disease forces physicians to allow the switch. An Intention-To-Treat analysis, comparing groups formed by randomization regardless of the treatment actually received, is often used; however, it provides valid causal estimates of the effect of assignment, but it does not give information about the effect of the actual receipt of the treatment. Other existing methods in the literature aim at reconstructing the outcome a unit would have had if he/she had not switched. The principal drawback of these methods is, however, to assume that there is no relation between patient's prognosis and switching behaviour. We instead provide estimates of the effect of the actual receipt of the new treatment versus the actual receipt of the standard treatment for (latent) subgroups of units, defined according to their switching behaviours.

2 Principal Stratification Approach to Treatment Switching

We use the potential outcome paradigm [2], which associates each unit-treatment pair with a potential outcome, defining the causal effects as the comparisons between potential outcomes. Consider for example a clinical trial, involving N units indexed by i, designed to test the effect of a new drug (the active treatment) with respect to an old one (the control treatment). Let $Y_i(0)$ and $Y_i(1)$ denote, respectively, the outcome under the control treatment and the outcome under the active treatment; the treatment level indicator we use is $Z = \{0, 1\}$. The *unit-level* causal effect is defined as a comparison between $Y_i(1)$ and $Y_i(0)$. The outcomes are referred to as potential, because only the one corresponding to the treatment actually taken will be realized and possible observed.

In order to identify subgroups of units not affected by the treatment, we rely on the principal stratification approach [1]. This approach consists of partitioning the units according to the potential values of a post-treatment intermediate variable, S, under standard and active treatment

$$\{i : S_i(0) = s_0, S_i(1) = s_1\}.$$

Hence, the couple of potential outcomes for S defines *strata*. Principal causal effects (PCEs), that is, effects defined on the common set of units with the same joint potential values of the intermediate variable are always causal.

We consider survival as the outcome of primary interest and the post-treatment variable S represents the switching behaviour. Specifically, we focus on situations where it is allowed to switch from the control to the active treatment only; thus, the strata will be functions only of the potential outcome of the switching time under the control treatment.

3 Identification Assumptions

Although survival and switching time are continuous variables, to ease explanation, both of them are transformed into categorical variables having 3 levels according to the interval on which the principal event (e.g. disease, death, etc.) for $Y(0)$ and $Y(1)$, and the switching for $S(0)$ happens. Assuming all units enter the study at the same time and the observation period is the same for all patients, denoted by Δ, the discretization consists of generating the following variables

$$Y_i^d(z) = \begin{cases} 1 & \text{if } 0 \le Y_i(z) < a \\ 2 & \text{if } a \le Y_i(z) < \Delta \\ 3 & \text{if } Y_i(z) \ge \Delta \end{cases} \qquad S_i^d(0) = \begin{cases} 1 & \text{if } 0 \le S_i(0) < a \\ 2 & \text{if } a \le S_i(0) < \Delta \\ 3 & \text{if } S_i(0) \ge \Delta \end{cases}$$

where $a \ll \Delta$.

Each unit can possibly be a *switcher at time 1* ($S_i^d(0) = 1$), *a switcher at time 2* ($S_i^d(0) = 2$) or a *non switcher* ($S_i^d(0) = 3$).

Based on this setting, the focus is on the causal effect of the new treatment on the survival beyond the end of study

$$\tau_s^3 = \frac{\sum\limits_{i \in S_i^d(0)=s} I(Y_i^d(1) = 3) - I(Y_i^d(0) = 3)}{N_s}$$

or after the time point a, i.e. for $a = 2$ we have

$$\tau_s^2 = \frac{\sum\limits_{i \in S_i^d(0)=s} I(Y_i^d(1) = 3) - I(Y_i^d(0) = 3)}{N_s} + \frac{\sum\limits_{i \in S_i^d(0)=s} I(Y_i^d(1) = 2) - I(Y_i^d(0) = 2)}{N_s}$$

for each type s of *switcher*.

Although the effects computed in each principal stratum have a causal meaning our interest is mainly for the principal stratum of *non switchers*, because they are patients complying for all the study period with the initial protocol.

However, due to the latency of the intermediary variable $S^d(0)$, without further assumptions it is not possible to identify the distribution of potential outcomes within principal strata. The latency is partial for the units assigned the control treatment and total for the units assigned the active treatment. Consider, for example, a patient randomized the control treatment who died at time 1 and had not switched; his/her switching time could have been at time 2 or he/she could have been a *non switcher*. For the units randomized to take the active treatment, we will never be able to observe their switching behaviour being them randomized to take the control, resulting on a total latency of $S^d(0)$.

To overcome the identifiability problem for the control group we stated the following assumption

Patients who die during the first or the second time interval and have not switched, are considered potential switchers in the next intervals.

This assumption excludes the possibility for patients dying without switching, to switch in the same interval allowing the total identifiability of the switching behaviour. However, it cannot be checked from the data but relies on knowledges about the behaviour of the units and the length of the intervals.

Regarding the units belonging to the active treatment group we make an assumption about the probabilistic dependency structure of the potential outcome $Y^d(1)$ and some selected pre-treatment covariates.

Recent interesting results have been developed (see Stanghellini et al. [4]) about identifying, under suitable conditional independence restrictions, the joint distribution of k + 1 random variables, one of which is hidden. These results are achieved by exploiting the dependence structure among the variables taken into account. A helpful representation of the association structure of the variables involved can be obtained by means of concentration graphs, undirected graphs where nodes represents random variables and the presence or absence of an arch represents conditional association o independence of two variables, given all the others in the graph. As outlined in Mealli et al. [3], we exploit these result in our causal framework by specifying the joint distribution of the potential outcome $Y^d(1)$, the (latent) switching time under control treatment $S^d(0)$ and a set of binary baseline covariates.

4 Partially Simulated Case Study

To illustrate the proposed method and evaluating the plausibility of the assumptions stated, we carry out an analysis based on a partially simulated dataset on a group of units with asymptomatic HIV infection randomized to immediate (the active treatment) or deferred (the standard treatment) Zidovudine, initially created by White et al. [6]. It was intended that patients in control group would not receive Zidovudine until they progressed to AIDS-related complex (ARC) or AIDS. However, on the basis of persistently low CD4 cell counts (a prognostic marker for disease progression), Zidovudine was allowed to patients in control group before presence of ARC or AIDS. The outcome used was time-to-disease progression or death. We refer to our data as partially simulated because, in order to present the method, three baseline covariates, labelled X_1, X_2, X_3, were simulated while the outcomes were used as originally created. In order to exploit the identification results discussed in the previous section, we assume a global conditional dependence structure according to identified concentration graph in Fig. 1. It represents a situation where the patients

Fig. 1 Concentration graph

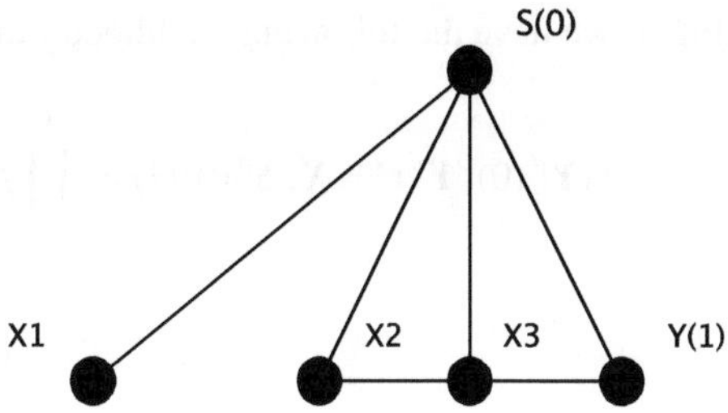

free-disease lifetime is independent of the covariates X_1 and X_2 conditionally to $S^d(0)$ and X_3. According to Fig. 1, the joint model, which is identified, factorizes to

$$\begin{aligned} f(X_1, X_2, X_3, Y^d(1), S^d(0)) =& f(X_3|X_2, S^d(0)) \cdot f(Y^d(1)|X_3, S^d(0)) \\ &\times f(X_1|S^d(0)) \cdot f(X_2|S^d(0)) \\ &\times f(S^d(0)). \end{aligned}$$

To show the flexibility of our method, we propose three different scenarios, according to the relation between the switching time and the potential outcomes. The first scenario resembles the common situation where patients switching sooner are in worse clinical conditions and consider the switching as a benefit possibly improving their conditions. The second scenario considers the switching time under active treatment being random. Interestingly we had a situation where the effect of the new treatment on the survival beyond the end of the study for non switchers is positive while that beyond time 2 is negative. The third scenario represents an extreme situation where switchers at time 1 have a greater probability of surviving beyond the end of the study if assigned to the control treatment. This can reflect a situation where the immediate Zidovudine causes unwanted side effects for weaker units.

4.1 Estimation Strategy

The process of estimation in a model-based approach appears to be rather straightforward from a Bayesian perspective. Potential outcomes are viewed as random variables and hence any function of them is also a random variable, including any causal estimand of interest. Considering the population of interest of size N, we have a N-vector $\boldsymbol{Z}$ and two $N \times 2$ matrices, $\boldsymbol{S}^d$ and $\boldsymbol{Y}^d$, for, respectively, the intermediate variable and the primary outcome. A $N \times 3$ matrix $\boldsymbol{X}$ is used to represent the observed values of the three baseline covariates of all units. Under row (unit) exchangeability of the matrices $\boldsymbol{Y}^d$, $\boldsymbol{S}^d$ and $\boldsymbol{X}$ and by appealing to the DeFinettis theorem, we can assume, with no loss of generality, that the rows of $(\boldsymbol{Y}, \boldsymbol{S}, \boldsymbol{X})$ are independent and identically distributed random variables given a parameter vector $\boldsymbol{\theta}$ with prior distribution $P(\boldsymbol{\theta})$. According to the dependence structure represented in

Fig. 1, we have the following likelihood function

$$
\begin{aligned}
f(\boldsymbol{Y}^d(0), \boldsymbol{Y}^d(1), \boldsymbol{X}, \boldsymbol{S}^d(0)|\boldsymbol{\theta}) = \prod_{i=1}^{N} & f(Y_i^d(0)|X_{1,i}, X_{2,i}, X_{3,i}, S_i^d(0), \theta) \\
& \times f(Y_i^d(1)|X_{3,i}, S_i^d(0), \theta) \\
& \times f(X_{1,i}|S_i^d(0), \theta) \cdot f(X_{2,i}|S_i^d(0), \theta) \\
& \times f(X_{3,i}|X_{2,i}, S_i^d(0), \theta) \\
& \times f(S_i^d(0)|\theta).
\end{aligned}
$$

As $S_i^d(0)$ takes values on $\{1, 2, 3\}$ and Z_i takes values on $\{0, 1\}$, we can decompose this *complete* likelihood function, into six parts each representing the likelihood $\mathscr{L}_{comp}^{z,s}$ of a subset S(z, s) of the sample such that $(Z = z,\ S^d(0) = s)$. All the different parts $\mathscr{L}_{comp}^{z,s}$ joint together constitute the *complete* likelihood function, namely the likelihood function we would have if, for each unit, the values of all the involved variables were observed. Unfortunately as we said previously, for some subsets of units there will be missing values in the variable $S^d(0)$, leading to a mixture of outcomes distributions generating the so called *observed* likelihood function.

In order to make posterior inference for $\boldsymbol{\theta}$, we rely on a Data Augmentation (DA) step [5]. A DA consists in drawing $S^d(0)$ from the posterior distribution of missing data conditional on parameters and observed data. Note that this is an iterative imputation method and thus relies on the current estimates of the parameters. After the DA step, Bayes' Rule is used to compute the posterior distribution of $\boldsymbol{\theta}$ and hence the posterior distribution of the causal estimands of interest.

4.2 *Results and Comments*

Because the simulation concerns only the baseline covariates and the switching time, the overall ITT estimate, defined as the difference between the proportion of surviving patients assigned the active treatment arm and surviving patients assigned the control arm, is the same for each scenario. The estimated ITT on the survival after the end of the study is $\widehat{ITT}^3 = 0.068$. This effect, however, does not take into account the switching from control to active treatment (Table 1).

We focus on the principal stratum of non switchers, being their outcomes not affected by the switching, computing the posterior distribution of the effect of the treatment on the survival beyond the end of the study, τ_3^3, and the effect on survival after time 1, τ_3^2. The results show that the effects can be highly heterogeneous depending on the latent switching behaviour and its relation with the outcome. Despite the fact that the switching behaviour is missing for some units, the Bayesian inferential analysis, conducted under the proposed structural assumptions, is able to recover the principal causal effects.

Table 1 Posterior estimates

Scenario	τ	Mean	$q^{.025}$	$q^{.975}$
1	τ_3^3	0.054	−0.090	0.201
	τ_3^2	0.002	−0.128	0.1445
2	τ_3^3	0.023	−0.133	0.187
	τ_3^2	−0.049	−0.169	0.063
3	τ_3^3	0.273	0.171	0.374
	τ_3^2	0.111	0.001	0.212

Table 2 True effects

Scenario	Effect	
	τ_3^3	τ_3^2
1	0.04	0.006
2	0.044	−0.003
3	0.173	0.005

Although the results are promising, repeated sample simulations would be required to check the frequentist properties of the Bayesian credibility intervals. For a significant improvement of the proposed method, assumptions and model specification should be generalized to allow continuous outcomes and switching time (Table 2).

References

1. Frangakis, C.E., Rubin, D.B.: Principal stratification in causal inference. Biometrics **58**(1), 21–29 (2002)
2. Imbens, G.W., Rubin, D.B.: Causal Inference in Statistics, Social, and Biomedical Sciences. Cambridge University Press, Cambridge (2015)
3. Mealli, F., Pacini, B., Stanghellini, E.: Identification of principal causal effects using additional outcomes in concentration graphs. J. Educ. Behav. Stat. (2016). http://dx.doi.org/10.3102/1076998616646199
4. Stanghellini, E., Vantaggi, B., et al.: Identification of discrete concentration graph models with one hidden binary variable. Bernoulli **19**(5A), 1920–1937 (2013)
5. Tanner, M.A., Wong, W.H.: The calculation of posterior distributions by data augmentation. J. Am. Stat. Assoc. **82**(398), 528–540 (1987)
6. White, I.R., Walker, S., Babiker, A., Darbyshire, J.: strbee: randomization-based efficacy estimator. Stata J. **2**(2), 140–150 (2002)

A Bayesian Joint Dispersion Model with Flexible Links

Rui Martins

Abstract The objective is to jointly model longitudinal and survival data taking into account their interdependence in a real HIV/AIDS dataset inside a Bayesian framework. We propose a linear mixed effects dispersion model for the CD4 longitudinal counts with a between-individual heterogeneity in the mean and variance, relaxing the usual assumption of a common variance for the longitudinal residuals. A hazard regression model is considered in addition to model the time since HIV/AIDS diagnostic until failure, where the coefficients accounting for the linking between the longitudinal and survival processes are time-varying. This flexibility is specified using penalized Splines and allows the relationship to vary in time. Because residual heteroscedasticity may be related with the survival, the standard deviation is considered as a covariate in the hazard model thus enabling to study the effect of the CD4 counts' stability on the survival. The proposed framework outperforms the traditional joint models, highlighting the importance in correctly taking account the individual heterogeneity for the measurement errors variance.

Keywords Repeated measurements · Variance model · Time-to-event · Penalized Splines · Time-dependent coefficients

1 Introduction

A joint model is, in simple terms, an approach to the construction and analysis of the joint distribution of a set of correlated response variables, which may be of different types. In this work we are interested in simultaneously analyse longitudinal and survival data, taking advantage of the information that flows between these two data sources collected on the same patients. The incorporation of survival information into the longitudinal process it is somewhat equivalent to take into account the

R. Martins (✉)
Centro de Investigação Interdisciplinar Egas Moniz (CiiEM),
Escola Superior de Saúde Egas Moniz, Monte de Caparica, 2829-511
Caparica, Portugal
e-mail: ruimartins@egasmoniz.edu.pt

R. Argiento et al. (eds.), *Bayesian Statistics in Action*, Springer Proceedings
in Mathematics & Statistics 194, DOI 10.1007/978-3-319-54084-9_5

effect of an informative missing-data mechanism in order to assess the trends of the repeated measures. On the other hand, including adjusted longitudinal characteristics into the survival model improves the fit of the survival regression. It is this latter feature of joint models that will permit this work to reach its objectives. We intend to perform a survival analysis of the time-to-event since the diagnostic of HIV/AIDS in infected individuals using simultaneously the longitudinal information coming from the $CD4^{+}T$ lymphocyte counts (CD4 counts for short) among other covariates (*see* Sect. 4 for a brief description of the dataset).

In a joint analysis context the dynamics of the longitudinal repeated measures are usually postulated to belong to the class of the linear mixed-effects models with Gaussian errors [2, 11]. Generally it is assumed that the residual variance for the individual longitudinal trajectory is common to all subjects. Few papers on joint modelling have been published considering a model for this source of variation, which means assuming different individual residual variances (dispersion modelling). The work of Gao et al. [3] is an example, where the within-subject variability is assumed to have a log-normal prior distribution. McLain et al. [12] presents a dispersion model in a frequentist framework, being the variation patterns modelled as a particular function of other variables (*vide* Sect. 2). In our study, we resorted to this strategy with promising results (Sect. 4.2). Instead, some authors assume a latent class model for that variability [8] and others model the error process with a skew-*t* distribution [1].

One of the investigators' biggest concerns in this field is how and what characteristics should be shared by the longitudinal and survival processes. Rizopoulos et al. [13] discuss this problem by means of general families of parametrisations describing the main features of the relationship between the two processes. The most used approach is to share a set of individual random effects believed to be the basis of a latent relationship. Generally those effects are used to explain the differences between the population mean longitudinal response and the individual-specific ones in a mixed model representation. We will extend this vision proposing the individual-specific longitudinal standard deviation as a covariate for the hazard model.

The most popular choices for the baseline hazard function specifications are the parametric forms (e.g. Weibull - [4]) or the piecewise constant approximations [15]. In seeking the most flexibility as possible we will use an approach rooted on penalized cubic B-Spline functions (P-Splines for short).

The majority of the joint analysis assume that the relationship between the longitudinal and survival processes has the same strength over time, i.e. the coefficients of the survival model accounting for the effect of a particular longitudinal characteristic are time-invariant. But, if we believe, for example, that an initial significant relationship will become non-significant some time later (or vice versa), what we are assuming is that the effect of the shared longitudinal characteristics on survival varies with time. We address this aspect using time-dependent coefficients approximated by P-Splines. Despite having a moderately use in survival analysis [6, 7] time-varying coefficients are not a first option in joint models. Fortunately there are some exceptions. For instance Yu et al. [17] use Splines to model these coefficients in a frequentist context; Hanson et al. [5] rely on a linear combination of Gaussian kernels and Song et al. [14] use first order Taylor expansions of the function representing the time-varying coefficients.

2 Model Specification

Let us consider a study where an individual should come to an health centre periodically in order to perform some blood tests (longitudinal process) and we follow him until he experiences the event of interest (failure) or be censored (lost to follow-up or end of the study). The longitudinal process for the ith individual, $m_i(t), i = 1, \ldots, N$, is measured with error. So, the observed one is $y_i(t) = m_i(t) + e_i(t), t > 0$.

The ith vector of the n_i observed repeated measures is defined as $\mathbf{y}_i = (y_i(t_{i1}), \ldots, y_i(t_{in_i})) \equiv (y_{i1}, \ldots, y_{in_i})$ being $\mathbf{t}_i = (t_{i1}, \ldots, t_{in_i})$ a vector of fixed individual times. This encompasses the possibility of having N different measurement schedules and follow-up times. We take T_i to be the observed (possibly right censored) time to event for the ith individual and δ_i is the failure indicator. The observed data without covariates for the N independent subjects is $\mathcal{D} = \{\mathcal{D}_i\}_{i=1}^N = \{(\mathbf{y}_i, \mathbf{t}_i, T_i, \delta_i)\}_{i=1}^N$.

2.1 Longitudinal Mixed Dispersion Model

In longitudinal models, the common assumption of residual homoscedasticity may not always be reasonable. For example, considering our dataset (Sect. 4), the plot of the individual $\sqrt{\text{CD4}}$ values versus the standard deviation (Fig. 1) suggests considerable within-subject variance heterogeneity. A higher individual mean appears

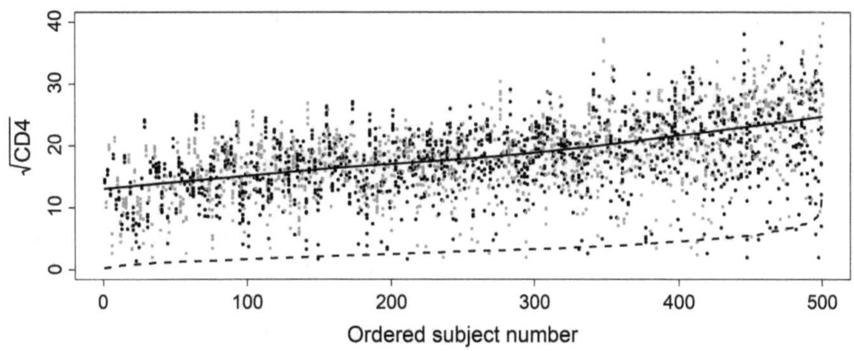

Fig. 1 Plot of the $\sqrt{\text{CD4}}$ values against the subject number ordered according to the standard deviation of their $\sqrt{\text{CD4}}$. The $\sqrt{\text{CD4}}$ values for one subject are plotted using the same colour and two adjacent subjects have different colours. The solid line represents the *lowess* smooth of the $\sqrt{\text{CD4}}$ values and the dashed one is simple the connection between the 500 standard deviations

to be associated with a higher variability. Few papers on joint modelling have been published considering a model for this source of variation (dispersion modelling). McLain et al. [12] presents a dispersion model in a frequentist framework but we can see it as an hierarchical model, easily implemented in Bayesian terms. Consider the longitudinal outcomes, y_{ij}'s, described by a mixed effects dispersion model,

$$y_{ij}|\boldsymbol{b}_i, \sigma_i^2 \sim \mathcal{N}\big(m_i(t_{ij}), \sigma_i^2\big), \quad j = 1, \ldots, n_i \tag{1}$$

$$m_i(t_{ij}) = \boldsymbol{\beta}_1^\top \mathbf{x}_{1i}(t_{ij}) + \boldsymbol{b}_{1i}^\top \mathbf{w}_{1i}(t_{ij}), \tag{2}$$

$$\sigma_i^2 = \sigma_0^2 \exp\{\, \boldsymbol{\beta}_2^\top \mathbf{x}_{2i}(t_{ij}) + \boldsymbol{b}_{2i}^\top \mathbf{w}_{2i}(t_{ij}) \,\}, \tag{3}$$

where $\mathbf{x}_{1i}$, $\mathbf{x}_{2i}$, $\mathbf{w}_{1i}$ and $\mathbf{w}_{2i}$ are appropriate subject-specific vectors of covariates (possibly time-dependent); $\boldsymbol{\beta}_1$ and $\boldsymbol{\beta}_2$ are vectors of population regression parameters; $(\boldsymbol{b}_{1i}^\top, \boldsymbol{b}_{2i}^\top) = \boldsymbol{b}_i|\Sigma \sim \mathcal{N}_p(\mathbf{0}, \Sigma)$ are time-independent subject-specific random effects capturing the inherent biological variability between individuals in the mean and variance, respectively.

Here the residual variance, σ_i^2, is assumed to be an individual property allowing for heterogeneity in the variance trends among the individuals. Modelling it and identify the covariates $\mathbf{x}_{2i}$ and $\mathbf{w}_{2i}$ influencing the within-subjects variance seems wise. Particularly, in many HIV/AIDS studies, where investigators are interested in understanding the trends of the variability, having an individual estimate of the subject-residual variance can be a plus in the assessment of whether individuals with different biomarker's stability have different prognosis. In addition, the specification in (3) can be considered as an extension to the dispersion model defined in Lin et al. [9] or the one defined in Gao et al. [3]. The former modelled the individual-specific measure of stability, σ_i^2, through an inverse gamma prior distribution, which is a special case of (3). The latter do not consider possible covariate effects. Indeed, if we consider that $\exp\{\, \boldsymbol{\beta}_2^\top \mathbf{x}_{2i}(t_{ij}) + \boldsymbol{b}_{2i}^\top \mathbf{w}_{2i}(t_{ij}) \,\} = 1$ in (3) we have $\sigma_i^2 = \sigma_0^2$ and we are back to the simple models where σ_i^2 accounts for the randomness in stability by using some prior distribution for this parameter. More details about the choose of this prior distribution are discussed in Sect. 4.1.

2.2 *Hazard Model with Time-Varying Coefficients*

Various approaches have been proposed to link the longitudinal and survival processes, namely sharing a set of individual random effects believed to be the basis of a latent relationship [13]. In this work, additionally to these random effects, we include the standard deviation, σ_i, as a covariate into the hazard model:

$$h_i(t \mid \boldsymbol{b}_i, \sigma_i) = h_0(t) \exp\{\, \boldsymbol{\beta}_3^\top \mathbf{x}_{3i} + \mathcal{C}_i\{\boldsymbol{b}_i, \sigma_i; \mathbf{g}(t)\} \,\} = h_0(t) \exp\{\psi_i(t)\}, \tag{4}$$

where $\mathbf{x}_{3i}$ is a subject-specific vector of baseline covariates and $\boldsymbol{\beta}_3$ is the respective population parameters vector. $\mathcal{C}_i\{.\}$ is a function specifying which components of the longitudinal process are directly related to $h_i(.)$. Finally, $\boldsymbol{g}(t)=(g_1(t),\ldots,g_L(t))$ is an appropriate vector of L smooth functions (approximated by P-Splines using a 1st order random-walk) representing the time-varying regression coefficients [6], which measure the effect of some characteristics of the longitudinal outcome to the hazard. These coefficients are particularly useful in explaining the effect of a time-independent covariate on survival when its impact is not constant throughout the time. Finally, the baseline hazard, $h_0(t)$, can have a parametric form (e.g. Weibull) or be specified using P-Splines or a piecewise constant function.

3 The Posterior Distribution

To form the contribution of the ith individual to the likelihood we assume: (i) a non-informative right censoring; (ii) events and censoring are independent of the process measurement schedule; (iii) longitudinal and survival processes are independent given the random effects, $\boldsymbol{b}_i$, and in addition (iv) the elements of the vector $\mathbf{y}_i$ are assumed independent given the same set of random effects. Thus, the posterior distribution, $\pi(\boldsymbol{\theta}|\mathcal{D})$, will be proportional to

$$\prod_{i=1}^{N}\{L_{1i}(\boldsymbol{\theta}\mid\mathcal{D}_i)\times L_{2i}(\boldsymbol{\theta}\mid\mathcal{D}_i)\}\times\pi(\boldsymbol{\theta})$$
$$=\prod_{i=1}^{N}\left\{\left[\prod_{j=1}^{n_i}f(y_{ij}|\boldsymbol{b}_i,\sigma_i^2)\right]\times\left[\,h(T_i|\boldsymbol{b}_i)^{\delta_i}\,S(T_i|\boldsymbol{b}_i)\,\right]\right\}\times\pi(\boldsymbol{\theta}), \tag{5}$$

where $\pi(\boldsymbol{\theta})$ denotes the prior distribution of the full parameters vector including the random effects, $\boldsymbol{b}_i$; $L_{1i}(\boldsymbol{\theta}|\mathcal{D}_i)$ is the ith individual longitudinal contribution, being $f(.)$ the Gaussian probability density function. $h(.)$ and $S(.)$ are, respectively, the hazard and the survival function for T_i. If one assumes that each T_i, $i=1,\ldots,N$, has a Weibull distribution, $\mathcal{W}(\rho,e^{\psi_i(t)})$, this implies the following ith individual survival contribution to the likelihood

$$L_{2i}(\boldsymbol{\theta}\mid\mathcal{D}_i)=\left\{\rho T_i^{\rho-1}e^{\psi_i(T_i)}\right\}^{\delta_i}\exp\left\{-e^{\psi_i(T_i)}\,T_i^{\rho}\right\}. \tag{6}$$

In case of a P-Spline approach to the reparametrized baseline hazard, $g_0(t)=\log[h_0(t)]$, we have

$$L_{2i}(\boldsymbol{\theta}\mid\mathcal{D}_i)=\left\{e^{g_0(T_i)+\psi_i(T_i)}\right\}^{\delta_i}\exp\left\{-\int_0^{T_i}e^{g_0(u)+\psi_i(u)}\,du\right\}. \tag{7}$$

Finally, if we define the baseline hazard to be a piecewise constant function over a finite partition of the time, $0=a_1,a_2,\ldots,a_K,a_{K+1}=\infty$, we have that $h_0(t)=$

$\sum_{k=1}^{K} \lambda_k \mathbb{1}_{[a_k, a_{k+1}[}(t)$, where $\boldsymbol{\lambda} = (\lambda_1, \ldots, \lambda_K)$ is a vector of positive but unknown parameters, each one representing the constant local hazard for $t \in [a_k, a_{k+1}[$, $k = 1, \ldots, K$, thus we can write

$$L_{2i}(\boldsymbol{\theta} \mid \mathcal{D}_i) = \left\{\lambda_{\mathring{k}} e^{\psi_i(T_i)}\right\}^{\delta_i} \exp\left\{-\left[\lambda_{\mathring{k}}(T_i - a_{\mathring{k}}) + \sum_{k=1}^{\mathring{k}-1} \lambda_k(a_{k+1} - a_k)\right] e^{\psi_i(T_i)}\right\}, \tag{8}$$

where $\mathring{k}$ is the largest integer such that $a_{\mathring{k}} \leq T_i$.

4 HIV/AIDS Data Analysis

The longitudinal and survival data were collected in a network of 88 laboratories located in every 27 states of Brazil during the years 2002–2006. CD4 counts and survival time were the responses collected in a random sample of $N = 500$ individuals. The explanatory variables were: age ($[15, 50[$, codded 0; ≥ 50, codded 1), sex (Female, codded 0; Male, codded 1) and PrevOI (previous opportunistic infection at study entry, codded 1; 0 otherwise). Survival time is the period between HIV/AIDS diagnosis and death, if happened before 31st December 2006. Otherwise it is censored. There were 34 deaths; 440 (88%) patients were between 15 and 49 years old; 298 (59.6%) were males of whom 23 died. 302 (60.4%) individuals had no previous infection. The initial median CD4 count was 269 cells/mm^3 (men - 250 cells/mm^3; women - 295 cells/mm^3) and patients made on average 5.51 CD4 exams resulting in a total of 2757 observations. The mean time of a censored patient in the study was 930 days and for those which had the event, that mean time was 862 days. For more about this dataset and for a modelling strategy which is similar in some aspects to the one implemented in Sect. 4.1, we refer the reader to [11] for comparisons.

4.1 *Fitted Models*

Considering the four assumptions refereed at the beginning of Sect. 3, Table 1 shows the form of the 33 joint models fitted to the data and the respective WAIC values (Widely Applicable Information Criterion [16]) for comparison. The adjusted longitudinal mean response, $m_i(t_{ij})$, will always have the representation

$$m_i(t_{ij}) = \beta_{10} + \beta_{11} t_{ij} + \beta_{12}\texttt{sex}_i + \beta_{13}\texttt{age}_i + \beta_{14}\texttt{PrevOI}_i + b_{1i,1} + b_{1i,2} t_{ij}. \tag{9}$$

The dispersion model (3) will assume several formulations. Namely,

$$\sigma_i^2 = \sigma_0^2 \exp\{\beta_{21}\texttt{sex}_i + \beta_{22}\texttt{age}_i + \beta_{23}\texttt{PrevOI}_i + b_{2i}\}; \quad \sigma_0^2 \sim \pi(\boldsymbol{\theta}_{\sigma_0^2}) \tag{10}$$

$$\sigma_i^2 = \sigma_0^2 \exp\{b_{2i}\}; \quad \sigma_0^2 \sim \pi(\boldsymbol{\theta}_{\sigma_0^2}) \tag{11}$$
$$\sigma_i^2 \sim \pi(\boldsymbol{\theta}_{\sigma_i^2}), \tag{12}$$
$$\sigma_i^2 = \sigma_0^2 \sim \pi(\boldsymbol{\theta}_{\sigma_0^2}), \quad i = 1, \ldots, N \tag{13}$$

where $\pi(.)$ is a suitable prior distribution. In the survival part we always consider

$$\boldsymbol{\beta}_3^\top \mathbf{x}_{3i} = \beta_{31}\texttt{sex}_i + \beta_{32}\texttt{age}_i + \beta_{33}\texttt{PrevOI}_i, \tag{14}$$

conjugated with four forms for the function $\mathcal{C}_i(.)$:

$$\mathcal{C}_i(.) = g_1(t)b_{1i,1} + g_2(t)b_{1i,2} + g_3(t)b_{2i}, \tag{15}$$
$$\mathcal{C}_i(.) = g_1(t)b_{1i,1} + g_2(t)b_{1i,2} + g_3(t)\sigma_i, \tag{16}$$
$$\mathcal{C}_i(.) = g_1(t)b_{1i,1} + g_2(t)b_{1i,2}, \tag{17}$$
$$\mathcal{C}_i(.) = g_1 b_{1i,1} + g_2 b_{1i,2}. \tag{18}$$

The P-Spline approach to the time-varying coefficients always considers 21 knots, corresponding to a knot every 3 months during the 5 years. In the case of a piecewise constant baseline-hazard, we used 20 subintervals with equal lengths, corresponding also to a length of 3 months each during the 5 years.

Assuming independence for all the elements in $\boldsymbol{\theta}$, model fitting were performed in WinBUGS. The following prior specifications were considered: all the elements of the vectors $\boldsymbol{\beta}_1$, $\boldsymbol{\beta}_2$ and $\boldsymbol{\beta}_3$ are independent of each other and Gaussian distributed, $\mathcal{N}(0, 100)$; $\Sigma^{-1} \sim \mathcal{Wish}(\text{diag}(100), 4)$; when needed $\lambda_k \sim \mathcal{G}(0.001, 0.001)$, $k = 1, \ldots, K$; for the common/individual residual variance, we considered $\log(\sigma_0) \sim \mathcal{U}(-100, 100)$ and $\log(\sigma_i) \sim \mathcal{U}(-100, 100)$, respectively.

4.2 *Results*

Based on the smallest WAIC value (Table 1), a P-Spline approach to the baseline hazard seems to be the best strategy. The chosen model ▲ assumes a heterogeneous within subject variability as in (12) considering $\log(\sigma_i) \sim \mathcal{U}(-100, 100)$. Additionally, because we use the linking structure (16), the patient-specific random intercept, slope and standard deviation are considered as covariates for the hazard model. The inclusion of the subject-specific variability as a covariate to explain the survival outcome improves the results (lower WAIC) compared to the models without this feature ($\mathcal{C}_i = (15)$ or $\mathcal{C}_i = (17)$). This is scientifically appealing, because the CD4's stability might contain extra information about health-related risks, complementing that provided by the mean-level trajectory. Although, for this particular case, our initial feeling – a possible effect of some set of covariates related to the individual heterogeneity – did not show up (models ◇; $\sigma_i^2 = (10)$).

Table 1 WAIC values for the 33 Bayesian joint dispersion models with flexible links

Longitudinal model		Hazard model			
m_i	σ_i^2	$\psi_i(t) = \boldsymbol{\beta}_3^\top \mathbf{x}_{3i} + \mathcal{C}_i$	h_0		
			Weibull	P-Spline	Piecewise
(9)	(10)	(14) + (15)	14671	12573 ◇	14317
(9)	(11)		14700	12848	14483
(9)	(10)	(14) + (16)	14307	12605 ◇	13365
(9)	(11)		14452	12917	13571
(9)	(12)		13134 △	**12104 ▲**	12921 △
(9)	(13)		13956 □	12887 □	13533 □
(9)	(10)	(14) + (17)	14811	13334 ◇	14463
(9)	(11)		14923	13688	14599
(9)	(12)		14314	13144	13968
(9)	(13)		14627 ○	13553 ○	14355 ○
(9)	(13)	(14) + (18)	16984 ⬠	15779 ⬟	16383 ⬠

The "traditional model" to jointly analyse longitudinal and survival data [2] considers: (i) a common residual variance as in (13), (ii) a set of two shared random effects ($b_{1i,1}$ and $b_{1i,2}$) believed to drive the relation between the two processes and (iii) the coefficients accounting for the effects of these latent variables on the hazard are time-independent, i.e., g_1 and g_2 do not depend on t, being the function $\mathcal{C}_i$ that specifies which components of the longitudinal process are shared with the survival process defined as in (18). In most of the literature, the baseline hazard has been tackled considering a Weibull form, but approaches considering splines or a piecewise-constant approximation also have its place [13]. Last row of Table 1 has been added so we can compare the performance of these "traditional" approaches (models ⬠) with our proposed models, namely △, □ and ○. Considering only the situations where h_0 is adjusted using P-Splines, we note that the traditional model ⬟ is the worst in terms of its WAIC value.

Table 2 shows the posterior mean estimates together with the 95% Credible Interval (CI) for a set of parameters for the best model ▲ compared to its counterparts for the traditional model ⬟. One can see that the variables `sex`, `age` and `PrevOI` are all significant in explaining the different longitudinal trajectories and survival times. Older males with opportunistic infections at the study entry have lower $\sqrt{\text{CD4}}$ values (negative estimates). Younger females without opportunistic infections at the study entry have a lower hazard of death (positive estimates). Model ▲ has been able to capture the effect of the `sex` on the survival outcome as opposed to the model ⬟. There is another important thing coming out of these estimates. The unique elements, σ_{11}^b, σ_{12}^b and σ_{22}^b, of the random-effects covariance matrix, Σ, for the chosen model, are lowered compared to its counterparts for the model ⬟, which means that we have been able to shrink more the individual random-effects under the presence of

Table 2 Posterior parameters estimates and 95% Credible Intervals (CI) for the selected model ▲ and for the traditional joint model ⬟ with a baseline adjusted with P-Splines

Parameter	Model ▲		Model ⬟	
	Mean	95% CI	Mean	95% CI
Longitudinal:				
Intercept (β_{10})	17.26	(16.59, 17.94)	17.3	(16.57, 18.02)
Time (β_{11})	1.74	(1.46, 2.04)	1.92	(1.66, 2.19)
sex (β_{12})	−0.66	(−1.42, −0.02)	−0.60	(−1.42, 0.21)
Age (β_{13})	−1.34	(−2.51, −0.41)	−1.59	(−2.58, −0.58)
PrevOI (β_{14})	−1.62	(−2.23, −0.99)	−1.80	(−2.77, −0.81)
σ_{11}^{b}	20.85	(17.83, 24.22)	24.58	(21.32, 28.24)
σ_{22}^{b}	4.71	(3.80, 5.78)	6.12	(5.03, 7.36)
σ_{12}^{b}	−3.07	(−4.46, −1.77)	−4.15	(−5.77, −2.67)
σ_0	–	–	2.59	(2.50, 2.67)
Survival:				
sex (β_{31})	0.74	(0.49, 1.01)	0.49	(−0.30, 1.29)
Age (β_{32})	0.99	(0.72, 1.27)	0.93	(0.04, 1.76)
PrevOI (β_{33})	1.04	(0.80, 1.28)	0.95	(0.19, 1.67)

a dispersion model for the individual-specific standard deviation than considering a common within-individual variance. This happens because the dispersion model is now explaining some of that additional variability. Some authors, namely Lyles et al. [10] have already noted this feature – random within-subject variances appears to describe a collection of longitudinal biomarker measurements well, as compared to a broad range of alternative models for the correlation structure.

5 Discussion

The three new aspects presented in this work seem to improve the traditional framework, *videlicet*: the time-dependent coefficients that account for the linking between the two processes, and estimated via P-Splines, allow us to understand the influence of the biomarker's values and its variability throughout the time in the survival-related outcome; the dispersion model is capturing some extra-variability in the individual repeated measures implying a reduction in the variability of the individual-specific random effects; the use of the individual level standard deviation as a covariate in the hazard model is scientifically appealing, highlighting that the biomarker's stability might contain extra information about health-related risks, complementing that provided by the mean-level trajectory. Unlike the random effects, $\boldsymbol{b}_i$, the biomarker variability, σ_i, has an intuitive meaning for the physicians and respective patients. In a joint model context, with the linking function (16), its interpretation is straightforward, because its contribution to understand the time-to-event can be readily quantified as a hazard ratio (HR). For instance, considering two subjects only differing in their repeated measures variability, everything else being equal for a specific time, the hazard ratio is $\text{HR} = \exp\{g_3(t)\}$, which is familiar to the clinical staff. Obviously, all these results need more studies to support them, but they seem to be encouraging. Beyond AIDS framework, there is an enormous potential to apply the methods developed in this work to other epidemiological situations.

Acknowledgements The author wish to thank Maria Goretti Fonseca and Cláudia Coeli for the database. This work was partially funded by Egas Moniz, Cooperativa de Ensino Superior.

References

1. Chen, J., Huang, Y.: A Bayesian mixture of semiparametric mixed-effects joint models for skewed-longitudinal and time-to-event data. Stat. Med. **34**(20), 2820–2843 (2015). doi:10.1002/sim.6517
2. Faucett, C.L., Thomas, D.C.: Simultaneously modelling censored survival data and repeatedly measured covariates: a gibbs sampling approach. Stat. Med. **15**(15), 1663–1685 (1996). doi:10.1002/(SICI)1097-0258(19960815)15:15<1663::AID-SIM294>3.0.CO;2-1
3. Gao, F., Miller, J., Xiong, C., Beiser, J., Gordon, M.: A joint-modeling approach to assess the impact of biomarker variability on the risk of developing clinical outcome. SMA **20**, 83–100 (2011). doi:10.1007/s10260-010-0150-z
4. Guo, X., Carlin, P.: Separate and joint modelling of longitudinal and event time data using standard computer packages. Am. Stat. **58**(1), 16–24 (2004)
5. Hanson, T., Branscum, A., Johnson, W.: Predictive comparison of joint longitudinal-survival modeling: a case study illustrating competing approaches. Lifetime Data Anal. **17**, 3–28 (2011). doi:10.1007/s10985-010-9162-0
6. Hennerfeind, A., Brezger, A., Fahrmeir, L.: Geoadditive survival models. JASA **101**, 1065–1075 (2006). doi:10.1198/016214506000000348
7. Hofner, B., Kneib, T., Hartl, W., Kuchenhoff, H.: Building cox-type structured hazard regression models with time-varying effects. Stat. Model. **11**(1), 3–24 (2011). doi:10.1177/1471082x1001100102
8. Jiang, B., Elliott, M.R., Sammel, M.D., Wang, N.: Joint modeling of cross-sectional health outcomes and longitudinal predictors via mixtures of means and variances. Biometrics **71**(2), 487–97 (2015). doi:10.1111/biom.12284
9. Lin, X., Raz, J., Harlow, S.: Linear mixed models with heterogeneous within-cluster variances. Biometrics **53**(3), 910–923 (1997). doi:10.2307/2533552
10. Lyles, R.H., Munõz, A., Xu, J., Taylor, J.M.G., Chmiel, J.S.: Adjusting for measurement error to assess health effects of variability in biomarkers. Stat. Med. **18**, 1069–1086 (1999). doi:10.1002/(SICI)1097-0258(19990515)18:9<1069::AID-SIM97>3.0.CO;2-7
11. Martins, R., Silva, G.L., Andreozzi, V.: Bayesian joint modeling of longitudinal and spatial survival AIDS data. Stat. Med. **35**(19), 3368–3384 (2016). doi:10.1002/sim.6937
12. McLain, A.C., Lum, K.J., Sundaram, R.: A joint mixed effects dispersion model for menstrual cycle length and time-to-pregnancy. Biometrics **68**(2), 648–656 (2012). doi:10.1111/j.1541-0420.2011.01711.x
13. Rizopoulos, D., Ghosh, P.: A bayesian semiparametric multivariate joint model for multiple longitudinal outcomes and a time-to-event. Stat. Med. **30**(12), 1366–1380 (2011). doi:10.1002/sim.4205

14. Song, X., Wang, C.: Semiparametric approaches for joint modeling of longitudinal and survival data with time-varying coefficients. Biometrics **64**(2), 557–566 (2008). doi:10.1111/j.1541-0420.2007.00890.x
15. Tang, N.S., Tang, A.M., Pan, D.D.: Semiparametric bayesian joint models of multivariate longitudinal and survival data. Comput. Stat. Data Anal. **77**, 113–129 (2014). doi:10.1016/j.csda.2014.02.015
16. Watanabe, S.: Asymptotic equivalence of Bayes cross validation and widely applicable information criterion in singular learning theory. J. Mach. Learn. Res. **11**(455), 3571–3591 (2010)
17. Yu, Z., Liu, L., Bravata, D.M., Williams, L.S.: Joint model of recurrent events and a terminal event with time-varying coefficients. Biom. J. **56**(2), 183–197 (2014). doi:10.1002/bimj.201200160

Local Posterior Concentration Rate for Multilevel Sparse Sequences

Eduard Belitser and Nurzhan Nurushev

Abstract We consider empirical Bayesian inference in the many normal means model in the situation when the high-dimensional mean vector is *multilevel sparse*, that is, most of the entries of the parameter vector are some fixed values. For instance, the traditional *sparse* signal is a particular case (with one level) of multilevel sparse sequences. We apply an empirical Bayesian approach, namely we put an appropriate prior modeling the multilevel sparsity and make data-dependent choices of certain parameters of the prior. We establish local (i.e., with rate depending on the "true" parameter) posterior contraction and estimation results. Global adaptive minimax results (for the estimation and posterior contraction problems) over sparsity classes follow from our local results if the sparsity level is of polynomial order. The results are illustrated by simulations.

Keywords Local posterior concentration rate · Multilevel sparse sequences · Empirical Bayesian approach

1 Introduction

Suppose we observe $X = X^{(n)} = (X_1, \dots, X_n) \in \mathbb{R}^n$, with

$$X_i = \theta_i + \xi_i, \quad i \in \mathbb{N}_n = \{1, \dots, n\}, \tag{1}$$

where $\theta = (\theta_1, \dots, \theta_n) \in \mathbb{R}^n$ is an unknown high-dimensional parameter of interest, the noise variables ξ_i's are independent standard Gaussian. In what follows, we let $n \geq 3$. The general goal is to make inference about θ based on the observed data X by using a Bayesian approach: in particular, recovery of the parameter θ and the

E. Belitser · N. Nurushev (✉)
Department of Mathematics, VU Amsterdam, Amsterdam, The Netherlands
e-mail: n.nurushev@vu.nl

E. Belitser
e-mail: e.n.belitser@vu.nl

R. Argiento et al. (eds.), *Bayesian Statistics in Action*, Springer Proceedings in Mathematics & Statistics 194, DOI 10.1007/978-3-319-54084-9_6

derivation of the local contraction rate of the empirical Bayes posterior. We consider mainly the non-asymptotic results, which imply asymptotic assertions if needed.

In this canonical high-dimensional problem, useful inference is clearly not possible without some structure on the parameter. One of the popular structural assumptions is *sparsity*. In this paper, we are concerned with a more generalized version of sparsity, namely, *multilevel sparsity*. The vector $\theta = (\theta_1, \ldots, \theta_n)$ is assumed to be a *multilevel sparse*, i.e., the large proportion of the entries of θ consist of some values $a_1, \ldots, a_m$. These values are known, but the proportions and the entries of θ at which these are taken are unknown. If $m = 1$ and $a_m = 0$, we obtain the traditional sparse signal.

One can extend the traditional sparsity class of *nearly black* vectors to multilevel sparsity class, but, to the best of our knowledge, multilevel sparsity structure is not considered in the literature and the minimax rate for this structure is not studied. For the traditional one-level ($m = 1$) sparsity structure, there is a variety of estimation methods and results are available in the literature: Donoho and Johnstone [7], Birgé and Massart [4], Johnstone and Silverman [9], Abramovich, Benjamini, Donoho and Johnstone [1], Abramovich, Grinshtein and Pensky [2], Castillo and van der Vaart [5], van der Pas, Kleijn and van der Vaart [10].

We pursue the novel local approach, namely, the posterior contraction (and estimation) rate $r^2(\theta)$ is allowed to be a function of θ, i.e., it is a local quantity. The local approach is more flexible than the global one; more on this is in Appendix 2. The point is that we do not need to impose any specific sparsity structure on θ, because the proposed local approach automatically exploits the "effective" sparsity of each underlying θ. For instance, if θ happens to lie in a sparsity class (say, $\ell_0([p_n]$ or $m_s[p_n]$, see Appendix 2) and the sparsity level p_n is of polynomial order, then the adaptive (global) minimax results (in fact, for the two problems: estimation and posterior contraction rate) over the sparsity class follow from the local results. In particular, our local results imply the same type of certain (global) minimax estimation results over sparsity classes as in Johnstone and Silverman [9], and the same type of global minimax (over sparsity classes) results on contraction posterior rates as in Castillo and van der Vaart [5].

The paper is organized as follows. In Sect. 2 we introduce the notation and describe the empirical Bayes procedure for multilevel sparse sequences. Section 3 contains the main results: the local (oracle) posterior contraction and estimation results for the constructed empirical Bayes posterior and the corresponding empirical Bayes posterior mean estimator, respectively, in terms of the local rate $r^2(\theta_0)$ uniformly over $\theta_0 \in \mathbb{R}^n$. If the sparsity level is of polynomial order, then the global adaptive minimax (over sparsity classes) results on contraction posterior and estimation rates follow as a consequence of our local results. The proofs, implications of the local results and the simulation study are presented in the three Appendix sections.

2 Preliminaries

First we introduce some notation, then construct an empirical Bayes posterior.

2.1 Notation

Denote the probability measure of X from the model (1) by $\mathrm{P}_\theta = \mathrm{P}_\theta^{(n)}$. For the notational simplicity we often skip the dependence on n of this quantity and many others. Denote by $1\{s \in S\} = 1_S(s)$ the indicator function of the set S, by $|S|$ the cardinality of the set S, the difference of sets $S\backslash S_0 = \{s \in S : s \notin S_0\}$, $\mathbb{N}_k = \{1, \dots, k\}$ for $k \in \mathbb{N} = \{1, 2, \dots\}$. For $I \subseteq \mathbb{N}_n$ define $I^c = \mathbb{N}_n \backslash I$. Denote by I_{a_k} be the index set of coordinates with a value a_k, by I be the set of index coordinates with values which are not equal to $a_1, \dots, a_m$, so that $\mathrm{I} = (I^{a_1}, \dots, I^{a_m}, I)$ forms the partition of $\mathbb{N}_n = \{1, \dots, n\}$. Without loss of generality, we assume that $a_1 = 0$. Let M_n^m be the family of all possible partitions I, except for the partitions with $I = \varnothing$. Then $|\mathrm{M}_n^m| = (m+1)^n - m^n$. Throughout $Z \sim \mathrm{N}(0, 1)$ will denote a generic standard normal random variable, with distribution function $\Phi(z) = \mathrm{P}(Z \le z)$ and density $\phi(z) = \Phi'(z)$. Let $\phi(x, \mu, \sigma^2)$ be the density of $\mu + \sigma Z \sim \mathrm{N}(\mu, \sigma^2)$ at point x. By convention, $\mathrm{N}(\mu, 0) = \delta_\mu$ denotes a Dirac measure at point μ.

2.2 Empirical Bayes Posterior

As we mentioned in the introduction, we deal with the classical high-dimensional normal model $X = (X_i,\, i \in \mathbb{N}_n) \sim \mathrm{P}_\theta = \bigotimes_{i=1}^n \mathrm{N}(\theta_i, 1)$, $\theta = (\theta_i,\, i \in \mathbb{N}_n) \in \mathbb{R}^n$. We would like to design a prior that models multilevel sparse sequences θ with m levels. Namely, there are $m+1$ groups in vector $\theta = (\theta_{I^{a_1}}, \dots, \theta_{I^{a_m}}, \theta_I)$, where $\theta_{I^{a_1}} = (\theta_i = a_1, i \in I^{a_1}), \dots, \theta_{I^{a_m}} = (\theta_i = a_m, i \in I^{a_m}), \theta_I = (\theta_i, i \in I)$. It is reasonable to impose a prior on θ given the partition $\mathrm{I} = (I^{a_1}, \dots, I^{a_m}, I)$ as follows:

$$\pi_{\mathrm{I}} = \bigotimes_{i=1}^n \mathrm{N}(\mu_i(\mathrm{I}), \tau_i^2(\mathrm{I})) = \Big[\bigotimes_{i \in I^{a_1}} \delta_{a_1}\Big] \times \cdots \times \Big[\bigotimes_{i \in I^{a_m}} \delta_{a_m}\Big] \times \Big[\bigotimes_{i \in I} \mathrm{N}(\mu_{m+1,i}, K)\Big], \tag{2}$$

where $\mu_i(\mathrm{I}) = \sum_{j=1}^m a_j 1\{i \in I^{a_j}\} + \mu_{m+1,i} 1\{i \in I\}$ and $\tau_i^2(\mathrm{I}) = K 1\{i \in I\}$, for some fixed $K > 0$. Next, we introduce the prior λ on $\mathrm{I} \in \mathrm{M}_n^m$ as follows: for $\kappa \ge 1$,

$$\lambda(\mathscr{I} = \mathrm{I}) = \lambda_{\mathrm{I}} = c_n \exp\Big\{-\kappa\big[|I| + \sum_{j=2}^m |I^{a_j}|\big] \log n\Big\}, \quad \mathrm{I} \in \mathrm{M}_n^m. \tag{3}$$

Since $\sum_{\mathrm{I} \in \mathrm{M}_n^m} \lambda_{\mathrm{I}} = 1$ and $|I| > 0$, the normalizing constant is $c_n = c_n(\kappa) = 1/\big[(1 + n^{-\kappa} + (m-1)n^{-\kappa} 1\{m \ge 2\})^n - (1 + (m-1)n^{-\kappa} 1\{m \ge 2\})^n\big]$. Putting a prior λ on M_n^m yields the resulting mixture prior for θ:

$$\pi = \sum_{\mathrm{I} \in \mathrm{M}_n^m} \lambda_{\mathrm{I}} \pi_{\mathrm{I}}, \tag{4}$$

where π_I is defined by (2). This in turn leads to the marginal distribution of X

$$\mathrm{P}_X = \sum_{\mathrm{I}\in\mathrm{M}_n^m} \lambda_\mathrm{I} \mathrm{P}_{X,\mathrm{I}}, \quad \mathrm{P}_{X,\mathrm{I}} = \bigotimes_{i=1}^n \mathrm{N}(\mu_i(\mathrm{I}), \tau_i^2(\mathrm{I})+1). \tag{5}$$

It remains to choose the parameters $\mu_{m+1,i}$ in the prior and we do this by using an empirical Bayes approach. The marginal likelihood P_X is readily maximized with respect to $\mu_{m+1,i}$: $\hat{\mu}_{m+1,i} = X_i$. Then we obtain the empirical Bayes posterior

$$\hat{\pi}(\theta|X) = \sum_{\mathrm{I}\in\mathrm{M}_n^m} \hat{\pi}(\theta, \mathscr{I}=\mathrm{I}|X) = \sum_{\mathrm{I}\in\mathrm{M}_n^m} \hat{\pi}(\theta|X, \mathscr{I}=\mathrm{I})\hat{\pi}(\mathscr{I}=\mathrm{I}|X), \tag{6}$$

where the empirical Bayes conditional posterior (recall that $\mathrm{N}(0,0)=\delta_0$)

$$\hat{\pi}(\theta|X, \mathscr{I}=\mathrm{I}) = \hat{\pi}_\mathrm{I}(\theta|X) = \Big[\bigotimes_{i\in I^{a_1}} \delta_{a_1}\Big] \times \cdots \times \Big[\bigotimes_{i\in I^{a_m}} \delta_{a_m}\Big] \times \Big[\bigotimes_{i\in I} \mathrm{N}\Big(X_i, \tfrac{K}{K+1}\Big)\Big] \tag{7}$$

and the empirical Bayes posterior for $\mathrm{I}\in\mathrm{M}_n^m$

$$\hat{\pi}(\mathscr{I}=\mathrm{I}|X) = \frac{\lambda_\mathrm{I} \bigotimes_{i=1}^n \phi(X_i, \sum_{j=1}^m a_j 1\{i\in I^{a_j}\} + X_i 1\{i\in I\}, \tau_i^2(\mathrm{I})+1)}{\sum_{\mathrm{J}\in\mathrm{M}_n^m} \lambda_\mathrm{J} \bigotimes_{i=1}^n \phi(X_i, \sum_{j=1}^m a_j 1\{i\in J^{a_j}\} + X_i 1\{i\in J\}, \tau_i^2(\mathrm{J})+1)}. \tag{8}$$

Denoting $\hat{\mu}(\mathrm{I}) = \sum_{j=1}^m a_j 1\{i\in I^{a_j}\} + X_i 1\{i\in I\}$, $i\in\mathbb{N}_n$, we get an estimator based on $\hat{\pi}(\cdot|X)$, namely,

$$\hat{\theta} = \hat{\theta}(\mathrm{I}) = \hat{\mathrm{E}}(\theta|X) = \sum_{\mathrm{I}\in\mathrm{M}_n^m} \hat{\mu}(\mathrm{I})\hat{\pi}(\mathscr{I}=\mathrm{I}|X), \tag{9}$$

which is nothing else but the *empirical Bayes mean*, with respect to the empirical Bayes posterior $\hat{\pi}(\theta|X)$ defined by (6).

3 Main Results

In this section we introduce the local contraction rate for the empirical Bayes posterior $\hat{\pi}(\cdot|X)$. Notice that $\hat{\pi}(\cdot|X)$ is a random mixture over $\hat{\pi}_\mathrm{I}(\cdot|X)$, $\mathrm{I}\in\mathrm{M}_n^m$. From the P_{θ_0}-perspective, each $\hat{\pi}_\mathrm{I}(\cdot|X)$ contracts to the true parameter θ_0 with the local rate

$$\begin{aligned}\mathrm{R}^2(\mathrm{I},\theta_0) &= \mathrm{R}^2(\mathrm{I},\theta_0,n,a_2,\ldots,a_m)\\ &= \sum_{i\in I^{a_1}}\theta_{0,i}^2+\sum_{j=2}^m\sum_{i\in I^{a_j}}(\theta_{0,i}-a_j)^2+|I|,\quad \mathrm{I}\in\mathrm{M}_n^m.\end{aligned}\tag{10}$$

Indeed, since $\hat{\mu}(\mathrm{I})=\sum_{j=1}^m a_j 1\{i\in I^{a_j}\}+X_i 1\{i\in I\}$, $i\in\mathbb{N}_n$, the Markov inequality yields

$$\begin{aligned}\mathrm{E}_{\theta_0}\hat{\pi}_{\mathrm{I}}(\|\theta-\theta_0\|^2\geq M^2\mathrm{R}^2(\mathrm{I},\theta_0)|X) &\leq \frac{\mathrm{E}_{\theta_0}\big(\|\hat{\mu}(\mathrm{I})-\theta_0\|^2+\frac{K|I|}{K+1}\big)}{M^2\mathrm{R}^2(\mathrm{I},\theta_0)}\\ &= \frac{(1+\frac{K}{K+1})|I|+\sum_{i\in I^{a_1}}\theta_{0,i}^2+\sum_{j=2}^m\sum_{i\in I^{a_j}}(\theta_{0,i}-a_j)^2}{M^2\mathrm{R}^2(\mathrm{I},\theta_0)}\leq\frac{2}{M^2}.\end{aligned}$$

For each $\theta_0\in\mathbb{R}^n$, there exists the best choice $\mathrm{I}_o=\mathrm{I}_o(\theta_0,a_2,\ldots,a_m)$ of the partition $\mathrm{I}\in\mathrm{M}_n^m$ corresponding to the fastest local rate over the family of local rates $\mathrm{R}^2(\mathrm{M}_n^m)=\{\mathrm{R}^2(\mathrm{I},\theta_0),\mathrm{I}\in\mathrm{M}_n^m\}$: with $R^2(\mathrm{I},\theta_0)$ defined by (10),

$$\mathrm{R}^2(\theta_0)=\min_{\mathrm{I}\in\mathrm{M}_n^m}\mathrm{R}^2(\mathrm{I},\theta_0)=\sum_{i\in I_o^{a_1}}\theta_{0,i}^2+\sum_{j=2}^m\sum_{i\in I_o^{a_j}}(\theta_{0,i}-a_j)^2+|I_o|.\tag{11}$$

Ideally, we would like to have the quantity $\mathrm{R}^2(\theta_0)$ defined by (11) as the benchmark for the contraction rate of the empirical Bayes posterior $\hat{\pi}(\cdot|X)$ defined by (6). However, this turned out to be impossible, which is also confirmed by following estimation result of Donoho and Johnstone [6] for *sparse* signals:

$$\liminf_{n\to\infty}\frac{1}{\log n}\inf_{\hat{\theta}}\sup_{\theta\in\mathbb{R}_n}\left[\frac{\mathrm{E}_{\theta_0}\|\theta-\hat{\theta}\|^2}{1+\sum_{i=1}^n\min\{\theta_{0,i}^2,1\}}\right]\geq 2,$$

where the infimum is taken over all estimators, measurable functions of X. This shows that a reasonable benchmark for the contraction rate must contain a logarithmic factor, as was also shown by Birgé and Massart [4] for the estimation problem.

For a parameter $s>1$, introduce the family of the so called s-*local rates* with a logarithmic factor:

$$\begin{aligned}r_m^2(\mathrm{I},\theta_0) &= r_m^2(\mathrm{I},\theta_0,s,n,a_2,\ldots,a_m)\\ &= \sum_{i\in I^{a_1}}\theta_{0,i}^2+\sum_{j=2}^m\Big(\sum_{i\in I^{a_j}}(\theta_{0,i}-a_j)^2+s^{-1}|I^{a_j}|\log n\Big)+|I|\log n,\end{aligned}$$

where

$$a_k^2\leq K\log n,\quad k=1,\ldots,m,\quad for\ \ some\ \ K>0.\tag{12}$$

There exists the best choice $\mathrm{I}_o = \mathrm{I}_o(\theta_0, s, n, a_2, \ldots, a_m)$ of the partition $\mathrm{I} \in \mathrm{M}_n^m$ such that

$$r_m^2(\theta_0) = r_m^2(\mathrm{I}_o, \theta_0) = \min_{\mathrm{I} \in \mathrm{M}_n^m} r_m^2(\mathrm{I}, \theta_0)$$
$$= \sum_{i \in I_o^{a_1}} \theta_{0,i}^2 + \sum_{j=2}^{m} \Big(\sum_{i \in I_o^{a_j}} (\theta_{0,i} - a_j)^2 + s^{-1} |I_o^{a_j}| \log n \Big) + |I_o| \log n. \quad (13)$$

We call the quantity $r_m^2(\theta_0)$ and the choice I_o *oracle rate* and *oracle partition*, respectively. If $m = 1$, then the quantity $r_1^2(\theta_0)$ is nothing else but the oracle rate for *sparse* signals considered by Belitser and Nurushev [3]. It is easy to see that $r_m^2(\theta_0) \le r_1^2(\theta_0)$ for all $m > 1$. Indeed, this follows immediately, since $r_m^2(\theta_0) \le \min_{I \subseteq \mathbb{N}_n} r_m^2(\mathrm{I}_*(I), \theta_0) = \min_{\mathrm{I} \in \mathrm{M}_n^1} r_1^2(\mathrm{I}, \theta_0)$, where $\mathrm{I}_*(I) = (I^c, \varnothing, \cdots, \varnothing, I)$.

The following theorem establishes that the empirical Bayes posterior $\hat{\pi}(\theta|X)$ contracts to θ_0 with the oracle rate $r_m^2(\theta_0)$ from the P_{θ_0}-perspective, and the empirical Bayes posterior mean $\hat{\theta}$, defined by (9), converges to $\hat{\theta}_0$ with the oracle rate $r_m^2(\theta_0)$, uniformly over the entire parameter space.

Theorem 1 *Let the relation* (12) *be fulfilled and let the empirical Bayes posterior* $\hat{\pi}(\theta|X)$ *and the oracle rate* $r_m^2(\theta_0)$ *be defined by (6) and (13), respectively. Then there exist a constant* $C_{or} = C_{or}(K, \kappa, m, s) > 0$ *such that for any* $\theta_0 \in \mathbb{R}^n$ *and* $M > 0$,

$$\mathrm{E}_{\theta_0} \hat{\pi}\big(\|\theta - \theta_0\|^2 \ge M^2 r_m^2(\theta_0) | X\big) \le \frac{C_{or}}{M^2}, \quad \text{(i)}$$

$$\mathrm{E}_{\theta_0} \|\hat{\theta} - \theta_0\|^2 \le C_{or} r_m^2(\theta_0). \quad \text{(ii)}$$

The oracle interpretation of this result is as follows. A family of priors $\{\pi_{\mathrm{I}}, \mathrm{I} \in \mathrm{M}_n^m\}$ leads to the family of empirical Bayes posteriors $\{\hat{\pi}_{\mathrm{I}}(\theta|X), \mathrm{I} \in \mathrm{M}_n^m\}$. The best choice $\hat{\pi}_{\mathrm{I}_o}(\theta|X)$ (with the fastest (oracle) concentration rate $r_m^2(\theta_0)$) is not available to the observer, it can only be picked by the *oracle* who knows the true θ_0. We propose the mixture prior $\hat{\pi}(\theta|X)$ which does not use any knowledge of the oracle I_o. The above theorem says basically that the proposed empirical Bayes posterior $\hat{\pi}(\theta|X)$ *mimics the oracle* in the posterior contraction and estimation problems, i.e., $\hat{\pi}(\theta|X)$ performs as good as the oracle choice $\hat{\pi}_{\mathrm{I}_o}(\theta|X)$.

Remark 1 Notice that we can make the oracle rate (13) smaller by choosing bigger values of the parameter s, but then the resulting constants C_{or} and C_{est} (which depend on s) will become bigger.

Appendix 1

Proofs

We provide a couple of technical lemmas used in the proof of the main result. For a $\tau_0 > 0$ and $\theta_0 \in \mathbb{R}^n$, introduce the families of sets:

$$\mathcal{O}(\tau_0) = \mathcal{O}(\tau_0, \theta_0) = \big\{\mathrm{I} \in \mathrm{M}_n^m : r^2(\mathrm{I}, \theta_0) \le \tau_0 r^2(\theta_0)\big\}, \tag{14}$$
$$\mathcal{O}^c(\tau_0) = \mathcal{O}^c(\tau_0, \theta_0) = \big\{\mathrm{I} \in \mathrm{M}_n^m : r^2(\mathrm{I}, \theta_0) > \tau_0 r^2(\theta_0)\big\}, \tag{15}$$

where the oracle partition $\mathrm{I}_o = \mathrm{I}_o(\theta_0)$ is given by (13). The families $\mathcal{O}(\tau_0)$ and $\mathcal{O}^c(\tau_0)$ form a partition of M_n^m, as they do not intersect and $\mathrm{M}_n^m = \mathcal{O}(\tau_0) \cup \mathcal{O}^c(\tau_0)$. Denote for brevity $\hat{\pi}_\mathrm{I} = \hat{\pi}(\mathscr{I} = \mathrm{I}|X)$, where $\hat{\pi}(\mathscr{I} = \mathrm{I}|X)$ is defined by (8).

Lemma 1 *Let the measure $\hat{\pi}_\mathrm{I}$ be defined by (8), the oracle rate $r^2(\mathrm{I}_o, \theta)$ be defined by (13), $K > 0$, $s > 1$ and $\kappa > \max\big\{\frac{5}{9}\log 10, 0.9K\big\} + 2.49$. Then there exist positive constants $c_1 = c_1(\kappa) > 2$, c_2 and $c_3 = c_3(K, \kappa, s)$ such that*

$$\mathrm{E}_{\theta_0}\hat{\pi}_\mathrm{I} \le n^{-c_1(|I|+\sum_{j=2}^m |I^{a_j}|)} \exp\Big\{ - c_2\big(r^2(\mathrm{I}, \theta_0) - c_3 r^2(\theta_0)\big)\Big\}.$$

Proof Since $\hat{\pi}(X_1, \ldots, X_n|\mathscr{I} = \mathrm{I})$ is the product of distributions $\mathrm{N}\big(\hat{\mu}_i(\mathrm{I}), \tau_i^2(\mathrm{I}) + 1\big)$, $i = 1, \ldots, n$, where $\hat{\mu}_i(\mathrm{I}) = \sum_{j=1}^m a_j 1\{i \in I^{a_j}\} + X_i 1\{i \in I\}$, $i \in \mathbb{N}_n$ and $\tau_i^2(\mathrm{I}) = K1\{i \in I\}$, with densities $\phi\big(X_i, \hat{\mu}_i(\mathrm{I}), \tau_i^2(\mathrm{I}) + 1\big)$, we compute for any $h \in [0, 1)$ and $\mathrm{I}_o \in \mathrm{M}_n^m$,

$$\begin{aligned}
\mathrm{E}_{\theta_0}\hat{\pi}_\mathrm{I} &= \mathrm{E}_{\theta_0}\frac{\lambda_\mathrm{I} \otimes_{i=1}^n \phi\big(X_i, \hat{\mu}_i(\mathrm{I}), \tau_i^2(\mathrm{I})+1\big)}{\sum_{\mathrm{J}\in\mathrm{M}_n^m} \lambda_\mathrm{J} \otimes_{i=1}^n \phi\big(X_i, \hat{\mu}_i(\mathrm{J}), \tau_i^2(\mathrm{J})+1\big)} \le \mathrm{E}_{\theta_0}\Bigg[\frac{\lambda_\mathrm{I} \otimes_{i=1}^n \phi\big(X_i, \hat{\mu}_i(\mathrm{I}), \tau_i^2(\mathrm{I})+1\big)}{\lambda_{\mathrm{I}_o} \otimes_{i=1}^n \phi\big(X_i, \hat{\mu}_i(\mathrm{I}_o), \tau_i^2(\mathrm{I}_o)+1\big)}\Bigg]^h \\
&= \mathrm{E}_{\theta_0}\Big[\frac{\lambda_\mathrm{I}}{\lambda_{\mathrm{I}_o}}\Big]^h \exp\Big\{ - \sum_{j=1}^m \sum_{i\in I^{a_j}} \frac{h(X_i - a_j)^2}{2} + \sum_{k=1}^m \sum_{i\in I_o^{a_k}} \frac{h(X_i - a_k)^2}{2} + \frac{h(|I_o| - |I|)\log(K+1)}{2}\Big\} \\
&\le \mathrm{E}_{\theta_0}\Big[\frac{\lambda_\mathrm{I}}{\lambda_{\mathrm{I}_o}}\Big]^h \exp\Big\{ - \sum_{j=1}^m \sum_{i\in I^{a_j}\setminus\cup_{k=1}^m I_o^{a_k}} \frac{h(X_i - a_j)^2}{2} + \sum_{k=1}^m \sum_{i\in I_o^{a_k}\setminus\cup_{j=1}^m I^{a_j}} \frac{h(X_i - a_k)^2}{2}\Big\} \\
&\quad \times \exp\Big\{ \sum_{j,k=1, j\ne k}^m \sum_{i\in I_o^{a_k}\cap I^{a_j}} \frac{h(2X_i a_j - 2X_i a_k + a_k^2 - a_j^2)}{2} + \frac{h|I_o|\log(K+1)}{2}\Big\}.
\end{aligned} \tag{16}$$

Recall the elementary identity for $Y \sim \mathrm{N}(\mu, \sigma^2)$, $a, d \in \mathbb{R}$ and $b > -\sigma^2$:

$$\mathrm{E}[\exp\{-b(Y-a)^2/2\}] = \exp\left\{-\frac{(\mu-a)^2 b}{2(1+b\sigma^2)} - \frac{1}{2}\log(1+b\sigma^2)\right\}, \tag{17}$$

$$\mathrm{E}[\exp\{dY\}] = \exp\left\{\frac{d(d\sigma^2+2\mu)}{2}\right\}. \tag{18}$$

Now take $h = 0.9$ in (16). By using (17), we derive

$$\mathrm{E}_{\theta_0}\hat{\pi}_{\mathrm{I}} \leq \Big[\frac{\lambda_{\mathrm{I}}}{\lambda_{\mathrm{I}_o}}\Big]^{0.9} \exp\Big\{-\tfrac{9}{38}\sum_{j=1}^m \sum_{i\in I^{a_j}\setminus\cup_{k=1}^m I_o^{a_k}} (\theta_i - a_j)^2 + 0.45|I_o|\log(K+1)\Big\}$$
$$\times \exp\Big\{4.5\sum_{k=1}^m \sum_{i\in I_o^{a_k}\setminus\cup_{j=1}^m I^{a_j}} (\theta_i - a_k)^2 + 0.5|I|\log 10\Big\}\mathrm{E}_{\theta_0} e^{T(X)}, \tag{19}$$

where $T(X) = 0.45\sum_{j,k=1,j\neq k}^m \sum_{i\in I_o^{a_k}\cap I^{a_j}} (2X_i(a_j-a_k) + a_k^2 - a_j^2)$.

By using the relations (12) and (18), we obtain

$$\mathrm{E}_{\theta_0} e^{T(X)} = \mathrm{E}_{\theta_0}\exp\Big\{0.45\sum_{j,k=1,j\neq k}^m \sum_{i\in I_o^{a_k}\cap I^{a_j}} \big(2X_i(a_j-a_k)+a_k^2-a_j^2\big)\Big\}$$
$$= \exp\Big\{0.45\sum_{j,k=1,j\neq k}^m \sum_{i\in I_o^{a_k}\cap I^{a_j}} \big((\theta_i-a_k)^2-(\theta_i-a_j)^2+0.9(a_k-a_j)^2\big)\Big\}$$
$$\leq \exp\Big\{\sum_{j,k=1}^m \sum_{i\in I_o^{a_k}\cap I^{a_j}} \big(4.5(\theta_i-a_k)^2 - \tfrac{9}{38}(\theta_i-a_j)^2\big)\Big\}$$
$$\times \exp\Big\{0.81K\Big(\sum_{k=2}^m |I_o^{a_k}| + \sum_{j=2}^m |I^{a_j}|\Big)\log n\Big\}. \tag{20}$$

Denote the constants $c_0 = \max\{0.9K, \frac{5}{9}\log 10\} + \frac{5}{19} < \max\{0.9K, \frac{5}{9}\log 10\} + 2.49 < \kappa$ and $c_1 = 0.9(\kappa - c_0) > 2$. The definition (3) of λ_{I}, $n \geq 3$ and $s > 1$ entail that

$$\Big[\frac{\lambda_{\mathrm{I}}}{\lambda_{\mathrm{I}_o}}\Big]^{0.9} \exp\Big\{0.5|I|\log 10 + 0.81K\Big(\sum_{j=2}^m |I^{a_j}|\Big)\log n\Big\}$$
$$\leq \exp\Big\{-\big[(c_1+\tfrac{9}{38})|I| + (c_1+\tfrac{9}{38s})\Big(\sum_{j=2}^m |I^{a_j}|\Big)\big]\log n + 0.9\kappa\Big(|I_o| + \tfrac{1}{s}\sum_{k=2}^m |I_o^{a_k}|\Big)\log n\Big\}.$$

Using the relations (19), (20) and the last inequality, we derive that

$$\mathrm{E}_{\theta_0}\hat{\pi}_{\mathrm{I}} \le n^{-c_1(|I|+\sum_{j=2}^m |I^{a_j}|)} \exp\Big\{-\tfrac{9}{38}\Big(\sum_{j=1}^m \sum_{i\in I^{a_j}} (\theta_i - a_j)^2 + (|I| + \tfrac{1}{s}\sum_{j=2}^m |I^{a_j}|)\log n\Big)\Big\}$$
$$\times \exp\Big\{C\Big(\sum_{k=1}^m \sum_{i\in I_o^{a_k}} (\theta_i - a_k)^2 + (|I_o| + \tfrac{1}{s}\sum_{k=2}^m |I_o^{a_k}|)\log n\Big\},$$

where $C = C(K, \kappa, s) = s\max\{4.5, 0.9\kappa + 0.81K\}$. This completes the proof, with the constants $c_2 = \frac{9}{38}$ and $c_3 = c_3(K, \kappa, s) = \frac{38s}{9}\max\{4.5, 0.9\kappa + 0.81K\}$.

Lemma 2 *Let $\theta_0 \in \mathbb{R}^n$ and let $\hat{\theta}(\mathrm{I})$, $\mathrm{I} \in \mathrm{M}_n^m$, be defined by (9), and the set $\mathscr{O}(\tau_0)$ by (14). Then*

$$\mathrm{E}_{\theta_0}\Big[\sum_{\mathrm{I}\in\mathscr{O}(\tau_0)} \|\hat{\theta}(\mathrm{I}) - \theta_0\|^2 \hat{\pi}_{\mathrm{I}}\Big] \le 6\tau_0 r^2(\theta_0).$$

Proof Recall that $\xi_i = (X_i - \theta_0) \overset{ind}{\sim} \mathrm{N}(0,1)$, $i \in \mathbb{N}_n$, under $X \sim \mathrm{P}_{\theta_0}$. Write

$$\mathrm{E}_{\theta_0}\Big[\sum_{\mathrm{I}\in\mathscr{O}(\tau_0)} \|\hat{\theta}(I) - \theta_0\|^2 \hat{\pi}_{\mathrm{I}}\Big] = \mathrm{E}_{\theta_0}\Big[\sum_{\mathrm{I}\in\mathscr{O}(\tau_0)} \Big(\sum_{i\in I}\xi_i^2 + \sum_{j=1}^m \sum_{i\in I^{a_j}} (\theta_{0,i} - a_j)^2\Big)\hat{\pi}_{\mathrm{I}}\Big]$$
$$\le \mathrm{E}_{\theta_0}\Big[\sum_{\mathrm{I}\in\mathscr{O}(\tau_0)} \Big(\sum_{i\in I}\xi_i^2\Big)\hat{\pi}_{\mathrm{I}}\Big] + \tau_0 r^2(\theta_0). \tag{21}$$

It is known fact that

$$\exp\{t\mathrm{E}[\max_{1\le i\le n}\xi_i^2]\} \le \mathrm{E}\exp\{t\max_{1\le i\le n}\xi_i^2\} \le \sum_{i=1}^n \mathrm{E}\exp\{t\xi_i^2\} = \frac{n}{\sqrt{1-2t}}.$$

Here we used Jensen's inequality. Therefore $\mathrm{E}[\max_{1\le i\le n}\xi_i^2] \le \frac{\log n}{t} - \frac{\log(1-2t)}{2t}$. Taking $t = \frac{2}{5}$, we derive that for any $n \ge 3$

$$\mathrm{E}[\max_{1\le i\le n}\xi_i^2] \le \frac{5\log n}{2} + \frac{5\log 5}{4} \le \Big(\frac{5}{2} + \frac{5\log 5}{4\log 3}\Big)\log n < 5\log n. \tag{22}$$

Since $\mathrm{I} \in \mathscr{O}(\tau_0)$, it is not difficult to see that $|I| \le \frac{\tau_0 r^2(\theta_0)}{\log n}$. Applying this and (22), we obtain

$$\mathrm{E}_{\theta_0}\Big[\sum_{\mathrm{I}\in\mathscr{O}(\tau_0)} \Big(\sum_{i\in I}\xi_i^2\Big)\hat{\pi}_{\mathrm{I}}\Big] \le \mathrm{E}_{\theta_0}\Big[\max_{1\le i\le n}\xi_i^2 \sum_{\mathrm{I}\in\mathscr{O}(\tau_0)} \Big(|I|\hat{\pi}_{\mathrm{I}}\Big)\Big]$$
$$\le \frac{\tau_0 r^2(\theta_0)}{\log n}\mathrm{E}_{\theta_0}\Big[\max_{1\le i\le n}\xi_i^2\Big] \le 5\tau_0 r^2(\theta_0). \tag{23}$$

Combining the last relation with (21) completes the proof of the lemma.

Now we are ready to prove the main result, Theorem 1.

Proof Let $\hat{\mathrm{E}}$ and $\widehat{\mathrm{var}}$ denote the (random) expectation and variance with respect to $\hat{\pi}(\theta|X, \mathscr{I} = I)$ given by (7). Then from (7), it follows that

$$\hat{\mathrm{E}}\big(\|\theta-\theta_0\|^2|X,\mathscr{I}=I\big) = \sum_{i\in\mathbb{N}_n}\widehat{\mathrm{var}}(\theta_i|X,\mathscr{I}=I)+\sum_{i\in\mathbb{N}_n}\big(\hat{\mathrm{E}}(\theta_i|X,\mathscr{I}=I)-\theta_{0,i}\big)^2$$
$$=\frac{K|I|}{K+1}+\sum_{i\in I}\xi_i^2+\sum_{j=1}^m\sum_{i\in I^{a_j}}(\theta_{0,i}-a_j)^2\le r^2(I,\theta_0)+\sum_{i\in I}\xi_i^2,$$

where $\xi_i=(X_i-\theta_{0,i})\sim\mathrm{N}(0,1)$.

The last relation and the Markov inequality imply that

$$\mathrm{E}_{\theta_0}\hat{\pi}\big(\|\theta-\theta_0\|\ge Mr(\theta_0)|X\big)=\mathrm{E}_{\theta_0}\sum_{\mathrm{I}\in\mathrm{M}_n^m}\hat{\pi}\big(\|\theta-\theta_0\|\ge Mr(\theta_0)|X,\mathscr{I}=\mathrm{I}\big)\hat{\pi}_{\mathrm{I}}$$
$$\le\mathrm{E}_{\theta_0}\sum_{\mathrm{I}\in\mathrm{M}_n^m}\frac{\hat{\mathrm{E}}\big(\|\theta-\theta_0\|^2|X,\mathscr{I}=\mathrm{I}\big)}{M^2r^2(\theta_0)}\hat{\pi}_{\mathrm{I}}$$
$$\le\frac{\sum_{\mathrm{I}\in\mathrm{M}_n^m}r^2(\mathrm{I},\theta_0)\mathrm{E}_{\theta_0}\hat{\pi}_{\mathrm{I}}}{M^2r^2(\theta_0)}+\frac{\mathrm{E}_{\theta_0}\Big[\sum_{\mathrm{I}\in\mathrm{M}_n^m}\big(\sum_{i\in I}\xi_i^2\big)\hat{\pi}_{\mathrm{I}}\Big]}{M^2r^2(\theta_0)}. \tag{24}$$

Let the sets $\mathscr{O}(\tau_0)$ and $\mathscr{O}^c(\tau_0)$ be defined by (14) and (15), respectively. Let τ_0 be chosen in such a way that $\tau_0>c_3=(38s/9)\max\{4.5,0.9\kappa+0.81K\}$ is defined in the proof of Lemma 1 and $\kappa>\max\big\{0.9K,\frac{5}{9}\log 10\big\}+2.49$. For $\mathrm{I}\in\mathscr{O}^c(\tau_0)$, we evaluate

$$r^2(\mathrm{I},\theta_0)-c_3r^2(\theta_0)\ge\big(1-\tfrac{c_3}{\tau_0}\big)r^2(\mathrm{I},\theta_0). \tag{25}$$

Denote $B=B(K,\kappa,s,\tau_0)=\frac{c_2(\tau_0-c_3)}{2\tau_0}=\frac{9(\tau_0-c_3)}{76\tau_0}$, where $c_2=\frac{9}{38}$ is defined in the proof of Lemma 1. Using Lemma 1, (25) and the facts that $\max_{x\ge0}\{xe^{-cx}\}\le(ce)^{-1}$ (for any $c>0$) and $(1+mn^{-c_1/2})^n\le e^m$, we obtain that

$$\sum_{\mathrm{I}\in\mathscr{O}^c(\tau_0)}r^2(\mathrm{I},\theta_0)\big[\mathrm{E}_{\theta_0}\hat{\pi}_I\big]^{\frac12}$$
$$\le\sum_{\mathrm{I}\in\mathscr{O}^c(\tau_0)}r^2(\mathrm{I},\theta_0)n^{-c_1(|I|+\sum_{j=2}^m|I^{a_j}|)}\exp\Big\{-c_2\big(r^2(\mathrm{I},\theta_0)-c_3r^2(\theta_0)\big)\Big\}$$
$$\le\sum_{\mathrm{I}\in\mathscr{O}^c(\tau_0)}n^{-c_1(|I|+\sum_{j=2}^m|I^{a_j}|)}r^2(\mathrm{I},\theta_0)e^{-Br^2(\mathrm{I},\theta_0)}\le\frac{1}{Be}\sum_{\mathrm{I}\in\mathscr{O}^c(\tau_0)}n^{-c_1(|I|+\sum_{j=2}^m|I^{a_j}|)}$$
$$\le\frac{1}{Be}\sum_{k_1+k_2+\cdots+k_{m+1}=n}\binom{n}{k_1,k_2,\ldots,k_{m+1}}n^{-\frac{c_1\sum_{j=2}^{m+1}k_j}{2}}\le\frac{e^{m-1}}{B}, \tag{26}$$

where $c_1 = c_1(\kappa) = 0.9(\kappa - c_0) > 2$ is defined in the proof of Lemma 1.

If $\mathrm{I} \in \mathscr{O}(\tau_0)$, then $|I| \le \frac{\tau_0 r^2(\theta_0)}{\log n}$. Combining this with Lemma 2 yields

$$\mathrm{E}_{\theta_0}\Big[\sum_{\mathrm{I}\in\mathscr{O}(\tau_0)} \hat{\pi}_{\mathrm{I}} \sum_{i\in I} \xi_i^2\Big] \le \mathrm{E}_{\theta_0}\Big[\max_{1\le i\le n} \xi_i^2 \sum_{\mathrm{I}\in\mathscr{O}(\tau_0)} |I|\hat{\pi}_{\mathrm{I}}\Big]$$
$$\le \frac{\tau_0 r^2(\theta_0)}{\log n}\mathrm{E}_{\theta_0}\Big[\max_{1\le i\le n}\xi_i^2\Big] \le 5\tau_0 r^2(\theta_0). \tag{27}$$

We have $\mathrm{E}(\sum_{i\in I}\xi_i^2)^2 = |I|^2 + 2|I| \le 3|I|^2$. Using this, Cauchy–Schwarz inequality and (26), we evaluate

$$\mathrm{E}_{\theta_0}\Big[\sum_{\mathrm{I}\in\mathscr{O}^c(\tau_0)} \hat{\pi}_{\mathrm{I}} \sum_{i\in I}\xi_i^2\Big] \le \sum_{\mathrm{I}\in\mathscr{O}^c(\tau_0)} \Big[\mathrm{E}_{\theta_0}\Big(\sum_{i\in I}\xi_i^2\Big)^2\Big]^{\frac{1}{2}}\Big[\mathrm{E}_{\theta_0}\hat{\pi}_{\mathrm{I}}^2\Big]^{\frac{1}{2}}$$
$$\le \sqrt{3}\sum_{\mathrm{I}\in\mathscr{O}^c(\tau_0)} r^2(\mathrm{I},\theta_0)\Big[\mathrm{E}_{\theta_0}\hat{\pi}_{\mathrm{I}}\Big]^{\frac{1}{2}} \le \frac{\sqrt{3}e^{m-1}}{B}. \tag{28}$$

From (27) and (28), it follows that

$$\mathrm{E}_{\theta_0}\Big[\sum_{\mathrm{I}\in\mathrm{M}_n^m}\Big(\hat{\pi}_{\mathrm{I}}\sum_{i\in I}\xi_i^2\Big)\Big] = \mathrm{E}_{\theta_0}\Big[\sum_{\mathrm{I}\in\mathscr{O}(\tau_0)}\Big(\hat{\pi}_{\mathrm{I}}\sum_{i\in I}\xi_i^2\Big) + \sum_{\mathrm{I}\in\mathscr{O}^c(\tau_0)}\Big(\hat{\pi}_{\mathrm{I}}\sum_{i\in I}\xi_i^2\Big)\Big]$$
$$\le 5\tau_0 r^2(\theta_0) + \frac{\sqrt{3}e^{m-1}}{B}. \tag{29}$$

Recall that $\sum_{\mathrm{I}}\hat{\pi}_{\mathrm{I}} = 1$ and $r^2(I,\theta_0) \le \tau_0 r^2(\theta_0)$ for all $\mathrm{I} \in \mathscr{O}(\tau_0)$. Using these relations and (26), we have

$$\sum_{\mathrm{I}\in\mathrm{M}_n^m} r^2(\mathrm{I},\theta_0)\mathrm{E}_{\theta_0}\hat{\pi}_{\mathrm{I}} = \sum_{\mathrm{I}\in\mathscr{O}(\tau_0)} r^2(\mathrm{I},\theta_0)\mathrm{E}_{\theta_0}\hat{\pi}_{\mathrm{I}} + \sum_{\mathrm{I}\in\mathscr{O}^c(\tau_0)} r^2(\mathrm{I},\theta_0)\mathrm{E}_{\theta_0}\hat{\pi}_{\mathrm{I}}$$
$$\le \tau_0 r^2(\theta_0) + \sum_{\mathrm{I}\in\mathscr{O}^c(\tau_0)} r^2(\mathrm{I},\theta_0)\mathrm{E}_{\theta_0}\hat{\pi}_{\mathrm{I}} \le \tau_0 r^2(\theta_0) + \frac{e^{m-1}}{B}. \tag{30}$$

Finally, combining the relations (24), (29) and (30), and taking into account that $r^2(\theta_0) \ge 1$, we finish the proof of assertion (i):

$$\mathrm{E}_{\theta_0}\hat{\pi}\Big(\|\theta-\theta_0\|^2 \ge M^2 r^2(\theta_0)|X\Big) \le \frac{6\tau_0}{M^2} + \frac{(\sqrt{3}+1)e^{m-1}}{M^2 r^2(\theta_0)B} \le \frac{C_{or}}{M^2},$$

where $C_{or} = 6\tau_0 + \frac{(\sqrt{3}+1)e^{m-1}}{B} = 6\tau_0 + \frac{76\tau_0(\sqrt{3}+1)e^{m-1}}{9(\tau_0-c_3)}$, and we take, say, $\tau_0 = c_3 + 1$.

The proof of assertion (ii) is essentially contained in the proof of the first assertion (i). Indeed, notice from (24), (29) and (30) that we proved a slightly stronger bound

$$\mathrm{E}_{\theta_0}\hat{\mathrm{E}}(\|\theta-\theta_0\|^2|X)=\mathrm{E}_{\theta_0}\sum_{\mathrm{I}\in\mathrm{M}_n^m}\hat{\mathrm{E}}\big(\|\theta-\theta_0\|^2|X,\mathscr{I}=\mathrm{I}\big)\hat{\pi}_{\mathrm{I}}\le C_{or}r^2(\theta_0).$$

This bound and $\|\hat{\theta}-\theta\|^2=\|\hat{\mathrm{E}}(\theta|X)-\theta\|^2\le\hat{\mathrm{E}}(\|\theta-\theta_0\|^2|X)$ imply the second assertion (ii): $\mathrm{E}_{\theta_0}\|\hat{\theta}-\theta_0\|^2\le\mathrm{E}_{\theta_0}\hat{\mathrm{E}}(\|\theta-\theta_0\|^2|X)\le C_{or}r^2(\theta_0)$.

Appendix 2

Implications: The Minimax Results over Sparsity Classes

We elucidate the potential strength of the oracle approach for sparse signals (i.e., $m=1$). When applied appropriately, the local approach is more flexible than global in that local result imply a whole panorama of global minimax results for all sparsity scales (covered by the local rate) at once. Namely, suppose we have a sparsity scale $\{\Theta[p],\,p\in\mathscr{P}\}$ so that $\theta\in\Theta[p]$ with unknown sparsity parameter $p\in\mathscr{P}$. Next, suppose we established for some local rate $r(\theta)$ that

$$r(\theta)\le cR(\Theta[p])\qquad\text{for all }\;\theta\in\Theta[p],\;\;p\in\mathscr{P},\tag{31}$$

with some uniform $c>0$. Then, clearly, the local results (for the posterior contraction and estimation problems) with local rate $r(\theta_0)$ will imply the global adaptive results *simultaneously for all scales* $\{\Theta[p],\,p\in\mathscr{P}\}$ with global rate $R(\Theta[p])$ for which (31) is satisfied. We say that the local rate $r(\theta)$ *covers* these scales.

Let us consider a couple of examples of sparsity scales $\{\Theta[p],\,p\in\mathscr{P}\}$ which are covered by our local rate $r_m(\theta_0)\le r_1(\theta_0)$ defined by (13). Let the conditions of Theorem 1 be fulfilled.

Nearly black vectors with sparsity level $p_n=n^{\gamma}$, $\gamma\in(0,1)$ as $n\to\infty$,

$$\ell_0[p_n]=\{\theta\in\mathbb{R}^n:\#(1\le i\le n:\theta_i\ne 0)\le p_n\}.$$

It is a well-known fact that the minimax estimation rate over the class of nearly black vectors $\ell_0[p_n]$ is $R^2(\ell_0[p_n])=2p_n\log(n/p_n)(1+o(1))$ as $n\to\infty$ (see Donoho et al. [8]). For $p_n=n^{\gamma}$ with $\gamma\in(0,1)$, this reduces to $R^2(\ell_0[p_n])=2p_n\log(n/p_n)(1+o(1))=O(p_n\log n)$.

We relate this minimax rate to the one-level oracle rate $r_1^2(\theta_0)$ (i.e., $m=1$), $\theta_0\in\ell_0[p_n]$, by taking $\mathrm{I}_*=\mathrm{I}_*(\theta_0)=(I_*^c,I_*)$ with $I_*=I_*(\theta_0)=\{i\in\mathbb{N}_n:\theta_{0,i}\ne 0\}$:

$$\sup_{\theta_0\in\ell_0[p_n]}r_1^2(\theta_0)\le\sup_{\theta_0\in\ell_0[p_n]}r_1^2(\mathrm{I}_*,\theta_0)\le p_n\log n=O(R^2(\ell_0[p_n])).$$

We thus have the property (31) for $\Theta[p]=\ell_0[p_n]$. Hence, Theorem 1 immediately implies the adaptive minimax results on the estimation and contraction rate problems for the empirical Bayes posterior $\hat{\pi}(\theta|X)$. We summarize these results in the following corollary.

Corollary 1 *Let the empirical Bayes posterior $\hat{\pi}(\theta|X)$ be defined by (6) and $\hat{\theta}$ be defined by (9). Then there exist constants $C, c > 0$ (depending only on K, κ) such that for any $M > 0$,*

$$\sup_{\theta_0 \in \ell_0[p_n]} \mathrm{E}_{\theta_0} \|\hat{\theta} - \theta_0\|^2 \le c\, p_n \log n,$$

$$\sup_{\theta_0 \in \ell_0[p_n]} \mathrm{E}_{\theta_0} \hat{\pi}\big(\|\theta - \theta_0\|^2 \ge M p_n \log n | X\big) \le \frac{C}{M}.$$

Weak ℓ_s-balls for $s \in (0, 2)$ with sparsity level $p_n = n^\gamma$, $\gamma \in (0, 1)$ as $n \to \infty$,

$$m_s[p_n] = \Big\{\theta \in \mathbb{R}^n : \tfrac{1}{n} \max_{1 \le i \le n} i|\theta_{[i]}|^s \le \big(\tfrac{p_n}{n}\big)^s\Big\},$$

where $|\theta_{[1]}| \ge \cdots \ge |\theta_{[n]}|$ are ordered values of $(|\theta_i|,\ i \in \mathbb{N}_n)$.

Denote $j = O_\theta(i)$ if $|\theta_i| = |\theta_{[j]}|$, with the convention that in the case $|\theta_{i_1}| = \cdots = |\theta_{i_k}|$ for $i_1 < \cdots < i_k$ we let $O_\theta(i_{l+1}) = O_\theta(i_l) + 1$, $l = 1, \ldots, k-1$. The minimax estimation rate over this class is $R^2(m_s[p_n]) = n(\frac{p_n}{n})^s \big(\log(\frac{n}{p_n})\big)^{(2-s)/2}(1 + o(1))$ as $n \to \infty$ (see Donoho and Johnstone [7]). Since $p_n = n^\gamma$, $\gamma \in (0, 1)$, $R^2(m_s[p_n]) = n(\frac{p_n}{n})^s \big(\log(\frac{n}{p_n})\big)^{(2-s)/2}(1 + o(1)) = O(n^{1-s} p_n^s (\log n)^{(2-s)/2})$. Then, with $p_n^* = \big(\frac{p_n^2 n^{(2/s-2)}}{\log n}\big)^{s/2}$, $\mathrm{I}_* = \mathrm{I}_*(\theta_0) = (I_*^c, I_*)$, $I_* = I_*(\theta_0) = \{i \in \mathbb{N}_n : O_{\theta_0}(i) \le p_n^*\}$, we derive that for large enough n

$$\sup_{\theta_0 \in m_s[p_n]} r_1^2(\theta_0) \le \sup_{\theta_0 \in m_s[p_n]} r_1^2(\mathrm{I}_*, \theta_0) \le p_n^* \log n + p_n^2 n^{(2-2s)/s} \sum_{i=p_n^*+1}^{\infty} i^{-2/s}$$
$$\le p_n^* \log n + \tfrac{s}{2-s} p_n^2 n^{(2-2s)/s} p_n^{*(s-2)/s} \le c n^{1-s} p_n^s (\log n)^{(2-s)/2} = O(R^2(m_s[p_n])). \tag{32}$$

We established (31) for $\Theta[p] = m_s[p_n]$, thus Theorem 1 implies the next corollary.

Corollary 2 *Let the empirical Bayes posterior $\hat{\pi}(\theta|X)$ be defined by (6) and $\hat{\theta}$ be defined by (9). Then there exist constants $C, c > 0$ (depending only on K, κ) such that for any $M > 0$,*

$$\sup_{\theta_0 \in m_s[p_n]} \mathrm{E}_{\theta_0} \hat{\pi}\big(\|\theta - \theta_0\|^2 \ge M n^{1-s} p_n^s (\log n)^{(2-s)/2} | X\big) \le \frac{C}{M},$$

$$\sup_{\theta_0 \in m_s[p_n]} \mathrm{E}_{\theta_0} \|\hat{\theta} - \theta_0\|^2 \le c\, n^{1-s} p_n^s (\log n)^{(2-s)/2}.$$

Remark 2 Recall that $r_m^2(\theta_0) \leq r_1^2(\theta_0) \leq R^2(\Theta)$ with both $\Theta = \ell_0[p_n]$ or $\Theta = m_s[p_n]$, for all $m \geq 2$, $a_2, \ldots, a_m$. Therefore by using multilevel sparsity model, we always improve upon the traditional minimax results for sparsity classes.

Appendix 3

Simulation Study

We simulated data according to the model (1) with dimension $n = 500$. We used signals $\theta_0 = (\theta_{0,1}, \ldots, \theta_{0,n})$ of the form $\theta_0 = (a_1, \ldots, a_1, \ldots, a_m, \ldots, a_m, A, \ldots, A)$, where $a_1 = 0$ and the value A is assumed to be unknown. Denote the cardinality of a_j values in the signal θ_0 by N_{a_j}, $j = 1, \ldots, m$. When performing simulations for the empirical Bayes posterior $\hat{\pi}(\theta|X)$ and some posterior based quantities, we used the values of the parameters $K = 10$ and $\kappa = 0.55$.

First, we did a small simulation study for the four estimators based on $\hat{\pi}(\theta|X)$: empirical Bayes posterior (EBP) mean given by (9) for multilevel sparse sequences ($a_1, a_2, \ldots, a_m$ are known values in advance, i.e., $m > 1$), EBP mean for one-level sparsity (only $a_1 = 0$ is known, $a_2, \ldots, a_m$ are unknown, i.e., $m = 1$) and the estimator $\check{\theta}$ (to be defined later) for multilevel sparse sequences and one-level sparsity, respectively. The construction of the estimator $\check{\theta}$ is straightforward (basically, it can be reduced to a hard thresholding estimator with a certain threshold). Computation of EBP mean, which is a shrinkage estimator, is bit more involved. We provide some technical preliminaries for efficient computation of the mean and $\check{\theta}$ with respect to the empirical Bayes posterior $\hat{\pi}(\theta|X)$.

EBP mean. According to (9), the EBP mean $\hat{\theta} = \int \theta d\hat{\pi}(\theta|X)$ is a random vector in $\mathbb{R}^n$. We can compute its ith coordinate as follows:

$$\hat{\theta}_i = \sum_{j=2}^{m} \frac{a_j \phi(X_i, a_j, 1)}{n^\kappa Q_i} + \frac{X_i}{n^\kappa \sqrt{2\pi(K+1)} Q_i}, \quad i = 1, \ldots, n,$$

where

$$Q_i = \phi(X_i, 0, 1) + \sum_{j=2}^{m} \frac{\phi(X_i, a_j, 1)}{n^\kappa} + \frac{1}{n^\kappa \sqrt{2\pi(K+1)}}, \quad i = 1, \ldots, n.$$

Estimator $\check{\theta}$. By applying the empirical Bayes approach with respect to I, we obtain that

$$\check{\mathrm{I}} = \arg\max_{\mathrm{I}\in \mathrm{M}_n^m} \hat{\pi}(\mathscr{I} = \mathrm{I}|X) = \arg\max_{\mathrm{I}\in \mathrm{M}_n^m} \Big\{ -\sum_{j=1}^{m}\sum_{i\in I^{a_j}} \frac{(X_i-a_j)^2}{2} + \log\lambda_{\mathrm{I}} - \frac{|I|}{2}\log(K+1)\Big\}$$

$$= \arg\min_{\mathrm{I}\in \mathrm{M}_n^m} \Big\{\sum_{j=1}^{m}\sum_{i\in I^{a_j}}(X_i-a_j)^2 + 2\kappa\sum_{j=2}^{m}|I^{a_j}|\log n + |I|\big(2\kappa\log n + \log(K+1)\big)\Big\}. \quad (33)$$

Plugging in this into $\hat{\pi}_{\mathrm{I}}(\theta|X)$ defined by (7) gives the corresponding empirical (now "twice empirical": with respect to $\mu_{m+1,i}$ and with respect to I) Bayes estimator for θ:

$$\check{\theta} = \check{\theta}(\check{\mathrm{I}}) = \sum_{j=1}^{m} a_j 1\{i \in \check{I}^{a_j}\} + X_i 1\{i \in \check{I}\}, i \in \mathbb{N}_n. \quad (34)$$

Table 1 shows estimates of the mean square errors $\mathrm{E}_{\theta_0}\|\hat{\theta} - \theta_0\|^2$. These results are the average (square) error of 100 estimates $\hat{\theta}_1, \ldots, \hat{\theta}_{100}$ computed from 100 data vectors simulated independently from the model (1). Besides, we also simulate the classical hard-thresholding *HT*, hard-thresholding oracle *HTO* and the empirical Bayes mean EBM considered by Johnstone and Silverman (2004) with a standard Laplace prior. The hard-thresholding *HT* and hard-thresholding oracle *HTO*, given by $\hat{\theta}_i^{HT} = X_i 1\{|X_i| > \sqrt{2\log n}\}$ and $\hat{\theta}_i^{HTO} = X_i 1\{|X_i| > \sqrt{2\log(n/p_n)}\}$. Note that the last estimator uses the "oracle" value of the sparsity parameter p_n, all the other estimators do not.

According to the results of Table 1, our estimators based on the *empirical Bayes posterior* $\hat{\pi}(\theta|X)$ are competitive to the other ones.

For further illustration in Fig. 1 we visualize 95% credible intervals (gray bars) for θ_0 with parameters $a_1 = 0, a_2 = 5, A = 9, N_{a_1} = 45, N_{a_2} = 45, N_A = 10, n = 100$ and the empirical Bayes posterior means (red dots), by simulating 1000 draws from

Table 1 Average square errors of seven estimators computed on 100 data vectors X of length $n = 500$ simulated from model (1) with $\theta_0 = (a_1, \ldots, a_1, \ldots, a_m, \ldots, a_m, A, \ldots, A)$

Estimators	Average square errors	Average square errors
	$a_1 = 0, a_2 = 5, A = 9$ $N_{a_1} = 225, N_{a_2} = 225,$ $N_A = 50$	$a_1 = 0, a_2 = 3, a_3 = 6, A = 9$ $N_{a_1} = 150, N_{a_2} = 150,$ $N_{a_3} = 150, N_A = 50$
EBP mean, $m > 1$	187	449
EBP mean, $m = 1$	411	807
$\check{\theta}, m > 1$	214	598
$\check{\theta}, m = 1$	385	980
EBM	612	688
HT	607	1249
HTO	384	477

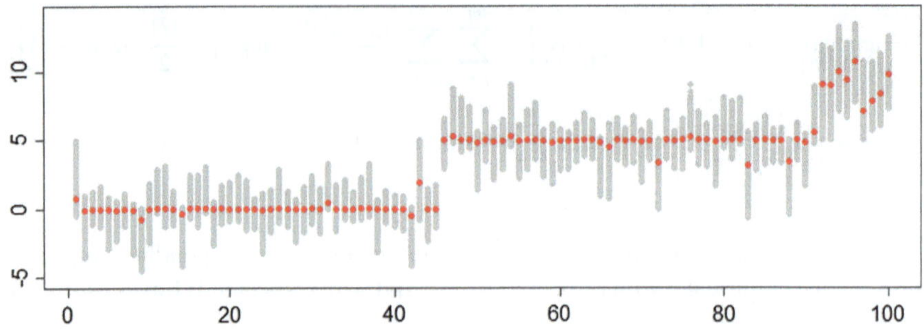

Fig. 1 Empirical Bayes posterior means (*red dots*) and 95% credible intervals (*gray bars*) for the signal $\theta_0 = (0, \ldots, 0, 5, \ldots, 5, 9, \ldots, 9)$ of length $n = 100$, where $N_0 = 45$, $N_5 = 45$ and $N_9 = 10$

the empirical Bayes posterior distribution $\hat{\pi}(\theta|X)$ and plotting the 95% draws out of the 1000 that are closest to the EBP mean. Note that this picture shows good coverage.

References

1. Abramovich, F., Benjamini, Y., Donoho, D.L., Johnstone, I.M.: Adapting to unknown sparsity by controlling the false discovery rate. Ann. Stat. **34**, 584–653 (2006)
2. Abramovich, F., Grinshtein, V., Pensky, M.: On optimality of Bayesian testimation in the normal means problem. Ann. Stat. **35**, 2261–2286 (2007)
3. Belitser, E., Nurushev, N.: Needles and straw in a haystack: empirical Bayes confidence for possibly sparse sequences. arXiv:1511.01803 (2015)
4. Birgé, L., Massart, P.: Gaussian model selection. J. Eur. Math. Soc. **3**, 203–268 (2001)
5. Castillo, I., van der Vaart, A.W.: Needles and Straw in a Haystack: posterior concentration for possibly sparse sequences. Ann. Stat. **40**, 2069–2101 (2012)
6. Donoho, D.L., Johnstone, I.M.: Ideal spatial adaptation by wavelet shrinkage. Biometrika **81**, 425–455 (1994)
7. Donoho, D.L., Johnstone, I.M.: Minimax risk over ℓ_p-balls for ℓ_q-error. Probab. Theory Related Fields. **99**, 277–303 (1994)
8. Donoho, D.L., Johnstone, I.M., Hoch, J.C., Stern, A.S.: Maximum entropy and the nearly black object(with Discussion). J. Roy. Stat. Soc. Ser. B **54**, 41–81 (1992)
9. Johnstone, I., Silverman, B.: Needles and straw in haystacks: empirical Bayes estimates of possibly sparse sequences. Ann. Stat. **32**, 1594–1649 (2004)
10. van der Pas, S.L., Kleijn, B.J.K., van der Vaart, A.W.: The horseshoe estimator: posterior concentration around nearly black vectors. Electron. J. Stat. **8**, 2585–2618 (2014)

Likelihood Tempering in Dynamic Model Averaging

Jan Reichl and Kamil Dedecius

Abstract We study the problem of online prediction with a set of candidate models using dynamic model averaging procedures. The standard assumptions of model averaging state that the set of admissible models contains the true one(s), and that these models are continuously updated by valid data. However, both these assumptions are often violated in practice. The models used for online tasks are often more or less misspecified and the data corrupted (which is, mathematically, a demonstration of the same problem). Both these factors negatively influence the Bayesian inference and the resulting predictions. In this paper, we propose to suppress these issues by extending the Bayesian update by a sort of likelihood tempering, moderating the impact of observed data to inference. The method is compared to the generic dynamic model averaging and to an alternative solution via sequential quasi-Bayesian mixture modeling.

Keywords Model averaging · Model uncertainty · Prediction · Sequential estimation · Tempered likelihood

1 Introduction

In many real-world applications of the statistical control theory we are interested in online prediction of process outcomes, evaluated by an a priori specified process model. In practice, it is mostly assumed that the adopted process model is sufficiently close to the true observations-generating model. However, this (surprisingly still prevailing) assumption is often violated due to various reasons, e.g., different operational regimes, noise heteroskedasticity or even unknown but complicated distribution, imperfect physical characterization of the observed process etc. If there

J. Reichl · K. Dedecius (✉)
Institute of Information Theory and Automation, Czech Academy of Sciences,
Pod Vodárenskou věží 4, 182 08 Prague 8, Czech Republic
e-mail: dedecius@utia.cas.cz

J. Reichl
e-mail: reichja3@fjfi.cvut.cz

R. Argiento et al. (eds.), *Bayesian Statistics in Action*, Springer Proceedings
in Mathematics & Statistics 194, DOI 10.1007/978-3-319-54084-9_7

exists a class of potentially admissible candidate models, maybe with different input explanatory data and valid through different periods of time, the model switching and averaging procedures represent an appealing way around the model uncertainty problem [7, 17]. Model switching is mostly based on the assumption of mutually exclusive models. Their probabilities are sequentially assessed and the model with the highest probability is used for a corresponding period of time. Model averaging virtually removes this exclusivity via concurrent predictive-performance-based assessment of uncertainty about the candidate models.

Although the model switching and averaging approaches admit that the adopted process models are not exact, they still ignore the possible misspecification issue in their inference. Indeed, even the best *available* models may be more or less misspecified, and provide only an approximation to the distribution of the observations. Although the consistency of the Bayesian inference guarantees the convergence of parameter estimates to the value minimizing the Kullback–Leibler divergence between the true distribution and its (imprecise) model [4], the results may be unacceptable for real applications. Among the typical demonstrations of such effects are negative estimates of strictly positive variables (fuel consumption, number of particles, object length etc.). Even worse, the standard Bayesian procedures are generally not robust to contamination by observations following other than the specified model, or to complete model misspecification [14].

This paper adheres to the dynamic model averaging (DMA) realm, providing parallel online assessment of candidate models probabilities. If the modeling goal lies in evaluating the predicted values, the DMA-point prediction is represented by a convex combination of individual models' predictions, whose contributions are proportional to the candidate model probabilities. The DMA method was originally proposed for linear models by Raftery et al. [17] as an online extension of the static Bayesian model averaging [12, 16], and later formulated for sequential logistic regression by McCormick et al. [13]. Since then, DMA attained a significant focus, mostly in econometrics and finance, e.g. [2, 3, 6, 9–11] to name a few.

More specifically, we focus on the issue of assimilation of model-incompatible information (observations) into the prior distribution during the online estimation process, where the prior distributions involved in DMA are updated regardless of the candidate models probability. It leads to unreliable estimates and may degrade the prediction quality for a long time period. Although the invalid information may be gradually eliminated by means of forgetting procedures (see [5] for an extensive overview), a rational argument is to prevent it from entering the prior distribution. For this reason, we study the possibilities of online likelihood tempering during the Bayesian inference and prediction. Our method is inspired by the weighted likelihood [20], recently appearing also in the c-posterior approach [14] aiming at the same objective – the robustness to model misspecification. The approach sketched in this paper is the first step towards extending of the c-posteriors to online modeling. The determination of tempering factors is based on the actual predictive performance of the particular models. In particular, we study two possibilities: (i) a simplified approach based on the (dis)similarity of the most likely observation and the true one, and (ii) a model-oriented approach where the model weights and tempering factors are estimated by means of the quasi-Bayesian framework for online mixture inference [8, 19].

For completeness we remark that the likelihood tempering may be found in the Markov chain Monte Carlo as the MC^3 likelihood tempering method [1], however, the motivation is different.

The paper is organized as follows: Sect. 2 overviews the principles of the dynamic model averaging method. It also sheds some light on the studied problem of the connection between model uncertainty and observations assimilation. Section 3 is devoted to the proposed tempered sequential Bayesian update. Section 4 studies an alternative approach inspired by the quasi-Bayesian mixture modeling. The ongoing Sect. 5 illustrates the effect of tempering on a simulation example. Finally, Sect. 6 concludes the paper.

2 On-Line Prediction with a Set of Admissible Models

In this section we describe the principles of the dynamic model averaging applied to a set of K admissible candidate models indexed by $k = 1, \ldots, K$. We consider discrete-time Bayesian modeling of a dynamic process with common observations y_t that are more or less determined by known possibly model-specific explanatory variables $x_{k,t}$, where $t = 0, 1, \ldots$ is the discrete time index. Assume, that some parametric models – probability density functions $p_k(y_t|x_{k,t}, \theta_k)$ parameterized by θ_k – are the admissible candidates for this purpose, and that (proper) prior distributions $\pi_k(\theta_k|x_{k,0:t-1}, y_{0:t-1})$ serve for their inference. The variables

$$x_{k,0:t-1} = \{x_{k,0}, \ldots, x_{k,t-1}\} \quad \text{and} \quad y_{0:t-1} = \{y_0, \ldots, y_{t-1}\}$$

express the statistical knowledge about θ_k up to time instant $t-1$, and $x_{k,0}$ and y_0 stand for pseudo-observations, expressing the initial prior knowledge before incorporation of the first observations. Let the modeling be performed under the uncertainty which model is true at the moment. The task is online prediction of y_t from $x_{k,t}$ and θ_k by means of model averaging.

2.1 Dynamic Model Averaging

As mentioned in the Introduction, the dynamic model averaging (DMA) methodology of Raftery et al. [17] extends Leamer's static Bayesian model averaging [12] to online problems, where the predictions are evaluated from sequentially acquired data.

The basic principle of DMA is that it assigns the candidate models $p_k(y_t|x_{k,t}, \theta_k)$ with probabilities (weights) $w_{k,t}$ taking values in the probabilistic $(K-1)$-simplex, that express the degree of evidence that these models are valid at the particular time instants. Recall, that the goal is the online prediction of the next observation y_t

given known explanatory variables $x_{k,t}$. Each of the models provides its own point prediction, mostly represented by the expected value

$$\widehat{y}_{k,t} = \mathbb{E}\left[y_t|x_{k,0:t,y_{0:t-1}}\right] = \int y_t p_k(y_t|x_{0:t}, y_{0:t-1})dy_t, \tag{1}$$

where

$$p_k(y_t|x_{0:t}, y_{0:t-1}) = \int p_k(y_t|x_t, \theta_k)\pi_k(\theta_k|x_{k,0:t-1}, y_{0:t-1})d\theta_k \tag{2}$$

is the predictive distribution connected with the kth model $p_k(y_t|x_{k,t}, \theta)$. The integrations are over the spaces of y_t and θ_k, respectively. The DMA point prediction reflects the uncertainty about the particular models by taking their probabilities into account, that is, it averages over all the available predictions,

$$\widehat{y}_t = \sum_{k=1}^{K} w_{k,t-1}\widehat{y}_{k,t}. \tag{3}$$

After acquiring the observation y_t, the update of the probabilities $w_{k,t-1}$ reflects the predictive performance of the particular models,

$$w_{k,t} \propto w_{k,t-1} \cdot p_k(y_t|x_{k,0:t}, y_{0:t-1}). \tag{4}$$

In practice, the distribution of model weights may be considerably influenced by outdated information. This issue can be resolved by an artificial increase of the uncertainty about the weights, e.g., by exponential forgetting, flattening $w_{1,t-1}, \ldots, w_{K,t-1}$ by a factor $\alpha \in [0, 1]$ as proposed by Peterka [15]. Furthermore, aberrant observations may get some $w_{k,t-1}$ too close to zero. This situation virtually eliminates the related models, as it is hard to recover from it. A workaround is to proceed with a stabilization additive constant during the weights update (4), e.g. $c = 10^{-3}/K$ proposed in [17]. The resulting equivalent of (4) then has the form

$$w_{k,t} \propto (w_{k,t-1}^{\alpha} + c) \cdot p_k(y_t|x_{k,0:t}, y_{0:t-1}). \tag{5}$$

The estimation of parameters $\theta_k, k = 1, \ldots, K$ is not influenced by DMA and has the form of the standard Bayes' theorem

$$\pi_k(\theta_k|x_{k,0:t}, y_{0:t}) \propto p_k(y_t|x_{k,t}, \theta_k)\pi_k(\theta_k|x_{k,0:t-1}, y_{0:t-1}). \tag{6}$$

We emphasize, that this is where the following question arises:

> *The dynamic model averaging (DMA) is used to assess the probability of several candidate models. Would it be possible and useful to take this probability into account in Eq.* (6)?

In other words, if one knows from $w_{k,t}$ that y_t are not well explained by $p_k(y_t|x_{k,t}, \theta_k)$, why should one violate the estimates provided by $\pi_k(\theta_k|\cdot)$? For instance, if the reality switches between two models, their prior distributions are updated regardless of which of the models is currently valid.

Below, we propose to solve this issue by likelihood tempering.

3 Tempered Bayesian Update

Let us drop the model indices k in this section. The standard Bayesian theory assumes that there is a true observations-generating model $q(y_t|x_t)$ which is approximated by the statistician using a parametric model $p(y_t|x_t, \theta)$, ideally as close as possible. Under certain assumptions, the posterior estimates then converge to the value minimizing the Kullback–Leibler divergence of the two models,

$$\widehat{\theta} = \arg\min_{\theta\in\Theta} \mathbb{D}\left(q(y_t|x_t)\big|\big|p(y_t|x_t, \theta)\right),$$

where Θ is the parameter space, see, e.g., [4]. That is, the classical Bayesian way of thinking admits, that there is a possible disagreement between the true but unknown model and the employed (approximate) model, and relying on the consistency of the Bayesian posterior distribution of θ, it updates the prior distribution via the Bayes' theorem

$$\pi(\theta|x_{0:t}, y_{0:t}) \propto p(y_t|x_t, \theta)\pi(\theta|x_{0:t-1}, y_{0:t-1}). \tag{7}$$

However, from the Bayesian asymptotic theory it is well known that (7) requires certain assumptions to provide "reasonable" results. Naturally, if these assumptions are not satisfied, e.g., some observations are not explained by the model, the posterior estimates are inappropriately influenced (biased) and unusable for prediction. This effect is pronounced in DMA, where the standard Bayesian update is used, too.

We propose to solve this issue by a weighted variant of the Bayesian update, suppressing the influence of such observations similarly as in the Miller and Dunson's version for static estimation [14]. It consists of a step to increase the uncertainty about the model using a tempering (flattening) factor $\zeta_t \in [0, 1]$,

$$\pi(\theta|x_{0:t}, y_{0:t}) \propto [p(y_t|x_t, \theta)]^{\zeta_t}\, \pi(\theta|x_{0:t-1}, y_{0:t-1}). \tag{8}$$

Miller and Dunson also propose a method for choosing a suitable value of ζ_t, however, it is not suitable for online cases.

A suboptimal solution of the first choice may be to base the factor on the predictive density and to compare the likelihood of the actually observed y_t with the expected $\widehat{y}_t$, i.e., the point estimate,

$$\zeta_t = \frac{p(y_t|x_{0:t}, y_{0:t-1})}{p(\widehat{y}_t|x_{0:t}, y_{0:t-1})}. \tag{9}$$

Naturally, it is possible to use the mode or other statistics in place of the mean value. Although this solution may lead to uncertainty underestimation, it could be easily counterbalanced by flattening of the posterior distribution, routinely used in engineering practice [5]. The predictive likelihood is analytically tractable in many practical cases, e.g. the linear regression models or Kalman filters, which allows for an easy computation of the factors in real time.

To summarize, the purpose of the proposed tempering update is to (i) penalize model misspecification, and (ii) to increase the robustness of estimation to contamination with other processes. The coefficient ζ_t can easily suppress the effect of unlikely observations (with respect to the predictive likelihood). If $\zeta_t \to 0$,

$$\pi(\theta|x_{0:t}, y_{0:t}) = \pi(\theta|x_{0:t-1}, y_{0:t-1}),$$

suppressing the influence of the extremely unlikely observation, while $\zeta_t \to 1$ recovers the standard Bayesian update. We suggest that this procedure should be used for estimation of parameters of models involved in DMA.

4 Mixture-Based Approach

Another possibility of model mixing[1] is to adopt the viewpoint that the class $\{p_k(y_t|x_{k,t}, \theta_k), k = 1, \ldots, K\}$ is a set of mixture components with weights $w_1, \ldots, w_K$, and to model these weights via the Dirichlet prior distribution whose hyperparameters $[\kappa_1, \ldots, \kappa_K]$ are updated in the sense of the quasi-Bayesian approach [19],

$$\pi(w|x_{1:K,0:t}, y_{0:t}) \propto \underbrace{\prod_{k=1}^{K} w_k^{\zeta_{k,t}}}_{\text{multinomial}} \underbrace{\prod_{k=1}^{K} w_k^{\kappa_{k,t-1}-1}}_{\text{Dirichlet prior}}, \tag{10}$$

where $\zeta_{k,t}$ is the estimate of the active component indicator,

$$\zeta_{k,t} \propto \frac{\kappa_{k,t-1}}{\sum_{l=1}^{K} \kappa_{l,t-1}} p_k(y_t|x_{k,0:t}, y_{0:t-1}).$$

Similarly to the Bayesian counterparts of the expectation-maximization algorithms, the quasi-Bayesian mixture estimation already assumes weighted Bayes' theorem for the estimation of component parameters,

$$\pi(\theta_k|x_{k,0:t}, y_{0:t}) \propto \left[p_k(y_t|x_{k,t}, \theta_k)\right]^{\zeta_{k,t}} \pi(\theta_k|x_{k,0:t-1}, y_{0:t-1}).$$

[1]Proposed in personal communication by Dr. Kárný (Institute of Information Theory and Automation, Czech Academy of Sciences).

The analytical updates of Dirichlet prior hyperparameters (10) are given by

$$\kappa_{k,t} = \kappa_{k,t-1} + \zeta_{k,t}.$$

Finally, the prediction of the upcoming observation is similarly to DMA a convex combination of component predictions weighted by component weights,

$$\widehat{y}_t = \frac{1}{\kappa_{0,t}} \sum_{k=1}^{K} \kappa_{k,t} \widehat{y}_{k,t}, \qquad \text{where} \qquad \kappa_{0,t} = \sum_{k=1}^{K} \kappa_{k,t}. \tag{11}$$

The key differences between this approach and the weighted DMA proposed above are apparent from the equations. First and foremost, the component weights are modelled as static (or slowly varying if a forgetting procedure is used). In the authors' viewpoint, this is where the main drawback lies. In many practical situations, the models switch rather abruptly, which needs to be reflected by quick changes of their weights. However, the forgetting procedures can be effective only in the cases of slow variations [5], and under abrupt changes lead to significantly biased weights estimation. This will be demonstrated in the following section. For this reason, this section is rather conceptual and included for completeness as another model-based dynamic approach to model mixing.

5 Simulation Results

In this section we present some simulation results comparing the studied strategies. The experiment is done on synthetic data generated by three models and contaminated with Laplacian noise. The performance is measured by the prediction mean squared error (MSE) and the mean absolute error (MAE).

The data set consists of 500 samples generated from the following normal models

$$\begin{aligned}
y_t &= -0.2 - 0.15 x_t^{(1)} + \varepsilon_t, && t = 150, \ldots, 300, \\
y_t &= 0.5 + 0.75 x_t^{(2)} + \varepsilon_t, && t = 1, \ldots, 150 \text{ and } t = 300, \ldots, 400, \\
y_t &= 0.95 x_t^{(3)} + \varepsilon_t, && t = 400, \ldots, 500,
\end{aligned}$$

with regressors $x_t^{(i)} \sim \mathcal{N}(\mu_i, 1)$, where $\mu_1 = 0$, $\mu_2 = 2$ and $\mu_3 = 3$. The normal noise variable ε_t is drawn from $\mathcal{N}(0, 1.25^2)$.

Two different scenarios are studied:

1. *'True' model scenario* where the three models are used to generate data and no misspecification is present. Effectively, it is a true model switching scenario.
2. *Misspecification scenario* where the data is additionally contaminated by a heavy-tailed Laplacian noise, namely by 200 samples from zero-mean Laplace distribution with a scale 2. These samples are randomly added to the data.

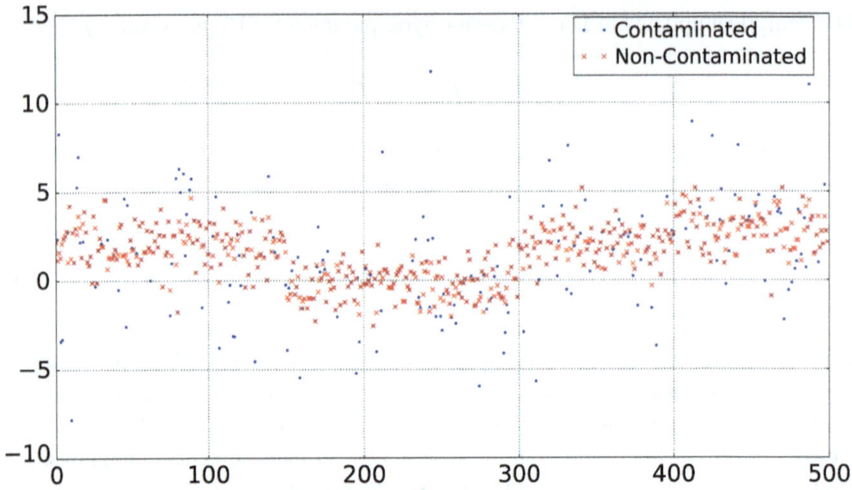

Fig. 1 Comparison of contaminated and non-contaminated data set

The resulting data sets depicts Fig. 1, where the red crosses indicate the 'true' model scenario data, while the blue dots show Laplacian noise-contaminated data for the misspecified scenario.

The process y_t is modeled with three normal linear models

$$y_t \middle| x_t, \beta, \sigma^2 \sim \mathcal{N}\left(x^\intercal \beta, \sigma^2\right), \tag{12}$$

where x_t represents 2-dimensional real regressors $x_t^{(1)}, \ldots, x_t^{(3)}$. Clearly, these models are not appropriate for the misspecification scenario with the Laplacian noise.

The prior placed on $\beta^{(i)}, \sigma^{2,(i)}$ is the normal-inverse gamma distribution in the compatible form conjugate to the exponential family form of the normal model (12) (more on this can be found, e.g., in [15]) with the same initialization for all considered models, namely with the sufficient statistic accumulating hyperparameter and the scalar degrees of freedom

$$\xi_0^{(i)} = \text{diag}(0.1, 0.01, 0.01) \quad \text{and} \quad \nu_0^{(i)} = 10,$$

respectively. The posterior predictive distribution is in this setting the Student's distribution [15].

Three averaging strategies are studied:

1. Basic dynamic model averaging (DMA) of Raftery et al. [17],
2. Tempered dynamic model averaging (t-DMA) proposed in Sect. 3, and
3. Quasi-Bayesian approach (q-B) inspired by mixture modelling and discussed in Sect. 4.

Table 1 Prediction MSEs and MAEs and averaged models weights $\widehat{\omega}$ of all tested strategies (DMA, tempered DMA and the quasi-Bayesian approach) and single models without switching/averaging. For comparison, MSE true and MAE true denote error statistics computed with respect to the actual non-contaminated observation

Model/Strategy	MSE	MAE	MSE true	MAE true	$\widehat{\omega}_1$	$\widehat{\omega}_2$	$\widehat{\omega}_3$
p_1	7.59	1.92	5.13	1.75	–	–	–
p_2	5.44	1.47	2.61	1.22	–	–	–
p_3	5.22	1.37	2.51	1.13	–	–	–
DMA	4.86	1.29	2.14	1.08	0.21	0.41	0.36
t-DMA	4.47	1.14	1.77	0.91	0.32	0.38	0.30
q-B	6.11	1.35	3.21	1.16	0.17	0.37	0.44

The initial setting of all three strategies is identical. Namely, the initial model weights are uniform, the initial prior for weights in q-B is the flat Dirichlet distribution, symmetric and uniform over the related simplex.

The predictive performance is measured in terms of the mean squared error (MSE) and the mean absolute error (MAE). Consistently with the forecasting theory, the goal is to minimize these measures.

The results are summarized in Table 1 for the individual models without switching, and for all three studied averaging approaches. Both the misspecification scenario (MSE and MAE) and the 'true model' scenario (MSE true and MAE true) are shown. From the results one may conclude that the tempered DMA performs best (in terms of MSE and MAE). That is, the weighted Bayesian update (8) effectively suppresses the influence of data that are not explained by the model(s) in use. The classical DMA strategy performs a bit worse as expected. The quasi-Bayesian approach leads to a prediction quality inferior to both t-DMA and DMA. Anyway, it can provide results better than certain isolated models.

The time evolution of the models weights is depicted in Fig. 2 for both the original DMA and the proposed tempered version. The evolution of q-B is omitted for its inferior quality. Apparently, the initial learning period is very similar in both DMA and t-DMA, however, the model switching after $t = 150$ is much better reflected by t-DMA (the weight of the corresponding model is mostly close to 1). Another model switches are better detected by t-DMA too, although the weights are not so pronounced. One can conclude that t-DMA is significantly more sensitive than the basic DMA.[2]

To summarize, the experimental results show that the dynamic model averaging strategy performs well even in complicated conditions, where the noise properties are different from the assumed and the models differ from the true ones. The proposed tempering strategy leads to better results than the pure DMA ignoring the fact that the observations are not well described by the models and fully assimilating them

[2]A thorough sensitivity analysis is postponed to further research.

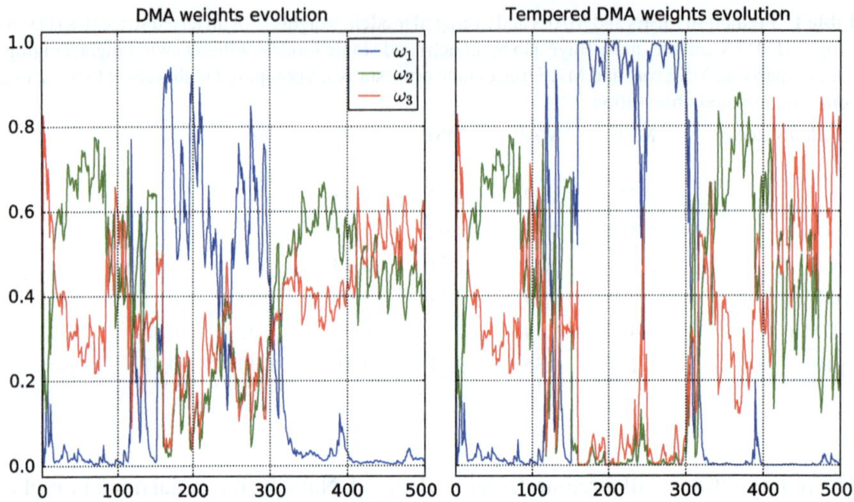

Fig. 2 Time evolution of model weights for the DMA and tempered-DMA. The correspondence between models and line types is the same in both plots

into the prior distributions. Our experience confirms that if the true model is present, its weight is dominant.

6 Conclusion

Applied statistical theory assumes that the models in use are relatively close to the true observations-generating models, and hence that the results (predictions, estimates) are close to the true values. However, the real-world phenomena are often rather roughly approximated by models. The standard dynamic model averaging provides a way around this issue by means of a concurrent assessment of models uncertainty, and by averaging over the results taking this uncertainty into account. However, it still neglects it at the level of the Bayesian update. In this paper, we focus specifically on this issue and propose to use a weighted – tempered – version of the Bayes' theorem, suppressing the impact of unlikely observations to the inference. The simulation example demonstrates that it provides an improvement of estimation quality.

Acknowledgements The project is supported by the Czech Science Foundation, project number 14–06678P. The authors thank to Dr. M. Kárný and Dr. L. Jirsa for the discussions about the model-based approach.

References

1. Altekar, G., Dwarkadas, S., Huelsenbeck, J.P., Ronquist, F.: Parallel Metropolis coupled Markov chain Monte Carlo for Bayesian phylogenetic inference. Bioinformatics **20**(3), 407–415 (2004)
2. Aye, G., Gupta, R., Hammoudeh, S., Kim, W.J.: Forecasting the price of gold using dynamic model averaging. Int. Rev. Financ. Anal. **41**, 257–266 (2015)
3. Bork, L., Møller, S.V.: Forecasting house prices in the 50 states using dynamic model averaging and dynamic model selection. Int. J. Forecast. **31**(1), 63–78 (2015)
4. Bunke, O., Milhaud, X.: Asymptotic behavior of Bayes estimates under possibly incorrect models. Ann. Stat. **26**(2), 617–644 (1998)
5. Dedecius, K., Nagy, I., Kárný, M.: Parameter tracking with partial forgetting method. Int. J. Adapt. Control Signal Process. **26**(1), 1–12 (2012)
6. Del Negro, M., Hasegawa, R.B., Schorfheide, F.: Dynamic prediction pools: An investigation of financial frictions and forecasting performance. J. Econom. **192**(2), 391–405 (2016)
7. Frühwirth-Schnatter, S.: Finite Mixture and Markov Switching Models. Springer, Heidelberg (2006)
8. Kárný, M., Kadlec, J., Sutanto, E.L.: Quasi-Bayes estimation applied to normal mixture. In: Proceedings of the 3rd European IEEE Workshop on Computer-Intensive Methods in Control and Data Processing, pp. 77–82. Praha (1998)
9. Koop, G., Korobilis, D.: Forecasting inflation using dynamic model averaging. Int. Econ. Rev. **53**(3), 867–886 (2012)
10. Koop, G., Korobilis, D.: A new index of financial conditions. Eur. Econ. Rev. **71**, 101–116 (2014)
11. Koop, G., Tole, L.: Forecasting the European carbon market. J. R. Stat. Soc. Ser. A (Stat. Soc.) **167**(3), 723–741 (2013)
12. Leamer, E.E.: Specification Searches: Ad hoc Inference with Nonexperimental Data. Wiley, New York (1978)
13. McCormick, T., Raftery, A., Madigan, D., Burd, R.: Dynamic logistic regression and dynamic model averaging for binary classification. Biometrics **68**(1), 23–30 (2012)
14. Miller, J.W., Dunson, D.B.: Robust Bayesian Inference via Coarsening pp. 1–31 (2015). http://arxiv.org/abs/1506.06101
15. Peterka, V.: Bayesian approach to system identification. In: Eykhoff, P. (ed.) Trends and Progress in System Identification, pp. 239–304. Pergamon Press, Oxford, U.K. (1981)
16. Raftery, A.E., Madigan, D., Hoeting, J.A.: Bayesian model averaging for linear regression models. JASA **92**(437), 179–191 (1997)
17. Raftery, A.E., Kárný, M., Ettler, P.: Online prediction under model uncertainty via dynamic model averaging: Application to a cold rolling mill. Technometrics **52**(1), 52–66 (2010)
18. Raiffa, H., Schlaifer, R.: Applied Statistical Decision Theory (Harvard Business School Publications). Harvard University Press, Cambridge (1961)
19. Titterington, D.M., Smith, A.F.M., Makov, U.E.: Statistical Analysis of Finite Mixture Distributions. Wiley, New York (1985)
20. Wang, X.: Approximating Bayesian inference by weighted likelihood. Cannadian J. Stat. **34**(2), 279–298 (2006)

Localization in High-Dimensional Monte Carlo Filtering

Sylvain Robert and Hans R. Künsch

Abstract The high dimensionality and computational constraints associated with filtering problems in large-scale geophysical applications are particularly challenging for the Particle Filter (PF). Approximate but efficient methods such as the Ensemble Kalman Filter (EnKF) are therefore usually preferred. A key element of these approximate methods is localization, which is a general technique to avoid the curse of dimensionality and consists in limiting the influence of observations to neighboring sites. However, while it works effectively with the EnKF, localization introduces harmful discontinuities in the estimated physical fields when applied blindly to the PF. In the present paper, we explore two possible local algorithms based on the Ensemble Kalman Particle Filter (EnKPF), a hybrid method combining the EnKF and the PF. A simulation study in a conjugate normal setup allows to highlight the trade-offs involved when applying localization to PF algorithms in the high-dimensional setting. Experiments with the Lorenz96 model demonstrate the ability of the local EnKPF algorithms to perform well even with a small number of particles compared to the problem size.

Keywords Particle Filter · Ensemble Kalman Filter · Data Assimilation · Filtering in High Dimension · Curse of Dimensionality · Localized Filter Algorithms

1 Introduction

Monte Carlo methods are becoming increasingly popular for filtering in large-scale geophysical applications, such as reservoir modeling and numerical weather prediction, where they are often called ensemble methods for data assimilation. The challenging (and interesting) peculiarity of this type of applications is that the state space is extremely high dimensional (the number of dimensions of the state x is typically of the order of 10^8 and the dimension of the observation y of the order of 10^6), while the computational cost of the time integration step limits the sample size to

S. Robert (✉) · H.R. Künsch
ETH Zürich Seminar for Statistics, Zürich, Switzerland
e-mail: robert@stat.math.ethz.ch

R. Argiento et al. (eds.), *Bayesian Statistics in Action*, Springer Proceedings in Mathematics & Statistics 194, DOI 10.1007/978-3-319-54084-9_8

less than a hundred. Because of those particularly severe constraints, the emphasis is on developing approximate but highly efficient methods, typically relying on strong assumptions and exploiting parallel architectures.

The Particle Filter (PF) provides a fully general Bayesian solution to filtering [1, 6, 13], but it is well-known that it suffers from sample degeneracy and cannot be applied to high-dimensional settings [16]. The most popular alternative to the PF in large-scale applications is the Ensemble Kalman Filter (EnKF) [2, 3], a successful but heuristic method, which implicitly assumes that the predictive distribution is Gaussian.

Three main routes for adapting the PF to high-dimensional settings can be identified. The first one is to use an adaptive PF with a carefully chosen proposal distribution [13, 18]. A second approach is to build hybrid methods between the EnKF and the PF, as for example the Ensemble Kalman Particle Filter (EnKPF) [4]. A third route is localization, as it is a key element of the success of the EnKF in practice and could avoid the curse of dimensionality [14, 16].

The first approach requires an explicit model for the transition probabilities, which is typically not available in practical applications. Furthermore [17] showed that even with the optimal proposal distribution the PF suffers from the curse of dimensionality. Therefore in the present paper we focus on the second and third approaches and explore some possible localized algorithms based on the PF and the EnKPF. In a simulation study, we extend an example of [16] to illustrate how localization seemingly overcomes the curse of dimensionality, but at the same time introduces some harmful discontinuities in the estimated state. In a second experiment we show how local algorithms can be applied effectively to a filtering problem with the Lorenz96 model [10]. The results from these numerical experiments highlight key differences between the algorithms and demonstrate that local EnKPFs are promising candidates for large-scale filtering applications.

2 Ensemble Filtering Algorithms

Consider a state space model with state process (x_t) and observations (y_t), where the state process evolves according to some deterministic or stochastic dynamics and the observations are assumed to be independent given the state process, with likelihood $p(x_t|y_t)$. The goal is to estimate the conditional distribution of x_t given $y_{1:t} = (y_1, \ldots, y_t)$, called the *filtering* distribution and which we denote by π_t^f. In general it is possible to solve this problem recursively by alternating between a *prediction* step where the filtering distribution at time $(t-1)$ is propagated into the predictive distribution π_t^p at time t, and an *update* step, also called assimilation, where the predictive distribution is updated with the current observation to compute π_t^f. The update step is done by applying Bayes' rule as $\pi_t^f(x) \propto \pi_t^p(x) \cdot p(x|y_t)$, where the predictive distribution is the prior and the filtering distribution the posterior to be estimated.

Sequential Monte Carlo methods [1] approximate the predictive and filtering distributions by finite samples, or ensembles of *particles*, denoted by $(x_t^{p,i})$ and

$(x_t^{f,i})$ respectively, for $i = 1, \dots, k$. The update step consists in transforming the predictive ensemble $(x_t^{p,i})$ into an approximate sample from the filtering distribution π_t^f. We briefly present the PF and EnKF in this context and give an overview of the EnKPF. Henceforth we consider the update step only and drop the time index t. Additionally for the EnKF and EnKPF we assume that the observations are linear and Gaussian, i.e. $p(x|y) = \phi(y;\ Hx, R)$, the Gaussian density with mean Hx and covariance R evaluated at y.

The Particle Fillter approximates the filtering distribution as a mixture of point masses at the predictive particles, reweighed by their likelihood. More precisely:

$$\hat{\pi}_{PF}^f(x) = \sum_{i=1}^{k} w_i\, \delta_{x^{p,i}}(x), \qquad w_i \propto \phi(y;\ Hx^{p,i}, R). \tag{1}$$

A non-weighted sample from this distribution can be obtained by resampling, for example with a balanced sampling scheme [9]. The PF is asymptotically correct (also for non-Gaussian likelihoods), but to avoid sample degeneracy it needs a sample size which increases exponentially with the size of the problem (for more detail see [16]).

The Ensemble Kalman Filter is a heuristic method which applies a Kalman Filter update to each particle with stochastically perturbed observations. More precisely it constructs $(x^{f,i})$ as a balanced sample from the following Gaussian mixture:

$$\hat{\pi}_{EnKF}^f(x) = \sum_{i=1}^{k} \frac{1}{k}\, \phi(x;\ x^{p,i} + \hat{K}(y - Hx^{p,i}), \hat{K}R\hat{K}'), \tag{2}$$

where $\hat{K}$ is the Kalman gain estimated with $\hat{\Sigma}^p$, the sample covariance of $(x^{p,i})$. The stochastic perturbations of the observations are added to ensure that the filter ensemble has the correct posterior covariance on expectation.

The Ensemble Kalman Particle Filter combines the EnKF and the PF by decomposing the update step into two stages as $\pi^f(x) \propto \pi^p(x) \cdot p(x|y)^\gamma \cdot p(x|y)^{1-\gamma}$, following the progressive correction idea of [11]. The first stage, going from $\pi^p(x)$ to $\pi^\gamma(x) \propto \pi^p(x) \cdot p(x|y)^\gamma$ is done with an EnKF. The second stage is done with a PF and goes from $\pi^\gamma(x)$ to $\pi^f(x) \propto \pi^\gamma(x) \cdot p(x|y)^{1-\gamma}$. The resulting posterior distribution can be derived analytically as the following weighted Gaussian mixture:

$$\hat{\pi}_{EnKPF}^f(x) = \sum_{i=1}^{k} \alpha^{\gamma,i}\, \phi(x;\ \mu^{\gamma,i}, \Sigma^\gamma), \tag{3}$$

where the expressions for the parameters of this distribution and more details on the algorithm can be found in [4]. To produce the filtering ensemble $(x^{f,i})$, one first samples the mixture components with probability proportional to the weights $\alpha^{\gamma,i}$, using for example a balanced sampling scheme, and then adds an individual noise term with covariance Σ^γ to each particle. The parameter γ defines a continuous interpolation between the PF ($\gamma = 0$) and the EnKF ($\gamma = 1$). In the present study the value of γ is either fixed, for the sake of comparison, or chosen adaptively. In the later case γ is chosen such that the equivalent sample size of the filtering ensemble is within some acceptable range. Alternative schemes for choosing γ such as minimizing an objective cost function are currently being investigated but are beyond the scope of this work.

3 Local Algorithms

Localization consists essentially in updating the state vector by ignoring long range dependencies. This is a sensible thing to do in geophysical applications where the state represents discretized spatially correlated fields of physical quantities. By localizing the update step and using local observations only, one introduces a bias, but achieves a considerable gain in terms of variance reduction for finite sample sizes. For local algorithms the error is asymptotically bigger than for a global algorithm, but it is not dependent on the system dimension anymore and therefore avoids the curse of dimensionality. Furthermore, local algorithms can be efficiently implemented in parallel and thus take advantage of modern computing architectures.

The Local EnKF (LEnKF) consists in applying a separate EnKF at each site, but limiting the influence of the observations to sites that are spatially close (there are different ways to accomplish this in practice, see for example [7, 8, 12]). Analogously, we define the Local PF (LPF) as a localized version of the PF, where the update is done at each location independently, considering only observations in a ball of radius ℓ. In order to avoid arbitrary "scrambling" of the particles indices, we use a balanced sampling scheme [9], and some basic ad-hoc methods to reduce the number of discontinuities, but we do not solve this problem optimally as it would greatly hinder the efficiency of the algorithm.

For the EnKPF we define two different local algorithms: the naive-local EnKPF (naive-LEnKPF), in which localization is done exactly as for the LEnKF, and the block-local EnKPF (block-LEnKPF), in which the observations are assimilated sequentially but their influence is restricted to a local area. The naive-LEnKPF does not take particular care of the introduced discontinuities beyond what is done for the PF, but it is straightforward to implement. The block-LEnkPF, on the other hand, uses conditional resampling in a transition area surrounding the local assimilation window, which ensures that there are no sharp discontinuities, but it involves more overhead computation. For more detail about the local EnKPF algorithms see [15].

4 Simulation Studies

We conducted two simulation studies: first a one-step conjugate normal setup where the effect of localization can be closely studied, and second a cycled experiment with the Lorenz96 model, a non-linear dynamical system displaying interesting non-Gaussian features.

4.1 Conjugate Normal Setup

We consider a simple setup similar to the one in [16], with a predictive distribution π^p assumed to be a N-dimensional normal with mean zero and covariance Σ^p. To imitate the kind of smooth fields that we encounter in geophysical applications, we construct the covariance matrix as $\Sigma^p_{ii} = 1$ and $\Sigma^p_{ij} = K_{GC}(d(i, j)/r)$, where K_{GC} is the Gaspari-Cohn kernel [5], $d(i, j)$ the distance between sites i and j on a one-dimensional domain with periodic boundary conditions, and the radius r in the denominator is chosen such that the covariance has a finite support of 20 grid points. From this process we generate observations of every component of x and standard Gaussian noise:

$$x \sim \mathcal{N}(0, \Sigma^p), \qquad y|x \sim \mathcal{N}(x, I). \tag{4}$$

In order to study the finite sample properties of the different algorithms, we compute the Mean Squared Error (MSE) of the ensemble mean in estimating the value x at each location, which we denote by MSE(x). Because the prior is conjugate to the likelihood, we can compute the MSE(x) of the posterior mean analytically for known Σ^p as the trace of the posterior covariance matrix and use this as a reference. For the simulation we use a sample size of $k = 100$ and average the results over 1000 runs. It should be noted that because the predictive distribution is normal, this setup is favorable to the EnKF and LEnKF, but the EnKPFs should still perform adequately. For the local algorithms the localization radius ℓ was set to 5, resulting in a local window of 11 grid points, which is smaller than the correlation length used to generate the data. Later on we study the effect of ℓ on the performance of the algorithms. For the EnKPF algorithms the parameter γ was fixed to 0.25, which means a quarter of EnKF and three-quarter of PF. In practice one would rather choose the value of γ adaptively, but the exact value does not influence the qualitative conclusions drawn from the experiments and fixing it in this way makes the comparison easier.

An example of a sample from the filtering distribution produced by different local algorithms is shown in Fig. 1, with each particle represented as a light blue line, the true state in dark and the observations in red. For more clarity the ensemble size is set to 10 and the system dimension to 40. While all algorithms manage to recover more or less the underlying state, it is clear that they vary in terms of quality. The LPF in particular suffers from sample depletion, even when applied locally, and displays strong discontinuities. If one looks closely at the naive-LEnKF ensemble, discontinuities can also be identified. The block-LEnKPF and the LEnKF, on the other hand, produce smoother posterior particles. This example is useful to highlight the behavior of the different local algorithms qualitatively, but we now proceed to a more quantitative assessment with a repeated simulations experiment.

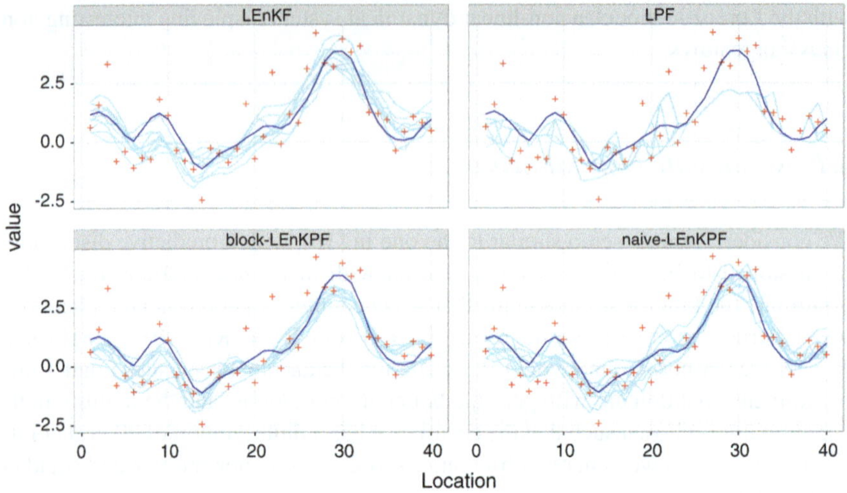

Fig. 1 Example of analysis ensemble with different local algorithms. Each particle is a *light blue line*, the true state in *dark* and the observations in *red*. The ensemble size is restricted to 10 and the domain size to 40 for better legibility

Beating the curse of dimensionality: In the first row of Fig. 2, the MSE(x) is plotted as a function of the system dimension N, for the global algorithms on the left and the local algorithms on the right. The values are normalized by the optimal MSE(x) to make them more interpretable. The PF degenerates rapidly, with an MSE(x) worse than using the prior mean (upper dashed line). The EnKF and the EnKPF suffer as well from the curse of dimensionality, although to a lesser extent. The local algorithms, on the other hand, are immune to the increase of dimensions N and their MSE(x) is constant and very close the optimum, which confirms that localization is working as expected. The LEnKF, naive-LEnKPF and block-LEnKPF make an error of less than 5% while the LPF is 20% worse than the optimum.

The cost of localization: As the old statistical adage goes, there is no free lunch: localization comes at a cost, particularly for PF algorithms. When doing the update locally with the EnKF, the filtering samples are relatively smooth fields, because the update applies spatially smoothly varying corrections to the predictive ensemble. However, for the LPF, when different particles are resampled at neighboring sites, arbitrarily large discontinuities can be created. While this might be discarded as harmless, it is not the case when the fields of interest are spatial fields of physical quantities used in numerical solvers of partial differential equations. One way to measure the impact of discontinuities is to look at the MSE in estimating the lag one increments Δx, which we denote as MSE(Δx). While the MSE(x) is computed for the posterior mean, the MSE(Δx) is computed for each particle separately and then averaged. We again compute the expected MSE(Δx) under the conjugate posterior distribution and use it as reference.

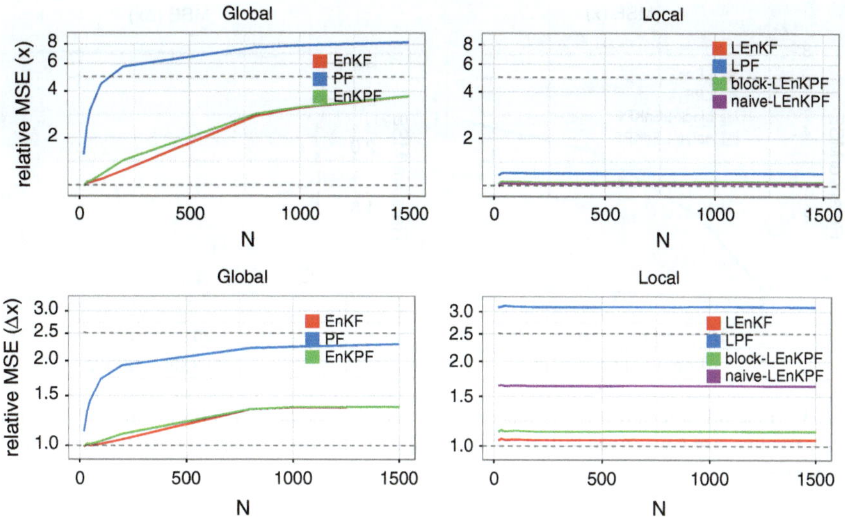

Fig. 2 Illustration of the relationship between the system dimension N and different quantities. In the first row is the MSE(x) for the global algorithms on the *left* and the local algorithms on the *right*. In the second row the same but for the MSE(Δx). All the values are given relative to the optimal one obtained with the true posterior distribution (*dashed line* at 1). The relative MSE of using the prior without using any observation is given by the second *dashed line*. Notice the log-scale on the y-axis

The plots in the second row of Fig. 2 show this quantity for the different algorithms averaged over 1000 simulation runs. The MSE(Δx) of the local algorithms is still constant as a function of N, as expected, but in the cases of the naive-LEnKPF and the LPF its value is worse than for the respective global algorithms. On the other hand, the LEnKPF and the block-LEnKPF improve on their global counterparts and have an error relatively close to the optimum.

Localization trade-off: In the previous experiment we fixed ℓ, the localization radius, to 5, and looked at what happens in terms of prediction accuracy with the MSE(x), and in terms of discontinuities with the MSE(Δx). In Fig. 3 we now look at MSE as a function of ℓ, fixing N to 200 and k to 100. For large values of ℓ the MSE(Δx) is smallest as discontinuities are avoided, but the MSE(x) is not optimal, particularly for the LPF. As ℓ is reduced the MSE(x) decreases for all methods, while MSE(Δx) is kept constant for a wide range of ℓ values. At some point, different for each algorithm, the localization is too strong and becomes detrimental, with both MSE(x) and MSE(Δx) sharply increasing. This behavior illustrates the trade-off at hand when choosing the localization radius: picking a too small value introduces a bias by neglecting useful information and creates too much discontinuities, while choosing a too large value does not improve MSE(Δx) but leads to poorer performance in terms of MSE(x).

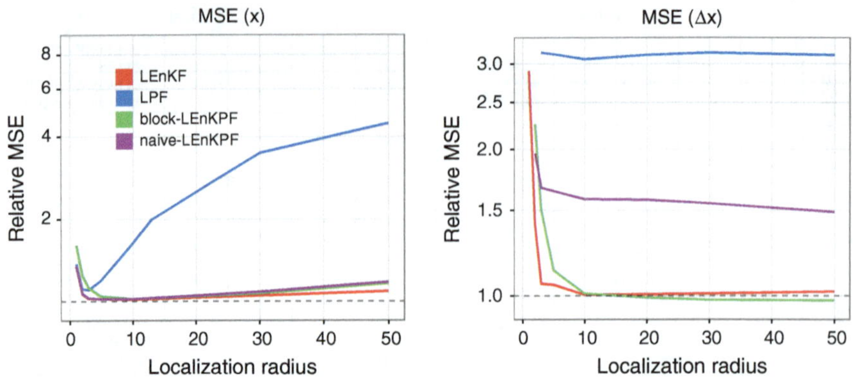

Fig. 3 Trade-off of localization: influence of ℓ on $\text{MSE}(x)$ and $\text{MSE}(\Delta x)$ for the local algorithms. The ensemble size k was fixed to 100 and the system dimension N to 200. Notice the log-scale on the y-axis

4.2 *Filtering with the Lorenz96 Model*

The Lorenz96 model [10] is a 40-dimensional non-linear dynamical system which displays a rich behavior and is often used as a benchmark for filtering algorithms. In [4] it was shown that the EnKPF outperforms the LEnKF in some setups of the Lorenz96 model, but the sample size required was of 400. In the present experiment we use the same setup as in [4] but with much smaller and realistic ensemble sizes. The data are assimilated at time intervals of 0.4, which leads to strong non-linearities and thus highlights better the relative advantages of the EnKPF. Each experiment is run for 1000 cycles and repeated 20 times, which provides us with stable estimates of the average performance of each algorithm. As in [4], the parameter γ of the EnKPFs is chosen adaptively such that the equivalent sample size is between 25 and 50% of the ensemble size. It should be noted that for local algorithms, a different γ is chosen at each location, which provides added flexibility and allows to adapt to locally non-Gaussian features of the distribution. We consider $\text{MSE}(x)$ only and denote it simply by MSE. It also takes errors in the estimation of increments into account through integration in time during the propagation steps.

In the left panel of Fig. 4 the MSE of the global algorithms is plotted against ensemble size. The PF is not represented as it diverges for such small values. The MSE is computed relative to the performance of the prior, which is simply taken as the $\text{MSE}(x)$ of an ensemble of the same size evolving according to the dynamical system equations but not assimilating any observations. With ensemble sizes smaller than 50, the filtering algorithms are not able to do better than the prior, which means that trying to use the observations actually makes them worse than not using them at all. Only for ensemble sizes of 100 and more do the global algorithms start to become effective. In practice we are interested in situations where the ensemble size

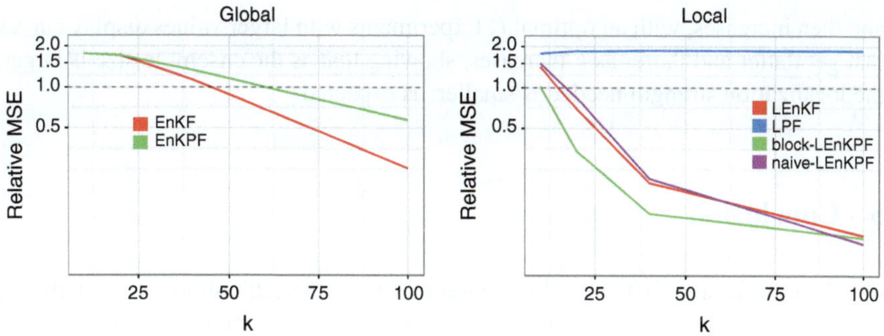

Fig. 4 Global and local algorithms results with the Lorenz96 model. On the y-axis the MSE and on the x-axis increasing ensemble sizes k. There is no global PF as the algorithm does not work for such small ensemble sizes. The MSE is computed relative to the MSE of an ensemble of the same size but which does not assimilate any observation. Notice the log-scale on the y-axis

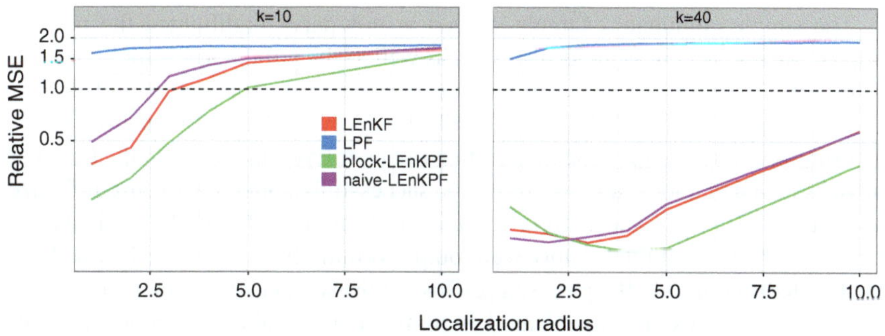

Fig. 5 Interplay of ensemble size k and localization radius ℓ for the the Lorenz96 model. The relative MSE is plotted as a function of ℓ for different value of k in the different panels. Notice the log-scale on the y-axis

is smaller than the system dimension (here 40), and thus the global methods are clearly not applicable.

On the right panel of Fig. 4 we show the same plot but for the local algorithms. For sample sizes as small as 20 or 30 the performances are already quite good. The LPF, however, does not work at all, probably because it still suffers from sample depletion and because the discontinuities it introduces have a detrimental impact during the prediction step of the algorithm. The block-LEnKPF clearly outperforms the other algorithms, particularly for smaller sample sizes. This indicates that it can localize efficiently the update without harming the prediction step by introducing discontinuities in the fields.

In order to better highlight the trade-off of localization, we plot similar curves but as a function of the localization radius ℓ in Fig. 5. One can see that for small k (left panel), the error is increasing with ℓ, which shows that localization is absolutely necessary for the algorithm to work. For $k = 40$ (right panel), the MSE first decreases and then increases, with an optimal ℓ. Experiments with larger values display curves that get flatter and flatter as k increases, showing that as the ensemble size is larger, the localization strength needed is smaller, as expected.

5 Conclusion

Localization is an effective tool to address some of the difficulties associated with high-dimensional filtering in large-scale geophysical applications. Methods such as the EnKF can be localized easily and successfully as they vary smoothly in space. At first sight, the LPF does seem to overcome the curse of dimensionality; however, looking more carefully, one notices that it introduces harmful discontinuities in the updated fields. The two localized EnKPFs both overcome the curse of dimensionality and handle better the problem of discontinuities.

The simple conjugate example studied in this paper highlighted the potential improvements coming from localization, as well as the pitfalls when applied blindly to the PF. The trade-off between the bias coming from localization and the gain coming from the reduced variance was illustrated by exploring the behavior of the algorithms as a function of the localization radius ℓ. Experiments with the Lorenz96 model showed that local algorithms can be successfully applied with ensemble sizes as small as 20 or 30, and highlighted the localization trade-off. In particular, the block-LEnKPF fared remarkably well, outperforming both the naive-LEnKPF and the LEnKPF in this challenging setup. This confirms other results that we obtained with more complex dynamical models mimicking cumulus convection [15] and encourages us to pursue further research with localized EnKPFs in a large-scale application in collaboration with Meteoswiss.

References

1. Doucet, A., De Freitas, N., Gordon, N.: Sequential Monte Carlo Methods in Practice. Springer New-York (2001)
2. Evensen, G.: Sequential data assimilation with a nonlinear quasi-geostrophic model using Monte Carlo methods to forecast error statistics. J. Geophys. Res. Oceans **99**(C5), 10143–10162 (1994)
3. Evensen, G.: The ensemble Kalman filter: theoretical formulation and practical implementation. Ocean Dyn. **53**(4), 343–367 (2003)
4. Frei, M., Künsch, H.R.: Bridging the ensemble Kalman and particle filters. Biometrika **100**(4), 781–800 (2013)
5. Gaspari, G., Cohn, S.E.: Construction of correlation functions in two and three dimensions. Q. J. R. Meteorol. Soc. **125**(554), 723–757 (1999)
6. Gordon, N., Salmond, D., Smith, A.: Novel approach to nonlinear non-Gaussian Bayesian state estimation. IEE Proc. Radar Signal Process. **140**(2), 107–113 (1993)
7. Houtekamer, P.L., Mitchell, H.L.: A sequential ensemble Kalman filter for atmospheric data assimilation. Mon. Weather Rev. **129**(1), 123–137 (2001)

8. Hunt, B.R., Kostelich, E.J., Szunyogh, I.: Efficient data assimilation for spatiotemporal chaos: a local ensemble transform Kalman filter. Physica D: Nonlinear Phenomena **230**(1–2), 112–126 (2007)
9. Künsch, H.R.: Recursive Monte Carlo filters: algorithms and theoretical analysis. Ann. Stat. **33**(5), 1983–2021 (2005)
10. Lorenz, E.N., Emanuel, K.A.: Optimal sites for supplementary weather observations: simulation with a small model. J. Atmos. Sci. **55**(3), 399–414 (1998)
11. Musso, C., Oudjane, N., Le Gland, F.: Improving regularised particle filters. In: A. Doucet, N. De Freitas, N. Gordon (eds.) Sequential Monte Carlo Methods in Practice, pp. 247–271. Springer, New-York (2001)
12. Ott, E., Hunt, B.R., Szunyogh, I., Zimin, A.V., Kostelich, E.J., Corazza, M., Kalnay, E., Patil, D.J., Yorke, J.A.: A local ensemble Kalman filter for atmospheric data assimilation. Tellus A **56**(5), 415–428 (2004)
13. Pitt, M.K., Shephard, N.: Filtering via simulation: auxiliary particle filters. J. Am. Stat. Assoc. **94**(446), 590–599 (1999)
14. Rebeschini, P., Handel, R.v.: Can local particle filters beat the curse of dimensionality? Ann. Appl. Prob. **25**(5), 2809–2866 (2015)
15. Robert, S., Künsch, H.R.: Local Ensemble Kalman Particle Filters for Efficient Data Assimilation. Tellus A **69**(1), 1–14 (2017)
16. Snyder, C., Bengtsson, T., Bickel, P., Anderson, J.: Obstacles to high-dimensional particle filtering. Mon. Weather Rev. **136**(12), 4629–4640 (2008)
17. Snyder, C., Bengtsson, T., Morzfeld, M.: Performance bounds for particle filters using the optimal proposal. Mon. Weather Rev. **143**(11), 4750–4761 (2015)
18. van Leeuwen, P.J.: Nonlinear data assimilation in geosciences: an extremely efficient particle filter. Q. J. R. Meteorol. Soc. **136**(653), 1991–1999 (2010)

Linear Inverse Problem with Range Prior on Correlations and Its Variational Bayes Inference

Ondřej Tichý and Václav Šmídl

Abstract The choice of regularization for an ill-conditioned linear inverse problem has significant impact on the resulting estimates. We consider a linear inverse model with on the solution in the form of zero mean Gaussian prior and with covariance matrix represented in modified Cholesky form. Elements of the covariance are considered as hyper-parameters with truncated Gaussian prior. The truncation points are obtained from expert judgment as range on correlations of selected elements of the solution. This model is motivated by estimation of mixture of radionuclides from gamma dose rate measurements under the prior knowledge on range of their ratios. Since we aim at high dimensional problems, we use the Variational Bayes inference procedure to derive approximate inference of the model. The method is illustrated and compared on a simple example and on more realistic 6 h long release of mixture of 3 radionuclides.

Keywords Linear inverse problem · Variational Bayes inference · Convex optimization · Uncertain correlations · Gamma dose rate measurements · Nuclide ratios

1 Introduction

Linear inverse problems are fundamental in many areas of science, signal processing, or machine learning. The conventional least squares method fails when the problem is ill-conditioned. In these cases, appropriate regularizations are beneficial to obtain desirable solution. Most commonly used regularizations are the Tikhonov [3] and LASSO [12] where different norms of the unknown vector are used, l_2 and l_1 respectively.

O. Tichý (✉) · V. Šmídl
Institute of Information Theory and Automation, Pod Vodarenskou Vezi 4, Prague, Czech Republic
e-mail: otichy@utia.cas.cz

V. Šmídl
e-mail: smidl@utia.cas.cz

R. Argiento et al. (eds.), *Bayesian Statistics in Action*, Springer Proceedings in Mathematics & Statistics 194, DOI 10.1007/978-3-319-54084-9_9

Both of these methods have Bayesian interpretation with different prior distribution of the unknown vector. However, parameters of these prior distributions are assumed to be known. More flexible models allow for estimation of the hyper-parameters, e.g. in the form of diagonal elements of the prior covariance matrix, which is known as the automatic relevance determination principle [14] since it favors sparse solutions. Theoretically, full covariance matrix can be also estimated using Wishart distribution [6, 13]. However, the problem is then over-parametrized and the influence of additional regularization is significant. In this contribution, we are concerned with models where some elements of the covariance matrix are vaguely known and need to be estimated from the data. We assume the knowledge of ranges of selected elements of the covariance matrix. We follow idea of Daniels and Pourahmadi [2] where modified Cholesky decomposition of the covariance matrix is used for longitudinal data. In our model, we restricted the possible interval for specific elements of the covariance matrix using truncated Gaussian distribution. These intervals are expert information and are considered as input of our algorithm.

The proposed approach is illustrated on simple synthetic example where comparison with Tikhonov and LASSO regularizations will be given. In addition, we apply the resulting algorithm on a problem of determination of the source term of an atmospheric release of radiation where ratios of the released nuclides are vaguely known. This scenario is relevant to the case of the Fukushima Daiichi nuclear power plant accident [8]. We aim for estimation of the time profile of the release using gamma dose rate (GDR) measurements, so our measurement vector does not contain nuclide-specific concentration activity measurements but bulk gamma dose rates from a mixture of nuclides. Particularly important are prior assumptions on the nuclide ratios and their treatment. These can be obtained, e.g., from physical analysis of the power plant state (reactor inventory combined with assumptions on the accident type) or from a few available nuclide-specific activity concentration samples downwind the release. In our simulated scenario, 6 h release of a mixture of 3 nuclides is considered and Austria monitoring network is used together with realistic meteorological data.

2 Mathematical Method

We study the following linear inverse problem

$$\mathbf{y} = M\mathbf{x} + \mathbf{e}, \tag{1}$$

where $\mathbf{y} \in \mathbf{R}^{p\times 1}$ is vector of measurements corrupted by error vector $\mathbf{e}$ of the same size, $M \in \mathbf{R}^{p\times n}$ is known matrix, and $\mathbf{x} \in \mathbf{R}^{n\times 1}$ is the unknown vector to be estimated. Solution of the noise-less problem via ordinary least square method is $\mathbf{x} = (M^T M)^{-1} M^T \mathbf{y}$, which is often infeasible due to ill-conditioned matrix M.

The problem is typically recast as an optimization problem

$$\mathbf{x}^* = \arg\min_{\mathbf{x}\in\mathcal{X}} \left\{ ||\mathbf{y} - M\mathbf{x}||_2^2 + \alpha g(\mathbf{x}) \right\}, \tag{2}$$

where $g(\mathbf{x})$ is a regularization term and α is its weight. Common regularization terms are Tikhonov regularization [3] or LASSO regularization [12]:

$$g_{\text{Tikhonov}}(\mathbf{x}) = ||\mathbf{x}||_2^2, \qquad g_{\text{LASSO}}(\mathbf{x}) = ||\mathbf{x}||_1, \tag{3}$$

however, the parameter α needs to be carefully selected or determined. The optimization approach (2) can be interpreted as a maximum a posteriori estimate of a Bayesian model. Many detailed analysis of Bayesian interpretations and also extensions are available, e.g. [7]. For the purpose of this text, we only note that the Tikhonov regularization is equivalent to MAP estimation of probabilistic model

$$\mathbf{x}^* = \arg\min_{\mathbf{x}\in\mathcal{X}} \left\{ -\log p(\mathbf{y}|M, \mathbf{x}) - \log p(\mathbf{x}|\alpha) \right\}, \tag{4}$$

with

$$p(\mathbf{y}|M, \mathbf{x}) = \mathcal{N}_{\mathbf{y}}\left(M\mathbf{x}, I_p\right), \qquad p(\mathbf{x}|\alpha) = \mathcal{N}_{\mathbf{x}}\left(\mathbf{0}, \alpha^{-1} I_n\right), \tag{5}$$

where $\mathcal{N}$ denotes Gaussian distribution and I_p denotes identity matrix with given size. For given α, the Bayesian model is fully equivalent to the optimization problem (2). However, the unknown parameters, α in this case, can be modeled using hierarchical priors and estimated within the model [1].

For problem specific tasks where assumption on same parameters arise such as non-negativity of $\mathbf{x}$, the optimization approach (2) can be supplemented using "subject to" condition. In Bayesian formulation, this condition can be enforced using truncated Gaussian prior denoted as $t\mathcal{N}$, see one dimensional example in Fig. 1 and Appendix for details.

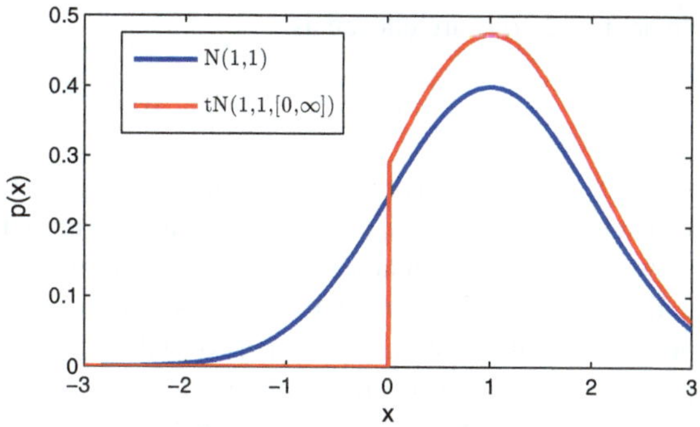

Fig. 1 Example of the Gaussian distribution $\mathcal{N}(1, 1)$, *blue line*, and the truncated Gaussian distribution $t\mathcal{N}(1, 1, [0, \infty])$, *red line*

2.1 Bayesian Hierarchical Model

Consider probabilistic formulation of linear inverse problem (1) with isotropic Gaussian noise

$$p(\mathbf{y}|\mathbf{x}, \omega) = \mathcal{N}_{\mathbf{y}}\left(M\mathbf{x}, \omega^{-1} I_p\right), \tag{6}$$

where ω is precision of noise. For unknown ω, we assume prior model in the form of Gamma $\mathcal{G}_\omega(\vartheta_0, \rho_0)$. All prior parameters (subscripted by 0) are set to non-informative values of 10^{-10}. We assume the unknown vector $\mathbf{x}$ to have Gaussian prior; however, with truncated support to positive values,

$$p(\mathbf{x}|\Omega) = t\mathcal{N}_{\mathbf{x}}\left(\mathbf{0}, \Omega^{-1}, [0, +\infty]\right). \tag{7}$$

We aim to model the precision matrix Ω in more detail; hence, we assume Ω in the form of modified Cholesky decomposition as

$$\Omega = L \Upsilon L^T, \tag{8}$$

where Υ is diagonal matrix with diagonal entries $\boldsymbol{\upsilon} = [\upsilon_1, \ldots, \upsilon_n]$ with prior Gamma model $\mathcal{G}_{\upsilon_j}(\alpha_0, \beta_0)$ for each element and L is lower triangular matrix

$$L = \begin{pmatrix} 1 & 0 & 0 & 0 \\ l_{2,1} & 1 & 0 & 0 \\ \vdots & \ddots & 1 & 0 \\ l_{n,1} & \ldots & l_{n,n-1} & 1 \end{pmatrix}, \tag{9}$$

with unknown off-diagonal elements forming column vectors $\mathbf{l}_i = \left[l_{i+1,i}, l_{i+2,i}, \ldots, l_{n,i}\right]^T \in \mathbf{R}^{(n-i)\times 1}$ for $i = 1, \ldots n-1$. We will introduce prior model for vectors $\mathbf{l}_i$ whose estimates together with estimate of vector $\boldsymbol{\upsilon}$ fully determine the covariance matrix decomposition (8). The prior model for each non-zero element of L, $l_{i,k}$, are chosen as

$$p\left(l_{i,k}|\psi_{i,k}\right) = t\mathcal{N}_{l_{i,k}}\left(0, \psi_{i,k}^{-1}, [a_{i,k}, b_{i,k}]\right), \tag{10}$$

where $\psi_{i,k}$ is unknown precision parameter with prior Gamma model $\mathcal{G}_{\psi_{i,k}}(\zeta_0, \eta_0)$ and with selected interval $[a_{i,k}, b_{i,k}]$ of truncated Gaussian distribution. These intervals allow us to select boundaries for each element of the covariance matrix.

Estimation of the model parameters is analytically intractable; hence, we employ the Variational Bayes method [10] to yield an approximate solution. The Variational Bayes method estimates the posterior solution in the form of conditionally independent distributions that minimize the Kullback–Leibler divergence to the true posterior. This minimization leads to a set of implicit equations which have to be solved iteratively. Here, shaping parameters of recognized posterior distributions

$$\tilde{p}(\mathbf{x}|\mathbf{y}) = t\mathcal{N}_{\mathbf{x}}\left(\mu_{\mathbf{x}}, \Sigma_{\mathbf{x}}, [0, +\infty]\right), \tag{11}$$
$$\tilde{p}(v_j|\mathbf{y}) = \mathcal{G}_{v_j}\left(\alpha_j, \beta_j\right), \tag{12}$$
$$\tilde{p}(l_{i,k}|\mathbf{y}) = t\mathcal{N}_{l_{i,k}}\left(\mu_{l_{i,k}}, \Sigma_{l_{i,k}}, [a_{i,k}, b_{i,k}]\right), \tag{13}$$
$$\tilde{p}(\psi_{i,k}|\mathbf{y}) = \mathcal{G}_{\psi_{i,k}}\left(\zeta_{i,k}, \eta_{i,k}\right) \tag{14}$$
$$\tilde{p}(\omega|\mathbf{y}) = \mathcal{G}_{\omega}\left(\vartheta, \rho\right), \tag{15}$$

are iteratively evaluated, see Algorithm 1. The algorithm will be denoted as the least square with the prior adaptive covariance with interval restrictions (LS-APCi) algorithm.

3 Experiments

To test and compare the studied LS-APCi algorithm, we first design a simple synthetic dataset. Second, we perform experiment on realistic gamma dose rate measurement with vaguely known ratios of selected radionuclides.

3.1 Toy Example

We select an ill-conditioned matrix $M \in \mathbf{R}^{6\times 3}$ with elements within 0 and 1 with eigenvalues $[2 \times 10^{-7}, 0.19, 0.23]$. The original vector $\mathbf{x}$ is selected as $\mathbf{x}_{\text{true}} = [1, 2, 3]^T$ and measurement vector is generated according to the assumed model (1) with $\mathbf{e} \sim \mathcal{N}(0, 0.1)$. The negative elements of $\mathbf{y}$ are cropped to 0. We will test two settings of the LS-APCi algorithm: (i) the space of possible solutions is restricted using fixed ratios of elements of vector $\boldsymbol{v}$: $\boldsymbol{v} = [v_1, 10v_1, 10v_1]$, and (ii) unrestricted $\boldsymbol{v}$. The prior intervals for the unknown elements of matrix L are

$$[a_{2,1}, b_{2,1}] = [-10; -1], \quad [a_{3,1}, b_{3,1}] = [-10; -1], \tag{22}$$

while the simulated are $l_{2,1} = -2$ and $l_{3,1} = -3$.

The results of the LS-APCi algorithm are given in Fig. 2. The results suggest that the restriction of the space of possible solutions are beneficial and the estimates converge to the true values, see Fig. 2a. On the other hand, estimation of full vector $\boldsymbol{v} = [v_1, v_2, v_3]$ results in over-parametrization of the problem and the estimates of the ratios in matrix L converge to the centers of the selected intervals. In result, the estimated vector $\mathbf{x}$ differs from the true vector, see Fig. 2b.

For comparison, we provide results of the LASSO algorithm, Fig. 3 left, and of the Tikhonov algorithm, Fig. 3 right. Since both algorithms need to preselect suitable regularization parameter, we run both algorithms for a wide range of the regularization parameters and select the best result for each algorithm. The key

Algorithm 1 The least square with the prior adaptive covariance with interval restrictions (LS-APCi) algorithm.

1. Initialization

   (a) Set all prior parameters (subscripted by 0) to 10^{-10}.
   (b) Set initial values: $\langle L\rangle = \langle \Upsilon\rangle = I_n$ and $\langle\omega\rangle = \frac{1}{\max(M^T M)}$.

2. Iterate until convergence or maximum number of iteration is reached:

   (a) Compute moments of $\langle \mathbf{x}\rangle$ using Appendix and shaping parameters of (11):

$$\Sigma_{\mathbf{x}} = \left(\langle\omega\rangle M^T M + \left\langle L\Upsilon L^T\right\rangle\right)^{-1}, \tag{16}$$

$$\mu_{\mathbf{x}} = \Sigma_{\mathbf{x}}\left(\langle\omega\rangle M^T\mathbf{y}\right), \tag{17}$$

   (b) Compute moment $\langle\Upsilon\rangle$ using shaping parameters of (12):

$$\boldsymbol{\alpha} = \alpha_0 + \frac{1}{2}\mathbf{1}_{n,1},\ \ \boldsymbol{\beta} = \beta_0 + \frac{1}{2}\text{diag}\left(\left\langle L^T\mathbf{x}\mathbf{x}^T L\right\rangle\right), \tag{18}$$

   (c) Compute moments of $\langle L\rangle$ with restricted ranges using Appendix and shaping parameters of (13):

$$\Sigma_{l_{i,k}} = \left(\langle\upsilon_i\rangle\left\langle x_{(i+1),k}x_{(i+1),k}^T\right\rangle + \text{diag}(\langle\psi_{i,k}\rangle)\right)^{-1}, \tag{19}$$

$$\mu_{l_{i,k}} = \Sigma_{l_{i,k}}\left(-\langle\upsilon_i\rangle\langle x_i x_{(i+1),k}\rangle\right), \tag{20}$$

   (d) Compute moment $\langle\omega\rangle$using shaping parameters of (15):

$$\vartheta = \vartheta_0 + \frac{p}{2},\ \ \rho = \rho_0 + \frac{1}{2}\text{tr}\left(\left\langle\mathbf{x}\mathbf{x}^T\right\rangle M^T M\right) - \mathbf{y}^T M\langle\mathbf{x}\rangle + \frac{1}{2}\mathbf{y}^T\mathbf{y}, \tag{21}$$

3. Report resulting estimated source term $\langle\mathbf{x}\rangle$

differences is in estimation of x_1. The LASSO algorithm estimates exact 0 which corresponds to its preference of a sparse solution. The Tikhonov algorithm estimates very similar result to the LS-APCi with unrestricted parameter $\boldsymbol{\upsilon}$. However, the LS-APCi with restriction is clearly closer to the true vector $\mathbf{x}$ as well as to the true covariance matrix and we will use this version of the algorithm in the next experiment.

3.2 *Realistic Example*

The linear inverse problem (1) is common in estimation of the source term of an atmospheric release. Here, the vector $\mathbf{y}$ contains gamma dose rate (GDR) measurements and the matrix M is a source-receptor-sensitivity matrix computed using an atmospheric transport model [9]. Note that the vector $\mathbf{y}$ does not contain any nuclide-specific information but only sum of GDR of a mixture of nuclides and the matrix

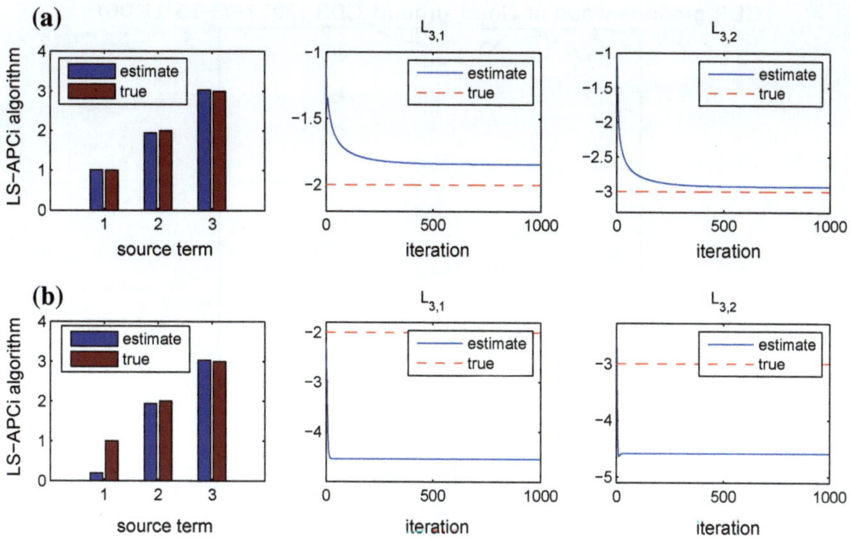

Fig. 2 The results of the LS-APCi algorithm with restricted (a) and unrestricted (b) parameter υ

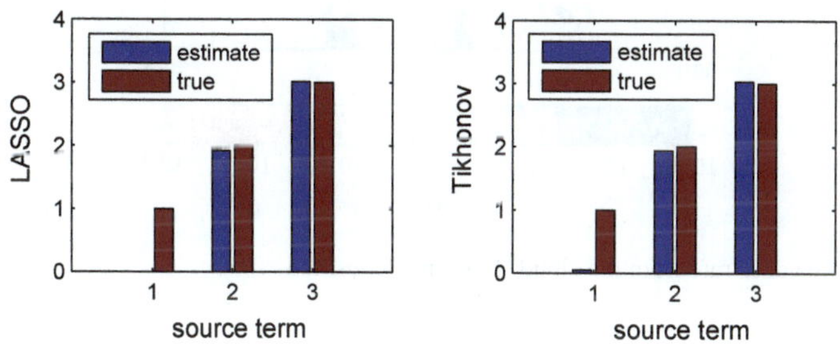

Fig. 3 The results of the LASSO algorithm (*left*) and Tikhonov algorithm (*right*)

M cumulates errors from atmospheric model including errors from the estimates of meteorological conditions (in this case, ECMWF Era-Interim data).

In this case, a 6 h long constant rate release is simulated using 3 nuclides: Cs-137, I-131, and Xe-133 from the Czech nuclear power plant Temelin. The Austrian radiation monitoring network is considered to provide measurements from more than 300 receptors implying $M \in R^{4032\times 18}$, see Fig. 4. To simulate realistic conditions, different meteorological data were used for generation matrix M and for generation of simulated measurements $\mathbf{y}$. The problem is critically ill-conditioned and classical optimization methods provide unsuitable results. For our algorithm, we use the following expert-defined intervals of nuclide ratios:

$$[a_{7:12,1}, b_{7:12,1}] = [-10, -3], \quad [a_{13:18,1}, b_{13:18,1}] = [-20, -50], \tag{23}$$

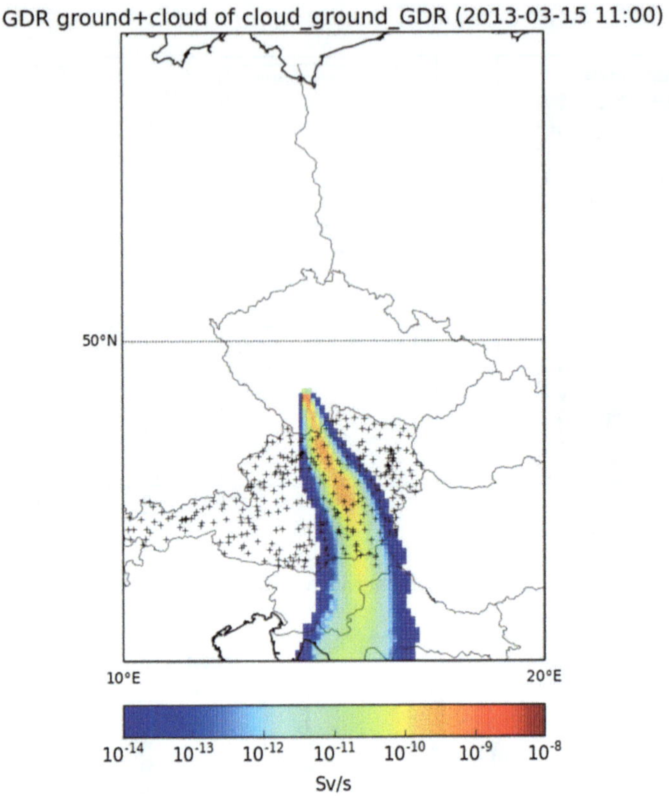

Fig. 4 Gamma dose rate from the cloud shine and deposition

covering the true (simulated) ratios $l_{7:12,1} = -3.8$ and $l_{13:18,1} = -31.3$ (which is, however, unknown in reality).

The results of the LS-APCi algorithm are given in Fig. 5 using subplot for each nuclide. We conclude that the results well correspond to the true releases. Note that in sums of the elements, $\mathbf{x}_{\text{true}}$ and the estimated $\mathbf{x}$ are almost equal. The dissimilarities can be caused by mismatch in the meteorological conditions as well as by uncertainty of the measurement. We perform also run of the LS-APCi algorithm with unrestricted υ with significantly worse results; hence, we conclude that the restriction of υ is crucial for the algorithm.

The results are compared with those of optimization approach with LASSO and Tikhonov regularization with the same ranges restrictions (23) as the LS-APCi algorithm. For this experiment, we used CVX toolbox [4, 5] where the optimization problem (2) can be formulated to respect the ranges given in (23). Since the crucial parameter of the optimization approach (2) is α, we run the LASSO and Tikhonov algorithms with $\alpha \in \left[10^{-5}, 10^{5}\right]$. Similarly, we identify as the most significant initial parameter of the LS-APCi algorithm as $\Upsilon = \alpha I_n$; hence, we compare these 3

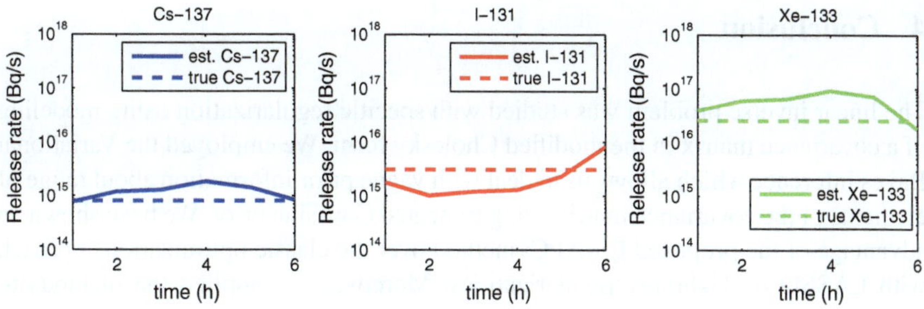

Fig. 5 The results of the source term estimation of 6 h constant release of 3 nuclides using LS-APCi algorithm

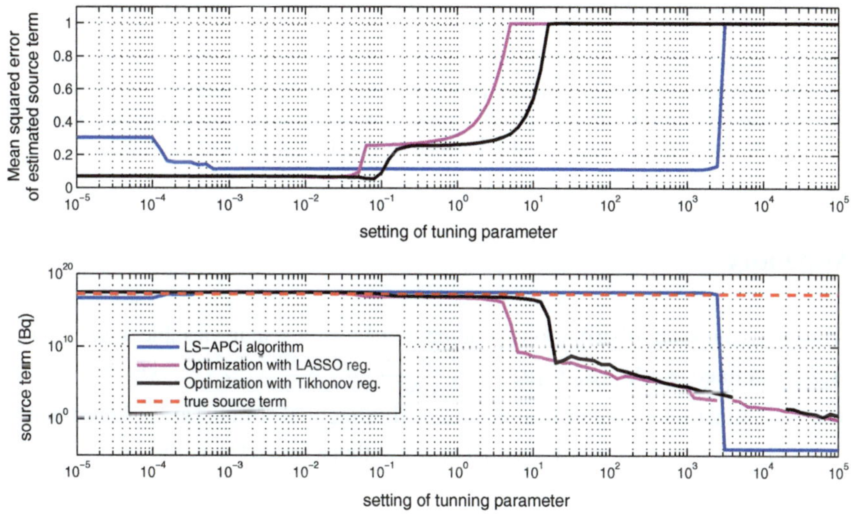

Fig. 6 **Top row** mean squared error between the true source term and the estimated source term for each tested algorithm and each parameter α. **Bottom row** sum of total activity of the source term for each algorithm accompanied by the true sum of the source term (*red dashed line*)

algorithm with respect to this parameter α. We normalize each nuclide activity to interval $[0, 1]$ and compute mean squared error (MSE) for each α and for each algorithm. The MSE depending on selected parameter α are given in Fig. 6, top, accompanied by the estimated sum of total activity of the source term. From these results, we can identify two main modes of the LS-APCi solution. Note that the natural choice $\Upsilon = I_n$, see Algorithm 1, lies in the correct mode of the solution, see Fig. 5, while the second mode of solution is clearly degenerate. Another situation is in the case of the optimization approaches where continuum of results are observed. Both optimization approaches were able to obtain slightly better results in terms of MSE for specific α; however, it would be difficult to select the correct parameter α without knowledge of the true solution.

4 Conclusion

The linear inverse problem was studied with specific regularization using modeling of a covariance matrix in the modified Cholesky form. We employed the Variational Bayes inference which allows us to deal with vague prior information about range of elements of the covariance matrix using truncated Gaussian prior. We have shown an advantage of the proposed LS-APCi method over the classic optimization approach with LASSO or Tikhonov regularizations. Moreover, we applied the methods to estimation of the source term of atmospheric release from realistic scenario where 6 h release of mixture of 3 nuclides is simulated. The results suggest that all methods are capable to reach a suitable solution using particular setting of parameters; however, LS-APCi method is much more robust to selection of the tuning parameters.

Acknowledgements The research leading to these results has received funding from the Norwegian Financial Mechanism 2009–2014 under Project Contract no. MSMT-28477/2014, Source-Term Determination of Radionuclide Releases by Inverse Atmospheric Dispersion Modelling (STRADI). The authors would like to thank to Radek Hofman for data for the realistic experiment.

Appendix

Truncated Gaussian distribution, denoted as $t\mathcal{N}$, of a scalar variable x on interval $[a; b]$ is defined as $t\mathcal{N}_x(\mu, \sigma, [a, b]) = \frac{\sqrt{2}\exp(-\frac{1}{2\sigma}(x-\mu)^2)}{\sqrt{\pi\sigma}(erf(\beta)-erf(\alpha))}\chi_{[a,b]}(x)$, where $\alpha = \frac{a-\mu}{\sqrt{2\sigma}}$, $\beta = \frac{b-\mu}{\sqrt{2\sigma}}$, function $\chi_{[a,b]}(x)$ is a characteristic function of interval $[a, b]$ defined as $\chi_{[a,b]}(x) = 1$ if $x \in [a, b]$ and $\chi_{[a,b]}(x) = 0$ otherwise. erf() is the error function defined as $\text{erf}(t) = \frac{2}{\sqrt{\pi}}\int_0^t e^{-u^2}du$.

The moments of truncated Gaussian distribution are $\langle x\rangle = \mu - \sqrt{\sigma}\frac{\sqrt{2}[\exp(-\beta^2)-\exp(-\alpha^2)]}{\sqrt{\pi}(\text{erf}(\beta)-\text{erf}(\alpha))}$ and $\langle x^2\rangle = \sigma + \mu\widehat{x} - \sqrt{\sigma}\frac{\sqrt{2}[b\exp(-\beta^2)-a\exp(-\alpha^2)]}{\sqrt{\pi}(\text{erf}(\beta)-\text{erf}(\alpha))}$. For multivariate case, see [11].

References

1. Bernardo, J.M., Smith, A.F.M.: Bayesian Theory. Wiley, New York (2009)
2. Daniels, M.J., Pourahmadi, M.: Bayesian analysis of covariance matrices and dynamic models for longitudinal data. Biometrika **89**(3), 553–566 (2002)
3. Golub, G.H., Hansen, P.C., O'Leary, D.P.: Tikhonov regularization and total least squares. SIAM J. Matrix Anal. Appl. **21**(1), 185–194 (1999)
4. Grant, M., Boyd, S.: Graph implementations for nonsmooth convex programs. In: Blondel, V., Boyd, S., Kimura, H. (eds.) Recent Advances in Learning and Control. Lecture Notes in Control and Information Sciences, pp. 95–110. Springer Limited, Heidelberg (2008)
5. Grant, M., Boyd, S.: CVX: Matlab Software for Disciplined Convex Programming, version 2.1. http://cvxr.com/cvx (2014)

6. Khare, K., Rajaratnam, B., et al.: Wishart distributions for decomposable covariance graph models. Ann. Stat. **39**(1), 514–555 (2011)
7. Kyung, M., Gill, J., Ghosh, M., Casella, G.: Penalized regression, standard errors, and Bayesian lassos. Bayesian Anal. **5**(2), 369–411 (2010)
8. Saunier, O., Mathieu, A., Didier, D., Tombette, M., Quélo, D., Winiarek, V., Bocquet, M.: An inverse modeling method to assess the source term of the Fukushima nuclear power plant accident using gamma dose rate observations. Atmos. Chem. Phys. **13**(22), 11403–11421 (2013)
9. Seibert, P., Frank, A.: Source-receptor matrix calculation with a lagrangian particle dispersion model in backward mode. Atmos. Chem. Phys. **4**(1), 51–63 (2004)
10. Šmídl, V., Quinn, A.: The Variational Bayes Method in Signal Processing. Springer, Heidelberg (2006)
11. Šmídl, V., Tichý, O.: Sparsity in Bayesian Blind Source Separation and Deconvolution. LNAI of Lecture Notes in Computer Science, vol. 8189, pp. 548–563 (2013)
12. Tibshirani. R.: Regression shrinkage and selection via the lasso. J. R. Stat. Soc. Ser. B (Methodol.) 267–288 (1996)
13. Tichý, O., Šmídl, V.: Non-parametric bayesian models of response function in dynamic image sequences. Comput. Vis. Image Underst. **151**, 90–100 (2016)
14. Tipping, M.E.: Sparse Bayesian learning and the relevance vector machine. J. Mach. Learn. Res. **1**, 211–244 (2001)

Part II
Applications and Case Studies

Bayesian Hierarchical Model for Assessment of Climate Model Biases

Maeregu Woldeyes Arisido, Carlo Gaetan, Davide Zanchettin and Angelo Rubino

Abstract Studies of climate change rely on numerical outputs simulated from Global Climate Models coupling the dynamics of ocean and atmosphere (GCMs). GCMs are, however, notoriously affected by substantial systematic errors (biases), whose assessment is essential to assert the accuracy and robustness of simulated climate features. This contribution focuses on constructing a Bayesian hierarchical model for the quantification of climate model biases in a multi-model framework. The method combines information from a multi-model ensemble of GCM simulations to provide a unified assessment of the bias. It further individuates different bias components that are characterized as non-stationary spatial fields accounting for spatial dependence. The approach is illustrated based on the case of near-surface air temperature bias over the tropical Atlantic and bordering regions from a multi-model ensemble of historical simulations from the fifth phase of the Coupled Model Intercomparison Project.

Keywords Bayesian hierarchical model · Climate bias · CMIP5 · Posterior inference · Spatial analysis

1 Introduction

During the last decades, GCMs generated numerical outputs data that deeply contributed to the understanding of climate dynamics and variability. The outputs provide

M.W. Arisido (✉) · C. Gaetan · D. Zanchettin · A. Rubino
Department of Environmental Sciences, Informatics and Statistics,
Ca' Foscari University of Venice, Via Torino 155, 30172 Venice, Italy
e-mail: maeregu.arisido@unive.it

C. Gaetan
e-mail: gaetan@unive.it

D. Zanchettin
e-mail: davido@unive.it

A. Rubino
e-mail: rubino@unive.it

R. Argiento et al. (eds.), *Bayesian Statistics in Action*, Springer Proceedings
in Mathematics & Statistics 194, DOI 10.1007/978-3-319-54084-9_10

quantitative estimates of many geophysical quantities such as atmospheric and oceanic temperatures and precipitation at discrete spatial locations and temporal intervals [4, 10]. Analyses of the outputs are based on the crucial assumption that a GCM represents the real climate system, hence, it is in agreement with observations. However, the current generation of GCMs is affected by substantial biases over extensive regions [11]. Accuracy and robustness of simulated climate features have to be carefully assessed when GCMs are used to understand climate variability and/or produce projections or predictions.

When analyzing GCMs outputs, often a multi-model approach is used as it allows to overcome the peculiarities of individual simulations, like those linked to the chosen initial conditions and applied external forcing, and the deficiencies of individual models, by combining the information into a multi-model consensus [10]. Multi-model intercomparisons have also demonstrated that there are common features in the bias generated by different models [11]. An elaborated statistical method is required to account for these relevant features. Here, we present a Bayesian hierarchical model to provide a unified assessment of climate model biases using spatially referenced data. We estimate an overall bias which is common for all GCMs and individual GCM biases, further characterizing each model contribution to the common bias component. Our approach demonstrates the improved quantification of the bias, culminating in interpretative results allowed by the posterior distributions.

Bayesian analysis is an attractive tool to provide probabilistic distributions for unknown quantities and related uncertainties [6]. The use of Bayesian hierarchical models for climate assessments using multi-model outputs is increasingly appealing [10]. These assessments often depend on spatially aggregated multi-model outputs [2]. Here, we define the overall climate model bias and other components in the Bayesian hierarchical model as spatial fields in which spatial dependencies are accounted for. We describe our methodology in Sect. 2; an application of the method and associated results are presented in Sect. 3. Final concluding remarks are given in Sect. 4.

2 Bayesian Hierarchical Approach for Climate Model Biases

Biases are quantified by comparing GCM outputs to observations. Let $\{M_j(s) : s \in D\}$ denote the bias process for GCM j at the spatial location $s \in D$ for the domain $D \subset \mathbb{R}^2$. Suppose that at n sites in D, we have observed biases on $M_j(s)$, namely $\{B_j(s_i), \ldots, B_j(s_n)\}$, and we consider the error model of the form

$$B_j(s) = M_j(s) + \varepsilon_j(s), \quad j = 1, \ldots, J, \tag{1}$$

where $\varepsilon_j(s) \sim \mathcal{N}(0, \sigma_\varepsilon^2)$ is Gaussian white noise, where σ_ε^2 is unknown. The white noise corresponds to the error associated with representing the true bias process by

$M_j(s)$. Spatial dependence for the bias of climate model j is captured by $M_j(s)$ which is independent of $\varepsilon_j(s)$. Using specification (1), the likelihood, which comprises the observed data is Gaussian conditional on the spatial process, $B_j(s)|M_j(s) \sim \mathcal{N}(M_j(s), \sigma_\varepsilon^2)$. Our interest is to model the true bias process $M_j(s)$. We assume the following low-dimensional deterministic specification for it

$$M_j(s) = \sum_{k=1}^{p} w_k(s)\beta_{j,k}, \quad j = 1, \ldots, J; k = 1, \ldots, p \tag{2}$$

where $w_k(s)$ is a weighting kernel with number of components $p << n$, and $\{\beta_{j,1}, \ldots, \beta_{j,p}\}'$ is a vector of coefficients with dimension p for climate model j. The weighting kernels are discussed in the next section. Further, let $\mu(s)$ represent the overall bias which is common for each $\{j : 1, \ldots, J\}$, and we specify as $\mu(s) = \sum_{k=1}^{p} w_k(s)\beta_k$, where $(\beta_1, \ldots, \beta_p)'$ is a vector of independent coefficients distributed as $\mathcal{N}(\mathbf{0}, \sigma_\beta^2\mathbf{I}_p)$, for which $\mathbf{I}_p$ is the $p \times p$ identity matrix. Because we assume that $\mu(s)$ represents a systematic average of $\{M_j(s), j = 1, \ldots, J\}$, the individual parameter $\beta_{j,k}$ is centered at the overall bias parameter β_k

$$\beta_{j,k} = \beta_k + \nu_{j,k}, \quad j = 1, \ldots, J; k = 1, \ldots, p \tag{3}$$

where $\boldsymbol{\nu}_j = \{\nu_{j,1}, \ldots, \nu_{j,p}\}'$ denotes a vector of independently distributed mean zero Gaussian processes, $\nu_j \sim \mathcal{N}(\mathbf{0}, \tau_j^2\mathbf{I}_p)$. The variance parameters τ_j^2 capture different amount of variability in various climate models after accounting for the underlying common overall bias signal. If we assume that $\tau_1^2 = \tau_2^2 = \cdots = \tau_J^2$, this corresponds to the common overall bias $\mu(s)$ being a simple average of $\{M_1(s), \ldots, M_J(s)\}$. Here allowing τ_j^2 to vary across the different climate models grants to determine the contribution of each climate model in predicting the common overall bias. Combining (2) and (3), the bias signal process $M_j(s)$ is expressed as

$$M_j(s) = \sum_{k=1}^{p} w_k(s)\big[\beta_k + \nu_{j,k}\big] = \mu(s) + \eta_j(s), \tag{4}$$

where $\eta_j(s) = \sum_{k=1}^{p} w_k(s)\nu_{j,k}$ describes the departure of each climate model bias from the common overall bias. Further, $\eta_j(s)$ follows the mean zero Gaussian distribution, $\eta_j(s) \sim \mathcal{N}\big(\mathbf{0}, \tau_j^2\mathbf{w}(s)\mathbf{w}(s)'\big)$, where $\mathbf{w}(s) = \{w_1(s), \ldots, w_p(s)\}$. For identifiability reason, we assume that $\sum_{j=1}^{J} \eta_j(s) = 0$. In general, model (4) suggests that the bias process $M_j(s)$ is a combination of the overall component $\mu(s)$, which captures large-scale variations, and the individual component $\eta_j(s)$, which describes the small-scale variations. In the parameter level of the hierarchical model, prior distributions for the variance components τ_j^2 and σ_β^2 are specified. We assign the inverse-gamma (IG) prior on these variance parameters, $\tau_j^2 \sim \text{IG}(a_1, b_1)$ independently for $j = 1 \ldots, J$ and $\sigma_\beta^2 \sim \text{IG}(a_2, b_2)$, where the a 's and the b 's are hyperparameters whose values are chosen. For our particular application given in the next section, we choose the same value, 0.001, for each hyperparameter so that the prior distributions are approximately non-informative.

3 Application to Temperature Bias in the Tropical Atlantic Region

In this section, we present an application of the model using near surface air temperature (unit: Kelvin) data for the Tropical Atlantic and bordering regions covering the domain 40W–20E longitude and 35S–15N latitude. Our database comprises information about observations and ensemble of climate output data. For observational data, we use reanalysis data, which are the output of a state-of-the-art analysis/forecast system with data assimilation using past data from 1948 to the present. The climate output data are an ensemble of *six historical* full-forcing climate simulations contributing to the fifth phase of the Coupled Model Intercomparison Project (CMIP5). An overview of the models' characteristics is provided in Table 1. These GCMs have been developed at different research institutes and have different major characteristics. We consider the employed ensemble to be representative of the range of model characteristics of current global climate models. The application covers the period of 1950–2005 CE in which we derive climatologies of annual-mean values starting from the monthly-mean time series of both observations and climate output.

3.1 *Choice of Weighting Functions*

Several types of functions have been used in the literature, including the B-splines [1], Gaussian kernels [9] and bisquare functions [8]. In this contribution, we consider the Gaussian kernel function, namely

$$w_k(s) \propto \exp\{-(s - c_k)'\Sigma^{-1}(s - c_k)/2\},\ k = 1, \ldots, p \tag{5}$$

Table 1 Six climate models and thir specific related information which are used in this study

Research centre	Country	Model ID	Resolution
National Center for Atmospheric Research	USA	CCSM4	288×192
Beijing Climate Center	China	BCC	128×64
NASA/Goddard Institute for Space Studies	USA	GISS	144×90
Institut Pierre Simon Laplace	France	IPSL	96×96
Max Planck Institute for Meteorology	Germany	MPI	192×96
Center for Climate System Research	Japan	MIROC	128×64

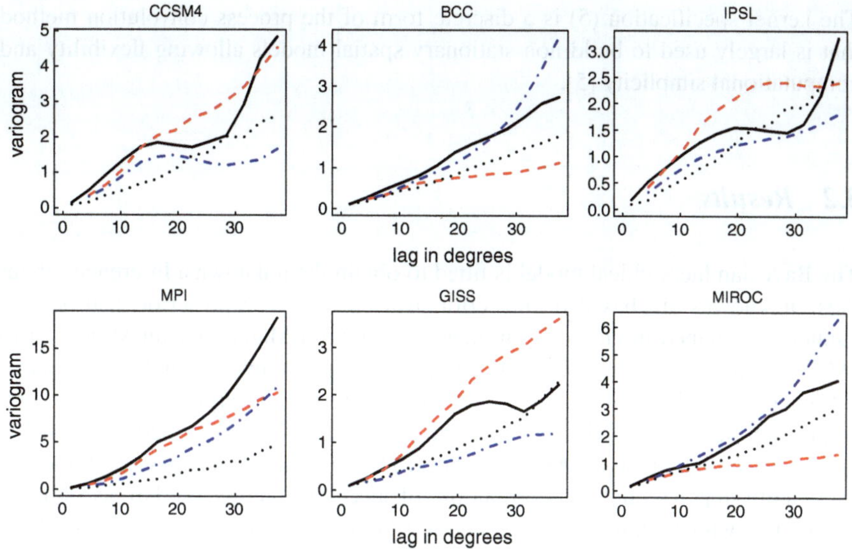

Fig. 1 Empirical variograms of the six GCM biases of near-surface temperature over the tropical Atlantic region for four different directions (*black solid* 0°, *red dash* 45°, *gray dotted* 90°, *blue dash* 135°). The variograms were analyzed using the robust estimator as given by [3]

where c_k is the center of the kernel and Σ determines the shape and smoothness of the kernel. The number of components p, Σ and the location of the kernels must be chosen. These choices are often based on the presence of prior information such as smoothness and spatial dependence related to the spatial process [9]. If we choose equally sized Gaussian kernels, i.e., spherically shaped kernels, we obtain an isotropic spatial process. Alternatively, a geometrically anisotropic process may be obtained if we choose non-spherical Gaussian kernels. One way to investigate whether the spatial biases are direction-dependent or not is to perform variogram analyses for different directions [3]. Figure 1 illustrates the empirical variograms of the six GCM biases for the directions: 0°, 45°, 90°, 135° (i.e. North, Northeast, East and Southeast direction, respectively). The variograms in the four directions do not reveal largely different spatial patterns, thus we assume an isotropic spatial bias process.

Furthermore, we assume Σ is the same for each kernel. Another advantage of the chosen kernel functions is that they allow to obtain non-stationary spatial covariance for $M_j(s)$. If $\mathscr{K}_j(s_i, s_l)$ is the covariance between two locations s_i and s_l, a simple algebraic manipulation lead to obtain a non-stationary covariance

$$\mathscr{K}_j(s_i, s_l) = (\sigma_\beta^2 + \tau_j^2) \sum_{k=1}^{p} w_k(s_i) w_k(s_l) \quad (6)$$

The kernel specification (5) is a discrete form of the process convolution method that is largely used to build non-stationary spatial models allowing flexibility and computational simplicity [5].

3.2 Results

The Bayesian hierarchical model is fitted to obtain the unknowns. Inferences about these unknowns are based on posterior distributions. The posterior distributions cannot be obtained in closed form, hence we rely on Markov Chain Monte Carlo (MCMC) simulation [7], with Gibbs sampler to draw samples from the posterior of full conditional distributions given the data. Posterior convergence is assessed by inspecting the simulation history using graphical tools. We performed 50000 simulations discarding the first 20000 as burn-in. The remaining samples were thinned at every tenth step to reduce autocorrelations of successive samples and to save storage space. The whole simulation takes about 3 hours on a 64-bit OS X 10.10.5 Intel Core i5 1.6 GHz. Once we were confident that the draws are from the stationary distribution of the Markov chain, we summarized these draws to make inferences.

Figure 2 summarizes the posterior results with respect to the overall climate model bias, $\mu(s)$, and departure of individual components from the overall estimate, $\{\eta_j(s) : j = 1, \dots, 6\}$, fixing $p = 12$ as in Fig. 2a. The overall common bias appears to be stronger in the southeast of the tropical Atlantic Ocean. Posterior estimates associated to $\eta_j(s)$ are widely heterogeneous across the six climate models from estimating the maximum warm bias in CCSM4 to the point of obtaining a cold bias of -2 Kelvin in MIROC. The individual bias component for CCSM4 essentially describes a generally warmer bias over the central southern tropical Atlantic Ocean. IPSL, MPI and MIROC produce warmer biases over the sub-Saharian Africa compared to the overall common bias.

To better understand the posterior distributional variability of the individual components, Fig. 3 depicts the posterior distribution of variance components $\{\tau_j^2 : j = 1, \dots, 6\}$. As pointed out in the previous section, these variance parameters represent the differences between the random effects $\boldsymbol{\beta}_j$ and the random effects of the overall common bias $\boldsymbol{\alpha}$. Hence, they are useful to assess how each climate model bias varies around the overall common bias. It is worth noting that there is marked difference across the random effects of the individual GCMs about the overall common random effects. CCSM4 varies the least, whereas IPSL and GISS vary the most about the overall common bias. Thus, in terms of weighting the contributions of each GCMS in synthesizing the overall common bias, CCSM4 is ranked first, whereas IPSL and GISS have smaller weights.

One way to check the adequacy of our modeling approach is to assess the robustness of the results with respect to the choices of the kernel weighting functions and hyperparameters. We investigated the sensitivity of $\mu(s)$ for different choices of p and Σ of the Gaussian kernels (results not shown). When we used smaller p, i.e., $p < 9$, $\mu(s)$ is oversmoothed and unable to capture the smaller details of the spatial

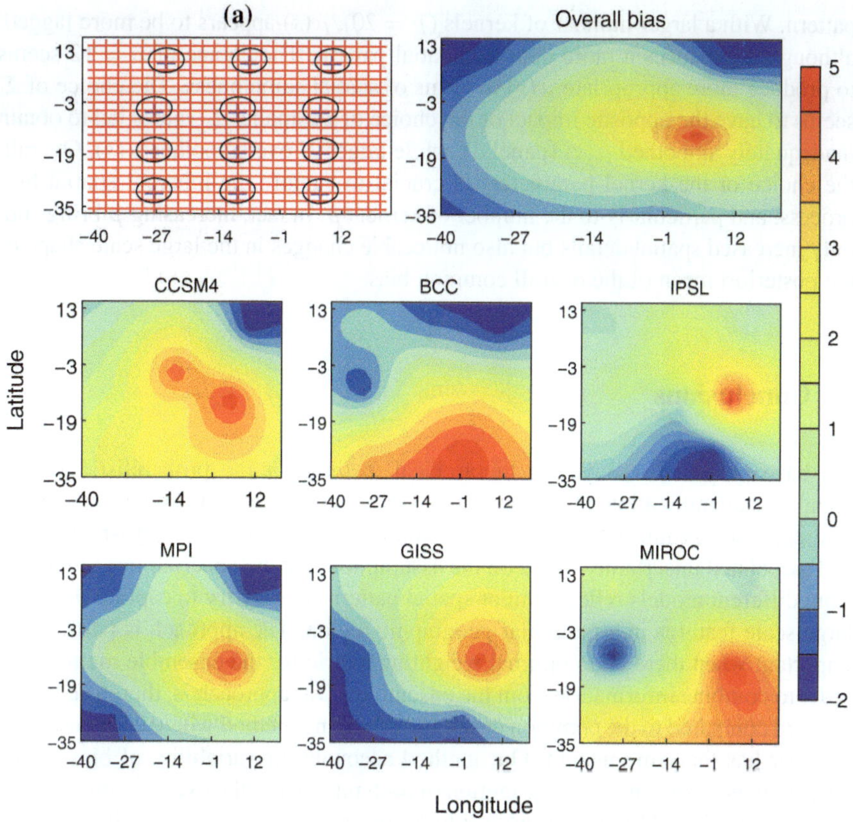

Fig. 2 **a** spherical shape kernels with $p = 12$. Posterior average of the spatially-varying overall bias $\mu(s)$ (*top right*) and departure of individual component from the overall mean $\eta_j(s)$, $j = 1, 2, \ldots, 6$ (rows 2 and 3). The color bar unit is in kelvin

Fig. 3 Boxplots of the posterior draws associated to the variance parameters τ_i for $j = 1, \ldots, 6$, where the solid horizontal bars denote the medians, the *lower* and *upper* ends of the *boxes* indicate the 25th and 75th percentiles respectively

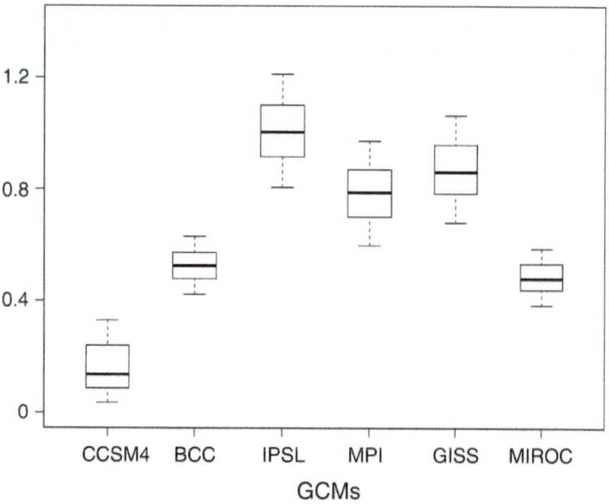

pattern. With a larger number of kernels ($p = 20$), $\mu(s)$ appears to be more jagged, although it produces a more detailed spatial pattern. The choice of $p = 12$ seems to produce more appropriate $\mu(s)$ in terms of spatial smoothness. The choice of Σ seems to have the opposite impact of the choice of p: smaller Σ values led to obtain inadequately smoothed $\mu(s)$ (panel d), while larger Σ oversmoothes $\mu(s)$. Overall, the choice of the kernel parameters is crucial to capture the inherent spatial bias process, and particularly to the number of kernels p. In fact, increasing p brings not only increased spatial details but also noticeable changes in the large scale shape of the posteriori mean of the overall common bias.

4 Conclusions

We have proposed a Bayesian hierarchical method for the probabilistic assessment and quantification of climate model biases using a multi-model ensemble. The approach synthesizes an overall common bias as non-stationary spatial field and associated uncertainty based on the assumption that the uncertainties of outputs from different models reflect similar spatial patterns as they try to capture the same large-scale features in a particular geographic region. The approach is particularly important when there is no optimal weighting metric for the ensemble members in order to combine information from the ensembles. Our approach is, therefore, a step forward compared to the common ensemble averaging method where each ensemble member has the same weight. Our method accounts for variability of bias across ensemble members and the contribution of each member to the overall common bias is determined based on the posterior inferences on each model's variability parameter. We have focused on a multi-model ensemble of six CMIP5 historical simulations. The basic idea is, however, generic and could be applied to a wider range of climate models, geographical locations, and geophysical variables.

Acknowledgements The research leading to these results has received funding from the European Union, Seventh Framework Programme (FP7/2007–2013) under Grant agreement n 603521 - PREFACE. The authors would like to thank the two anonymous reviewers.

References

1. Arisido, M.W.: Functional measure of ozone exposure to model short-term health effects. Environometrics **27**, 306–317 (2016)
2. Buser, C.M., Knsch, H.R., Lthi, D., Wild, M., Schr, C.: Bayesian multi-model projection of climate: bias assumptions and interannual variability. Clim. Dyn. **33**, 849–868 (2009)
3. Cressie, N.: Statistics for Spatial Data. Wiley-Interscience, New York (1993)
4. Furrer, R., Geinitz, S., Sain, S.R.: Assessing variance components of general circulation model output fields. Environmetrics **23**, 440–450 (2012)
5. Gelfand, A.E., Diggle, P., Guttorp, P., Fuentes, M. (eds.) Handbook of Spatial Statistics. CRC Press, Boca Raton (2010)

6. Gelman, A., Carlin, J.B., Stern, H.S., Rubin, D.B.: Bayesian Data Analysis. Chapman and Hall/CRC Texts in Statistical Science, London (2003)
7. Gilks, W.R., Richardson, S., Spiegelhalter, D.J.: Markov chain Monte Carlo in practice. Chapman and Hall, London (1996)
8. Kang, E.L., Cressie, N., Sain, S.R.: Combining outputs from the North American regional climate change assessment program by using a Bayesian hierarchical model. J. R. Stat. Soc. Ser. C (Appl. Stat.) **61**, 291–313 (2012)
9. Stroud, J.R., Muller, P., Sansó, B.: Dynamic models for spatiotemporal data. J. R. Stat. Soc. Ser. B. Stat. Methodol. **63**, 673–689 (2001)
10. Tebaldi, C., Smith, R.L., Nychka, D., Mearns, L.O.: Quantifying uncertainty in projections of regional climate change: a Bayesian approach to the analysis of multimodel ensembles. J. Clim. **18**, 1524–1540 (2005)
11. Wang, C., Zhang, L., Lee, S.K., Wu, L., Mechoso, C.R.: A global perspective on CMIP5 climate model biases. Nat. Clim. Change **4**, 201–205 (2014)

An Application of Bayesian Seemingly Unrelated Regression Models with Flexible Tails

Charles Au and S.T. Boris Choy

Abstract Seemingly unrelated regression (SUR) models are useful for capturing the correlation structure between different regression equations. While the multivariate normal distribution is a common choice for the random error term in an SUR model, the multivariate t-distribution is also popular for robustness considerations. However, the multivariate t-distribution is elliptical which leads to the limitation that the degrees of freedom of its marginal distributions are identical. In this paper, we consider a non-elliptical multivariate Student-t error distribution which allows flexible shape parameters for the marginal distributions. This non-elliptical distribution is constructed via a scale mixtures of normal form and therefore the Markov chain Monte Carlo (MCMC) algorithms are used for Bayesian inference of SUR models. In the empirical study of the capital asset pricing model (CAPM), we show that this non-elliptical Student-t distribution outperforms the multivariate normal and multivariate Student-t distributions.

Keywords Modified multivariate t-distribution · Scale mixtures of normal (SMN) · Markov chain Monte Carlo (MCMC) · Capital Asset Pricing Model (CAPM)

1 Introduction

The seemingly unrelated regression (SUR) model, proposed by [12], is useful for modelling the relationship amongst several variables. In particular, it captures the correlation structure between several linear regression equations in the model. To provide a better fit to the error terms, [13] proposed a Bayesian algorithm for the SUR model with the multivariate t error distribution. [1] showed that the univariate Student-t distribution has a scale mixtures of normal (SMN) representation, which is

C. Au (✉) · S.T.B. Choy
Discipline of Business Analytics, The University of Sydney,
Sydney, NSW 2006, Australia
e-mail: charles.au.32@gmail.com

S.T.B. Choy
e-mail: boris.choy@sydney.edu.au

R. Argiento et al. (eds.), *Bayesian Statistics in Action*, Springer Proceedings in Mathematics & Statistics 194, DOI 10.1007/978-3-319-54084-9_11

also valid for the multivariate t-distribution. However, the multivariate t-distribution belongs to the elliptical family and therefore all shape parameters of its marginal distributions must be identical. To overcome this deficiency, [3] proposed a non-elliptical multivariate t-distribution whose marginal distributions are allowed to have different degrees of freedom and they called this distribution the modified multivariate t-distribution (Mod-t distribution).

The Capital Asset Pricing Model (CAPM) [5–8, 10] postulates a relationship between the excess return of an individual asset and the excess return of the market portfolio. This model has been studied widely in econometrics and finance. For example, [9] fitted a copula regression model with Student-t marginals to the multivariate CAPM.

In this paper, we study the excess market returns under the CAPMs and the SUR models are used to explore the correlation structure amongst different CAPMs. Section 2 presents the Mod-t distribution and its statistical properties and the SUR model, along with the necessary Markov chain Monte Carlo (MCMC) algorithm for statistical inference. In Sect. 3, we compare the performance of various error distributions in the CAPM analysis. The paper is concluded in Sect. 4.

2 Bayesian SUR Model with Mod-t Distribution

2.1 Mod-t Distribution

In the SMN representation, the m-dimensional elliptical multivariate t-distribution with the degrees of freedom ν has the following probability density function (p.d.f.)

$$f_{\mathbf{X}}(\mathbf{x}) = \int_0^\infty \mathcal{N}_m(\mathbf{x}|\boldsymbol{\mu}, \lambda^{-1}\boldsymbol{\Sigma})\mathrm{Ga}\left(\lambda|\frac{\nu}{2}, \frac{\nu}{2}\right) \mathrm{d}\,\lambda, \tag{1}$$

where λ is the scale mixture variable, $\mathcal{N}_m(\mathbf{x}|\boldsymbol{c}, \boldsymbol{D})$ is the m-dimensional multivariate normal p.d.f. with location $\boldsymbol{c}$ and scale matrix $\boldsymbol{D}$, and $\mathrm{Ga}(\lambda|a, b)$ is the gamma p.d.f. given by

$$\mathrm{Ga}(\lambda|a, b) = b^a/\Gamma(a) \times \lambda^{a-1}\exp(-b\lambda). \tag{2}$$

Let $\boldsymbol{\lambda} = (\lambda_1, \ldots, \lambda_m)^\mathsf{T}$ be the vector of scale mixture variables, $\boldsymbol{\Lambda} = \mathrm{diag}(\boldsymbol{\lambda})$ and $\boldsymbol{\nu} = (\nu_1, \ldots, \nu_m)^\mathsf{T}$ be the vector of degrees of freedom. The p.d.f. of the non-elliptical Mod-t distribution, proposed by [3], is given by

$$f_{\mathbf{X}}(\mathbf{x}) = \underbrace{\int_0^\infty \cdots \int_0^\infty}_{m} \mathcal{N}_m(\mathbf{x}|\boldsymbol{\mu}, \boldsymbol{\Lambda}^{-\frac{1}{2}}\boldsymbol{\Sigma}\boldsymbol{\Lambda}^{-\frac{1}{2}}) \times \prod_{j=1}^m \mathrm{Ga}\left(\lambda_j|\frac{\nu_j}{2}, \frac{\nu_j}{2}\right) \mathrm{d}\,\lambda_1, \ldots, \mathrm{d}\,\lambda_m. \tag{3}$$

[3] revealed that the marginal distribution of X_j, $j = 1, \ldots, m$, is the Student-t distribution with location μ_j, scale Σ_{jj} (the jj-th element of $\boldsymbol{\Sigma}$) and degrees of freedom ν_j. Moreover, they also showed that the correlation coefficient between X_j and X_k ($j = 1, \ldots, m$; $k = 1, \ldots, m$) is given by

$$\mathrm{Corr}(X_j, X_k) = \frac{\rho_{jk}}{2}\sqrt{(\nu_j - 2)(\nu_k - 2)}\frac{\Gamma\left(\frac{\nu_j-1}{2}\right)\Gamma\left(\frac{\nu_k-1}{2}\right)}{\Gamma\left(\frac{\nu_j}{2}\right)\Gamma\left(\frac{\nu_k}{2}\right)}, \; \nu_j > 2, \nu_k > 2, \quad (4)$$

where ρ_{jk} is the correlation coefficient of X_j and X_k of the elliptical multivariate Student-t distribution.

The multivariate normal and multivariate Student-t distributions are special cases of the Mod-t distribution. For the normal distribution, all λ_{ij} degenerate to 1, whereas for the Student-t distribution, $\lambda_1 = \lambda_2 = \cdots = \lambda_m = \lambda$, where $\lambda \sim \mathrm{Ga}(\nu/2, \nu/2)$.

The contour plots for various standard bivariate t and standard Mod-t distributions are shown in Fig. 1. The bivariate t distributions are on the left column and the Mod-t distributions are on the right column. It is obvious that the contour plots of the Mod-t distributions are non-elliptical but they are getting closer to an elliptical shape as ν_1 and ν_2 increase.

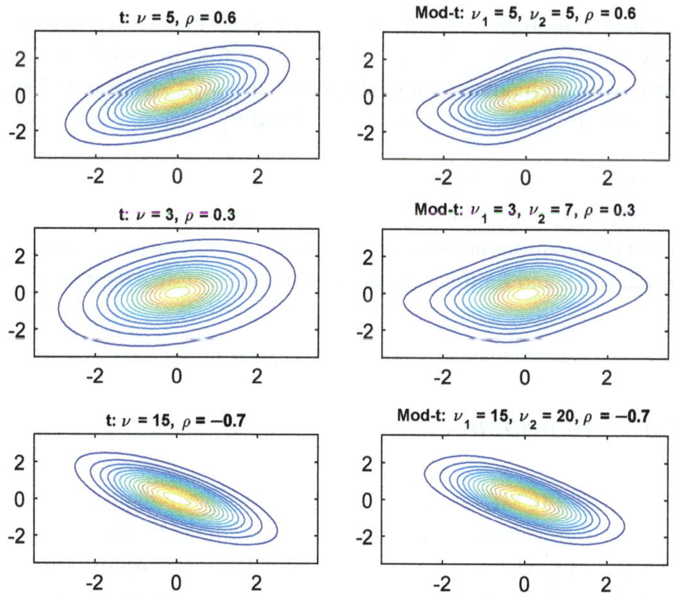

Fig. 1 Bivariate t and Mod-t distributions — Contour plots

2.2 SUR Model

The SUR model, proposed by [12], is given below.

$$y_{ij} = \mathbf{x}_{ij}^{\mathsf{T}}\boldsymbol{\beta}_j + \varepsilon_{ij} \quad (i = 1, \ldots, n, \quad j = 1, \ldots, m). \tag{5}$$

There are n observations in each of the m dimensions in the SUR model. In vector form, it is expressed as

$$\mathbf{y}_i = \mathbf{X}_i\boldsymbol{\beta} + \boldsymbol{\varepsilon}_i \quad (i = 1, \ldots, n), \tag{6}$$

where $\mathbf{y}_i = (y_{i1}, y_{i2}, \ldots, y_{im})^{\mathsf{T}}$ are the dependent variables, $\mathbf{X}_i$ are the independent variables ($m \times p$ matrix), $\boldsymbol{\beta}$ are the regression coefficients ($p \times 1$ vector) and $\boldsymbol{\varepsilon}_i = (\varepsilon_{i1}, \varepsilon_{i2}, \ldots, \varepsilon_{im})^{\mathsf{T}}$ are the error terms. The subscript i denotes the i-th observation in the model.

The errors $\boldsymbol{\varepsilon}_i$ follow the Mod-t distribution with location $\mathbf{0}$, scale matrix $\boldsymbol{\Sigma}$ and degrees of freedom vector $\boldsymbol{\nu}$. Expressing the Mod-t distribution in the SMN form, the error terms has the conditional distribution given below

$$\boldsymbol{\varepsilon}_i|\boldsymbol{\lambda}_i \sim \mathcal{N}_m(\mathbf{0}, \boldsymbol{\Lambda}_i^{-1/2}\boldsymbol{\Sigma}\boldsymbol{\Lambda}_i^{-1/2}), \tag{7}$$

where $\boldsymbol{\Lambda}_i$ is an $m \times m$ matrix of scale mixture variables in the SMN distribution with $\boldsymbol{\Lambda}_i = \text{diag}(\boldsymbol{\lambda}_i)$ and $\boldsymbol{\lambda}_i = (\lambda_{i1}, \lambda_{i2}, \ldots, \lambda_{im})^{\mathsf{T}}$. $\boldsymbol{\Sigma}$ is a positive-definite symmetrical $m \times m$ matrix with elements σ_j^2 on its diagonals and $\rho_{jk}\sigma_j\sigma_k$ on row j and column k ($\rho_{jk} = \rho_{kj}$) for all $j, k = 1, \ldots, m$. $\rho_{jk} = 1$ and $\sigma_j = \sigma_k$ if $j = k$, and $|\rho_{jk}| < 1$ otherwise.

For the Mod-t error distribution, $\lambda_{ij} \sim \text{Ga}(\nu_j/2, \nu_j/2)$, as the marginal t-distribution of each dimension j has ν_j degrees of freedom. The multivariate normal and t error distributions are special cases. For the multivariate t error distribution, $\lambda_{ij} \sim \text{Ga}(\nu/2, \nu/2)$, and for the multivariate normal error distribution, all λ_{ij} degenerate to 1.

2.3 MCMC Algorithm

A commonly used MCMC algorithm for performing simulation-based Bayesian inference is the Gibbs sampler. For an SUR model with the Mod-t error distribution, posterior samples of model parameters $\boldsymbol{\beta}$, $\boldsymbol{\Sigma}$ and $\boldsymbol{\nu}$ and scale mixture variables λ_{ij} are generated successively from their full conditional distributions. Starting values for all model parameters are provided. Before a Gibbs sampler is initiated, the following prior distributions are used in the empirical study of this paper.

- $\boldsymbol{\beta} \sim \mathcal{N}_p(\mathbf{0}, 100\boldsymbol{I}_p)$
- $\boldsymbol{\Sigma} \sim \mathcal{IW}_m(0.01\boldsymbol{I}_m, m)$ (inverse Wishart distribution)
- $\nu_j \sim \text{Exp}(0.1)I(2 \leq \nu_j \leq 30)$ for $j = 1, \ldots, m$

Posterior samples are generated in the following order using the Gibbs sampler.

1. Simulate the regression coefficients $\boldsymbol{\beta}$ from a multivariate normal distribution.
2. Simulate the matrix $\boldsymbol{\Sigma}$ from an inverse Wishart distribution.
3. Simulate the scale mixture variables λ_{ij} from a non-standard distribution for $i = 1, \ldots, n$ and $j = 1, \ldots, m$.
4. Simulate the degrees of freedom parameters ν_j for $j = 1, \ldots, m$ from a non-standard distribution.

The Metropolis–Hastings (M–H) algorithm is used within a Gibbs sampler to simulate random variates from a non-standard distribution. For the multivariate t error distribution with ν degrees of freedom, λ_{ij} are simulated independently from a gamma distribution and ν is simulated using the M–H algorithm. Furthermore, for the multivariate normal error distribution, only Steps 1 and 2 are required.

3 Application

3.1 *Capital Asset Pricing Model (CAPM)*

In this section, we compare the performance of SUR models with different multivariate error distributions in the CAPM, which, in the form of a regression model, is expressed as

$$y_{ij} = \beta_{1j} + \beta_{2j}x_{ij} + \varepsilon_{ij} \quad (i = 1, \ldots, n, \quad j = 1, \ldots, m), \tag{8}$$

where y_{ij} are the excess returns of security j at time i, x_{ij} are the excess returns on the market portfolio, β_{1j} are the intercepts, β_{2j} are the beta coefficients in the CAPM for security j and ε_{ij} are the error terms. If $\beta_{2j} > 1$ ($\beta_{2j} < 1$), there is a higher (lower) risk for security j than the market portfolio. If $\beta_{2j} = 1$, security j is as risky as the market portfolio.

[4] studied the CAPM for portfolios of stock in four regions: Asia Pacific (excluding Japan) (Region 1), Europe (Region 2), Japan (Region 3) and North America (Region 4). We extend their work by modelling the correlation structure between the excess returns of the portfolio of stocks in all four regions. The data used for statistical analysis is chosen from July 1990 to March 2016 (309 months) and is obtained from the Kenneth French data library (http://mba.tuck.dartmouth.edu/pages/faculty/ken.french/data_library.html). [4] divided the portfolios in these regions into five quintiles based on size and five quintiles based on the ratio of book equity to market equity (B/M) to obtain 25 portfolios. y_{ij} are the excess returns of the portfolio from

the third size quintile and first B/M quintile of region j ($j = 1, \dots, 4$) at month i ($i = 1, \dots, 309$) and x_{ij} are the excess returns on the market portfolio from region j at month i.

In the Bayesian inference, we adopt the prior distributions shown in Sect. 2.3 for all model parameters. The degrees of freedom parameters are restricted to the range between 2 and 30. The Gibbs sampler is run for a total of 20000 iterations. The first 10000 iterations are discarded as the burn-in period and the remaining 10000 iterations are used for statistical inference.

The performance of the three SUR models are compared using the Deviance Information Criterion (DIC) [11], and the model with the smallest DIC is preferred. However, for ease of calculation, the conditional DIC, DIC_7 in [2], is used instead. DIC_7 is given by

$$\text{DIC}_7 = -4\text{E}_{\boldsymbol{\theta},\boldsymbol{\lambda}}\left[\log f(\boldsymbol{y}|\boldsymbol{\theta}, \boldsymbol{\lambda})|\boldsymbol{y}\right] + 2\log f(\boldsymbol{y}|\widehat{\boldsymbol{\theta}}(\boldsymbol{y}), \widehat{\boldsymbol{\lambda}}(\boldsymbol{y})), \tag{9}$$

where $\boldsymbol{y} = (\boldsymbol{y}_1^{\mathsf{T}}, \dots, \boldsymbol{y}_n^{\mathsf{T}})^{\mathsf{T}}$ is the vector for the response variable in the SUR model, $\boldsymbol{\theta} = (\boldsymbol{\beta}, \boldsymbol{\Sigma}, \boldsymbol{\nu})$ and $\boldsymbol{\lambda}$ is the vector for the scale mixture variables. $(\widehat{\boldsymbol{\theta}}(\boldsymbol{y}), \widehat{\boldsymbol{\lambda}}(\boldsymbol{y}))$ is the joint maximum a posteriori (MAP) estimator. The conditional log-likelihood is given by $f(\boldsymbol{y}|\boldsymbol{\theta}, \boldsymbol{\lambda}) = \prod_{i=1}^{n} f(\boldsymbol{y}_i|\boldsymbol{\theta}, \boldsymbol{\lambda}_i) = \prod_{i=1}^{n} \mathcal{N}_m(\boldsymbol{y}_i|\boldsymbol{\mu}, \boldsymbol{\Lambda}_i^{-\frac{1}{2}}\boldsymbol{\Sigma}\boldsymbol{\Lambda}_i^{-\frac{1}{2}})$. Note that for the SUR model with multivariate normal error distribution, $\boldsymbol{\theta} = (\boldsymbol{\beta}, \boldsymbol{\Sigma})$, $f(\boldsymbol{y}|\boldsymbol{\theta}) = \prod_{i=1}^{n} \mathcal{N}_m(\boldsymbol{y}_i|\boldsymbol{\mu}, \boldsymbol{\Sigma})$ and $\text{E}_{\boldsymbol{\theta},\boldsymbol{\lambda}}\left[\log f(\boldsymbol{y}|\boldsymbol{\theta}, \boldsymbol{\lambda})|\boldsymbol{y}\right] = \text{E}_{\boldsymbol{\theta}}\left[\log f(\boldsymbol{y}|\boldsymbol{\theta})|\boldsymbol{y}\right]$.

3.2 Results

The posterior mean, posterior standard deviation and 95% credible interval (CI) for the model parameters in three different SUR models are given in Table 1. The error distributions used are the multivariate normal, multivariate t and Mod-t distributions. For each error distribution, the 95% CI is listed underneath the posterior mean and standard deviation.

All the slope parameters β_{21}, β_{22}, β_{23} and β_{24} are significant. Across all three error distributions, the beta coefficients for Asia Pacific (excluding Japan) (β_{21}), and Europe (β_{22}) are close to 1, meaning that the portfolio of stocks for these regions is just as risky as the market portfolio. For Japan, the parameter β_{23} is around 1.1, and the portfolio of stocks for Japan is riskier than the market portfolio in Japan. For North America, the beta coefficients are highest across the three error distributions, with β_{24} between 1.26 and 1.30. β_{11} is the only intercept parameter that is significant, whereas the other three parameters β_{12}, β_{13} and β_{14} are not.

The posterior means of $\sigma_1, \sigma_2, \sigma_3$ and σ_4 are the greatest for the multivariate normal error distribution, with 3.52, 2.81, 3.71 and 3.97 respectively. For the multivariate t error distribution, the posterior means of σ_1, σ_2, σ_3 and σ_4 are 2.97, 2.00, 3.11 and 2.74. For the Mod-t error distribution, these are 2.81, 1.83, 3.27 and 2.44, which are

Table 1 Posterior means, standard deviations and the 95% credible intervals for the parameters in SUR models with the multivariate normal, multivariate t and Mod-t error distributions

	Normal		t		Mod-t	
	Mean	S.D.	Mean	S.D.	Mean	S.D.
β_{11}	−0.5949	0.2022	−0.5565	0.1888	−0.6578	0.1826
	[−0.9907, −0.2067]		[−0.9212, −0.1837]		[−1.0187, −0.2974]	
β_{21}	1.0462	0.0334	1.0238	0.0354	1.0140	0.0344
	[0.9823, 1.1128]		[0.9540, 1.0943]		[0.9472, 1.0815]	
β_{12}	−0.1868	0.1598	−0.0773	0.1269	−0.1349	0.1272
	[−0.5054, 0.1228]		[−0.3311, 0.1712]		[−0.3855, 0.1153]	
β_{22}	1.0419	0.0309	0.9981	0.0277	0.9868	0.0270
	[0.9822, 1.1030]		[0.9432, 1.0520]		[0.9353, 1.0407]	
β_{13}	−0.1319	0.2122	−0.2737	0.1985	−0.2423	0.2046
	[−0.5461, 0.2890]		[−0.6615, 0.1200]		[−0.6395, 0.1641]	
β_{23}	1.1217	0.0353	1.1039	0.0356	1.1073	0.0356
	[1.0529, 1.1911]		[1.0352, 1.1740]		[1.0389, 1.1774]	
β_{14}	−0.0679	0.2287	−0.1107	0.1749	−0.1577	0.1725
	[−0.5153, 0.3821]		[−0.4559, 0.2299]		[−0.4946, 0.1797]	
β_{24}	1.3025	0.0513	1.2875	0.0429	1.2635	0.0400
	[1.2007, 1.4031]		[1.2045, 1.3729]		[1.1845, 1.3425]	
σ_1	3.5213	0.1421	2.9700	0.1525	2.8116	0.2256
	[3.2629, 3.8161]		[2.6790, 3.2819]		[2.3779, 3.2643]	
σ_2	2.8095	0.1140	2.0027	0.1059	1.8338	0.1312
	[2.5991, 3.0465]		[1.8059, 2.2196]		[1.5909, 2.0995]	
σ_3	3.7113	0.1506	3.1142	0.1576	3.2656	0.2154
	[3.4311, 4.0238]		[2.8178, 3.4385]		[2.8408, 3.6770]	
σ_4	3.9691	0.1602	2.7414	0.1465	2.4427	0.1761
	[3.6737, 4.2959]		[2.4693, 3.0401]		[2.1100, 2.7967]	
ρ_{12}	0.2259	0.0539	0.2244	0.0584	0.2642	0.0590
	[0.1177, 0.3295]		[0.1080, 0.3352]		[0.1456, 0.3767]	
ρ_{13}	−0.0177	0.0574	−0.0178	0.0632	−0.0155	0.0607
	[−0.1288, 0.0953]		[−0.1405, 0.1064]		[−0.1338, 0.1043]	
ρ_{14}	0.1652	0.0559	0.1817	0.0591	0.2122	0.0608
	[0.0552, 0.2735]		[0.0646, 0.2967]		[0.0903, 0.3309]	
ρ_{23}	0.2012	0.0556	0.1587	0.0608	0.1675	0.0623
	[0.0891, 0.3106]		[0.0385, 0.2776]		[0.0439, 0.2873]	
ρ_{24}	0.4139	0.0474	0.3184	0.0574	0.3608	0.0580
	[0.3171, 0.5049]		[0.2022, 0.4281]		[0.2438, 0.4705]	
ρ_{34}	0.0540	0.0577	0.0391	0.0622	0.0517	0.0644
	[−0.0598, 0.1653]		[−0.0869, 0.1608]		[−0.0750, 0.1779]	
ν	—	—	5.2095	0.7516	—	—
			[3.9168, 6.8434]			

(continued)

Table 1 (continued)

	Normal		t		Mod-t	
	Mean	S.D.	Mean	S.D.	Mean	S.D.
ν_1	—	—	—	—	5.9478	2.7313
					[2.9664, 12.9272]	
ν_2	—	—	—	—	3.6109	0.7725
					[2.4263, 5.3731]	
ν_3	—	—	—	—	10.6512	5.2546
					[4.2580, 25.0797]	
ν_4	—	—	—	—	3.5181	0.7379
					[2.4193, 5.3183]	

Table 2 Deviance Information Criterion (DIC) for the three SUR models

Distribution	**Normal**	t	**Mod-t**
DIC	6517.62	6171.27	6095.94

the smallest out of all three error distributions with the exception of the posterior mean of σ_3.

The 95% CIs for ρ_{13} and ρ_{34} include 0, whereas those for ρ_{12}, ρ_{14}, ρ_{23}, and ρ_{24} do not include 0. This indicates that the excess returns of portfolios between Asia Pacific (excluding Japan) and Japan, and between Japan and North America are not correlated, whereas the other combinations of the four regions are correlated.

For the Mod-t model, the degrees of freedom for the excess returns on the portfolio in Asia Pacific (Region 1), Europe (Region 2), Japan (Region 3) and North America (Region 4) are 5.95, 3.61, 10.65 and 3.52 respectively. The excess returns on the portfolio in Japan (Region 3) are relatively lighter-tailed than those of the other three regions, as the degrees of freedom parameter for this region is higher than those of the other three. For the t model, the only degrees of freedom parameter has a posterior mean 5.21, posterior standard deviation 0.75, and the 95% CI [3.92, 6.84].

Table 2 presents the DIC values of the SUR models with multivariate normal, Student-t and Mod-t error distributions. It is observed that the multivariate t and Mod-t error distributions achieve a lower DIC than the multivariate normal error distribution, and this is unsurprising because financial data are always heavy-tailed. However, the Mod-t distribution outperforms the t distribution in data fitting.

Figure 2 shows the posterior distribution of ν for the SUR model with the multivariate t error distribution. The posterior distribution is skewed slightly to the right. Similarly, from Fig. 3, the posterior distributions for ν_1, ν_2, ν_3 and ν_4 in the SUR model with the Mod-t error distribution are also obviously skewed to the right to different extents.

Figure 4 shows the plots of the correlation matrices for the SUR models with multivariate normal, multivariate t and Mod-t error distributions. The colours indicate the strength and direction of the correlation between any two of the four regions.

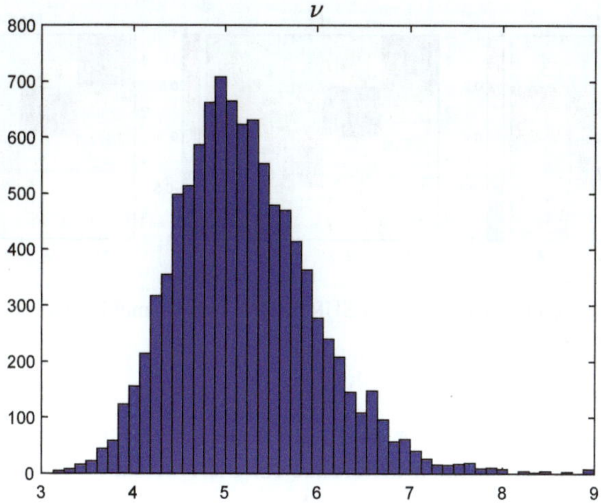

Fig. 2 Histogram of the posterior samples for ν in the multivariate t error distribution

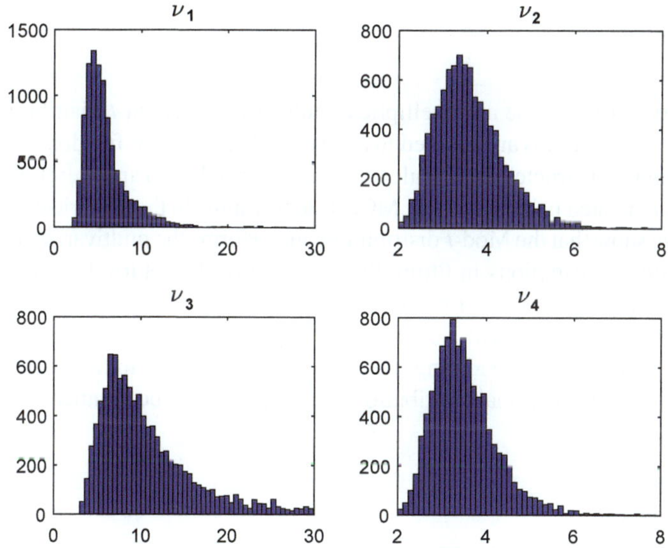

Fig. 3 Histograms of the posterior samples for ν_1, ν_2, ν_3 and ν_4 in the Mod-t error distribution

Colours closer to brown, green and blue indicate correlation coefficients closer to 1, 0 and -1 respectively. For the multivariate normal and t-distributions, the off-diagonal elements of the correlation matrices are the posterior means of ρ_{jk} for all $j, k = 1, \ldots, m$ and $j \neq k$. For the Mod-t distribution, the off-diagonal elements are calculated by the formula for $\mathrm{Corr}(X_j, X_k)$ shown in Eq. 4, using the posterior means of ρ_{jk}, ν_j and ν_k for all $j, k = 1, \ldots, m$ and $j \neq k$. The correlation coefficients between Regions 2 and 3 and between Regions 2 and 4 are furthest from 0

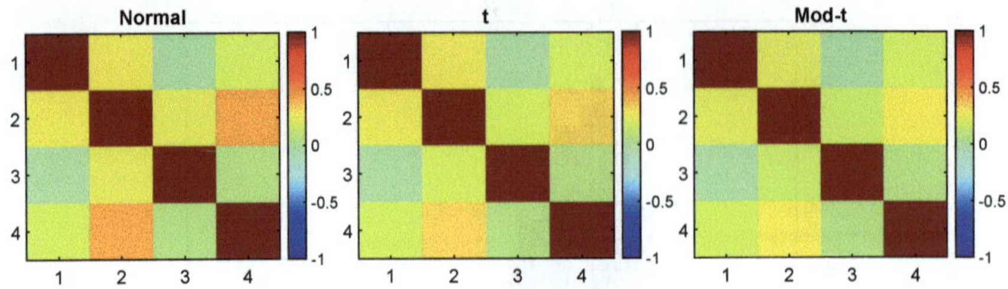

Fig. 4 Plots of the correlation matrices for SUR models with the multivariate normal, multivariate t and Mod-t error distributions

for the multivariate normal error distribution and closest to 0 for the Mod-t error distribution. There is not much difference for the correlation coefficients between any of the other regions.

4 Conclusion

In this paper, we propose a non-elliptical multivariate Student-t distribution whose marginal t-distributions are allowed to have flexible degrees of freedom. This Mod-t distribution is constructed via a scale mixture of normal form and therefore, it can be easily implemented using Bayesian MCMC algorithms. In the empirical study of the CAPM, we show that the Mod-t distribution outperforms the multivariate normal and multivariate t-distributions in fitting the excess portfolio return data. The standard deviations in the covariance matrix of the error term are smaller under the Mod-t distribution than the Student-t distribution, confirming a better fit being obtained. Therefore, it is worth considering the Mod-t distribution in multivariate statistical analysis where the marginal distributions are less likely to be identical.

References

1. Andrews, D.F., Mallows, C.L.: Scale mixtures of normal distributions. J. Royal Stat. Soc. Ser. B (Methodol.) **36**, 99–102 (1974)
2. Celeux, G., Forbes, F., Robert, C.P., Titterington, D.M.: Deviance information criteria for missing data models. Bayesian Anal. **1**, 651–674 (2006)
3. Choy, S.T.B., Chen, C.W.S., Lin, E.M.H.: Bivariate asymmetric GARCH models with heay tails and dynamic conditional correlations. Quant. Finance **14**, 1297–1313 (2014)
4. Fama, E.F., French, K.R.: Size, value and momentum in international stock returns. J. Financ. Econ. **105**, 457–472 (2012)
5. Lintner, J.: The valuation of risk assets and the selection of risky investments in stock portfolios and capital budgets. Rev. Econ. Stat. **47**, 13–37 (1965)
6. Markowitz, H.: Portfolio selection. J. Finance **7**, 77–91 (1952)
7. Markowitz, H.: Foundations of portfolio theory. J. Finance **46**, 469–477 (1991)

8. Markowitz, H.: The early history of portfolio theory. Finance Anal. J. **55**(4), 5–16 (1999)
9. Pitt, M., Chan, D., Kohn, R.: Efficient Bayesian inference for Gaussian copula regression models. Biom. **93**, 537–554 (2006)
10. Sharpe, W.F.: Capital asset prices: a theory of market equilibrium under conditions of risk. J. Finance **19**, 425–442 (1964)
11. Spiegelhalter, D.J., Best, N.G., Carlin, B.P., van der Linde, A.: Bayesian measures of model complexity and fit. J. Royal Stat. Soc. Ser. B (Methodol.) **64**, 583–639 (2002)
12. Zellner, A.: An efficient method of estimating seemingly unrelated regressions and tests for aggregation bias. J. Am. Stat. Assoc. **57**, 348–368 (1962)
13. Zellner, A.: Bayesian and non-Bayesian analysis of the regression model with multivariate student-t error terms. J. Am. Stat. Assoc. **71**, 400–405 (1976)

Bayesian Inference of Stochastic Pursuit Models from Basketball Tracking Data

Harish S. Bhat, R.W.M.A. Madushani and Shagun Rawat

Abstract We develop a Metropolis algorithm to perform Bayesian inference for models given by coupled stochastic differential equations. A key challenge in developing practical algorithms is the computation of the likelihood. We address this problem through the use of a fast method to track the probability density function of the stochastic differential equation. The method applies quadrature to the Chapman–Kolmogorov equation associated with a temporal discretization of the stochastic differential equation. The inference method can be adapted to scenarios in which we have multiple observations at one time, multiple time series, or observations with large and/or irregular temporal spacing. Computational tests show that the resulting Metropolis algorithm is capable of efficient inference for an electrical oscillator model.

Keywords Bayesian inference · Stochastic differential equations · Nonlinear systems · Sports analytics

1 Introduction

In 2010, the National Basketball Association (NBA) began to install a camera system to track the positions of the players and the ball as a function of time. For the ball and for each of the 10 players on the court, the system records an (x, y) position 25 times per second. Ultimately, this wealth of data should enable us to answer a number of questions regarding basketball strategy that would have seemed intractable just a few years ago. To bring this vision to reality, we must develop new algorithms that can efficiently use the data for inference of appropriate models.

H.S. Bhat (✉) · R.W.M.A. Madushani · S. Rawat
University of California Merced, N. Lake Rd., Merced, CA 5200, USA
e-mail: hbhat@ucmerced.edu

R.W.M.A. Madushani
e-mail: rmadushani@ucmerced.edu

S. Rawat
e-mail: srawat2@ucmerced.edu

R. Argiento et al. (eds.), *Bayesian Statistics in Action*, Springer Proceedings
in Mathematics & Statistics 194, DOI 10.1007/978-3-319-54084-9_12

In this work, we focus on so-called "fast break" situations where an offensive player races towards the basket in an attempt to score before the defensive team has time to set up their defense. In many such situations, it is relatively easy to identify from the data a runner and a chaser. This motivates the following question that is central to the present paper: using the NBA's spatial tracking data, how can we infer a stochastic model for the chaser's pursuit of the runner?

To answer this question, we first formulate a stochastic version of the classical pursuit model. Our model consists of a set of coupled, nonlinear stochastic differential equations with time-dependent coefficients. To perform Bayesian inference for this stochastic model, we develop a Markov Chain Monte Carlo (MCMC) algorithm. The MCMC algorithm is derived using a Metropolis scheme; our innovation is to evaluate the log likelihood efficiently using a novel, deterministic method called density tracking by quadrature (DTQ). The DTQ method applies quadrature to the Chapman–Kolmogorov equation associated with a time-discretization of the original stochastic differential equation (SDE). For the case of scalar SDE, we have established conditions under which the DTQ approximation converges to the true density of the SDE at a rate that is linear in the time step [3].

The MCMC algorithm developed here can be applied to Bayesian inference for a class of two-dimensional SDE, not just the pursuit model considered here. Note that inference of SDE models is a challenging problem, due to the fact that a closed-form likelihood function is generally unavailable [4, 5, 9]. Most existing parametric inference methods for discretely observed SDE require inter-observation times to be small. As a way to facilitate approximation of the transition density for parametric inference for large inter-observation times, Bayesian methods are used to simulate missing values of the observations to form a high-frequency data set. In situations where the likelihood function is either analytically unavailable or computationally prohibitive to evaluate, Bayesian inference of SDE makes use of likelihood-free methods such as Approximate Bayesian Computation [7], variational methods [2, 10], and/or Gaussian processes [1, 8]. In ongoing and future work, we will conduct a careful comparison of our method against these other methods. For the purposes of the present paper, we are more interested in establishing the appropriateness of a stochastic pursuit model for basketball fast breaks.

2 Derivation of the Model and Inference Method

Let the *runner* be the player (on offense) who has the ball and is running toward the basket. Let the *chaser* be the player (on defense) who is trying to prevent the runner from scoring. Let the current spatial coordinates of the runner and chaser be, respectively, $(x^r(t), y^r(t))$ and $(x^c(t), y^c(t))$.

Since the chaser is moving towards the runner, the velocity vector of the chaser points toward the runner's current position. Let $\boldsymbol{\phi} = (x^r - x^c, y^r - y^c)$. Then the unit vector that points toward the runner from the chaser is $\phi/\|\phi\|$. The velocity of the chaser, $(\dot{x}^c, \dot{y}^c)$, can thus be given as

$$(\dot{x}^c, \dot{y}^c) = \gamma(t)\phi/\|\phi\|, \tag{1}$$

where $\gamma(t) = \|(\dot{x}^c, \dot{y}^c)\|$, the instantaneous speed of the chaser. Note that (1) is a coupled system of nonlinear ordinary differential equations known as the pursuit model—classically, one assumes that $\gamma(t)$ and $(x^r(t), y^r(t))$ are given, in which case one typically solves an initial-value problem for $(x^c(t), y^c(t))$. To generalize the classical model to the real data context considered here, we multiply both sides of (1) by $\mathrm{d}t$ and then add noise to each component:

$$\mathrm{d}(x^c, y^c) = \gamma(t)\,[\boldsymbol{\phi}/\|\boldsymbol{\phi}\|]\,\mathrm{d}t + (\nu_1\,\mathrm{d}W_t^1, \nu_2\,\mathrm{d}W_t^2) \tag{2}$$

Here $W_{1,t}$ and $W_{2,t}$ denote two independent Wiener processes with $W_{1,0} = W_{2,0} = 0$ almost surely. We refer to this model as the stochastic pursuit model.

Given time-discrete observations of (x^c, y^c) and (x^r, y^r), how do we infer $\gamma(t)$ together with ν_1 and ν_2? Consider (2) as a particular example of:

$$\mathrm{d}X_{1,t} = f_1(t, \mathbf{X}_t, \theta)\,\mathrm{d}t + g_1(t, \mathbf{X}_t, \theta)\,\mathrm{d}W_{1,t} \tag{3a}$$
$$\mathrm{d}X_{2,t} = f_2(t, \mathbf{X}_t, \theta)\,\mathrm{d}t + g_2(t, \mathbf{X}_t, \theta)\,\mathrm{d}W_{2,t}. \tag{3b}$$

Here $\mathbf{X}_t = (X_{1,t}, X_{2,t})$ is a two-dimensional stochastic process. For $j = 1, 2$, we refer to f_j and g_j as, respectively, drift and diffusion functions. Both drift and diffusion functions may depend on a parameter vector $\theta \in \mathbb{R}^N$.

For the stochastic pursuit model (2), we take $\mathbf{X}_t = (x^c(t), y^c(t))$. We treat $\gamma(t)$ as piecewise constant. Each constant value of $\gamma(t)$ is one component of the parameter vector θ; the final two components of θ are ν_1 and ν_2. If we treat $(x^r(t), y^r(t))$ as given, then we can identify the time-dependent drift functions f_1 and f_2 as the two components of $\gamma(t)\phi/\|\phi\|$.

Our goal is to infer θ from discrete-time observations of $\mathbf{X}_t$. Suppose that at a sequence of times $0 = t_0 < t_1 < \cdots < t_M = T$, we have observations $\mathbf{x} := \{(x_{1,m}, x_{2,m})\}_{m=0}^M$. Here $\mathbf{x}_m = (x_{1,m}, x_{2,m})$ is a sample of $\mathbf{X}_{t_m}$. In this paper, we will assume equispaced temporal observations, i.e., $t_m = m\Delta t$ for fixed step size $\Delta t > 0$. We make this assumption purely for notational simplicity; the method we describe can be easily adapted for nonequispaced temporal observations.

We discretize the SDE (3) in time using the Euler–Maruyama scheme:

$$X_1^{n+1} = X_1^n + f_1(t_n, X_1^n, X_2^n, \theta)h + g_1(t_n, X_1^n, X_2^n, \theta)\sqrt{h}Z_1^{n+1} \tag{4a}$$
$$X_2^{n+1} = X_2^n + f_2(t_n, X_1^n, X_2^n, \theta)h + g_2(t_n, X_1^n, X_2^n, \theta)\sqrt{h}Z_2^{n+1}. \tag{4b}$$

Here $h > 0$ is a fixed time step, the time step of our numerical method. We shall choose h to be a fraction of Δt, i.e., $Fh = \Delta t$ for integer $F \geq 2$. The random variables X_i^n for $i = 1, 2$ are approximations of $X_{i,nh}$. The Z_i^n are independent and identically distributed random variables, normally distributed with mean 0 and variance 1, i.e., $Z_i^n \sim \mathcal{N}(0, 1)$.

The posterior density of the parameter vector given the observations is $p(\theta \mid \mathbf{x}) \propto p(\mathbf{x} \mid \theta) p(\theta)$, where $p(\mathbf{x} \mid \theta)$ is the likelihood and $p(\theta)$ is the prior. Let $\widetilde{p}(\mathbf{x} \mid \theta)$ denote the likelihood under the discrete-time model (4), an approximation to the true likelihood $p(\mathbf{x} \mid \theta)$. Note that (4) describes a discrete-time Markov chain. By the Markov property, the likelihood $\widetilde{p}(\mathbf{x} \mid \theta)$ factors and we can write:

$$p(\mathbf{x} \mid \theta) \approx \widetilde{p}(\mathbf{x} \mid \theta) = \prod_{m=0}^{M-1} \widetilde{p}(\mathbf{x}_{m+1} \mid \mathbf{x}_m, \theta). \tag{5}$$

The term $\widetilde{p}(\mathbf{x}_{m+1} \mid \mathbf{x}_m, \theta)$ is the transition density for (4), from state $\mathbf{x}_m$ at time t_m to state $\mathbf{x}_{m+1}$ at time t_{m+1}. This suggests a numerical method for computing this density, which we explore in the next subsection.

2.1 Density Tracking by Quadrature (DTQ)

Equation (4) describes a Markov chain over a continuous state space. If we let $\widetilde{p}^n(x_1, x_2 \mid \theta)$ denote the joint probability density function of X_1^n and X_2^n given θ, then the Chapman–Kolmogorov equation associated with (4) is

$$\widetilde{p}^{n+1}(x_1, x_2 \mid \theta) = \int_{y_1, y_2 \in \mathbb{R}^2} K(x_1, x_2, y_1, y_2, t_n; \theta) \widetilde{p}^n(y_1, y_2 \mid \theta)\, dy, \tag{6}$$

where

$$K(x_1, x_2, y_1, y_2, t_n; \theta) = \widetilde{p}^{n+1|n}(x_1, x_2 | y_1, y_2, \theta)$$
$$= (2\pi\sigma_1^2)^{-1/2} \exp\left[-(x_1 - \mu_1)^2/(2\sigma_1^2)\right] (2\pi\sigma_2^2)^{-1/2} \exp\left[-(x_2 - \mu_2)^2/(2\sigma_2^2)\right].$$

Here $\mu_1 = y_1 + f_1(t_n, y_1, y_2; \theta)h$, $\mu_2 = y_2 + f_2(t_n, y_1, y_2; \theta)h$, $\sigma_1^2 = g_1^2(t_n, y_1, y_2; \theta)h$ and $\sigma_2^2 = g_2^2(t_n, y_1, y_2; \theta)h$. That is, $K(x_1, x_2, y_1, y_2, t_n; \theta)$ is the conditional density of X_1^{n+1} and X_2^{n+1} given $X_1^n = y_1$, $X_2^n = y_2$ and a fixed θ, evaluated at the point (x_1, x_2). The fact that the conditional density is a product of normal distributions with means μ_1, μ_2 and variances σ_1^2, σ_2^2 can be shown using (4) together with the fact that X_1^{n+1} and X_2^{n+1} are conditionally independent given X_1^n and X_2^n. This conditional independence is a direct consequence of having two independent random variables Z_1^n and Z_2^n in (4).

The crux of the DTQ method is to apply quadrature to (6) to evolve an initial density forward in time. We use the trapezoidal rule for quadrature because, for analytic drift and diffusion functions, this leads to exponential convergence of $\hat{p}$ to $\widetilde{p}$ [3]. While other quadrature rules may yield improvements for non-analytic drifts/diffusions, the overall error between p and $\hat{p}$ will still be dominated by the $O(h)$ error between $\widetilde{p}$ and p stemming from the Euler–Maruyama discretization.

Consider a $(2M+1) \times (2M+1)$ spatial grid with fixed spacing $k > 0$ and grid points $x_1^i = ik$, $x_2^j = jk$, $y_1^{i'} = i'k$, and $y_2^{j'} = j'k$, where $i, i', j, j' \in \{-M, M\}$. Then we apply the trapezoidal rule in both the y_1 and y_2 variables to obtain:

$$\hat{p}^{n+1}(x_1^i, x_2^j; \theta) = k^2 \sum_{i'=-\infty}^{\infty} \sum_{j'=-\infty}^{\infty} K(x_1^i, x_2^j, y_1^{i'}, y_2^{j'}, t_n; \theta) \hat{p}^n(y_1^{i'}, y_2^{j'}; \theta) \quad (7)$$

It is unnecessary to sum over all of $\mathbb{Z}^2$. We know that a two-dimensional Gaussian decays to zero far from its mean. Since the mean (μ_1, μ_2) is approximately (y_1, y_2), we sum only from $y_1 = x_1 - \zeta k$ to $y_1 = x_1 + \zeta k$ and similarly for y_2:

$$\hat{p}^{n+1}(x_1^i, x_2^j; \theta) = k^2 \sum_{i'=i-\zeta}^{i+\zeta} \sum_{j'=j-\zeta}^{j+\zeta} K(x_1^i, x_2^j, y_1^{i'}, y_2^{j'}, t_n; \theta) \hat{p}^n(y_1^{i'}, y_2^{j'}; \theta) \quad (8)$$

We choose ζ manually to ensure the accuracy of the computation. We now have our method to evaluate $\widetilde{p}(\mathbf{x}_{m+1} \,|\, \mathbf{x}_m, \theta)$. Let us take $n = 0$ in (8) to correspond to the time t_m. We start with the deterministic initial condition $\mathbf{X}^0 = \mathbf{x}_m$, corresponding to the density $\widetilde{p}^0(\mathbf{x}) = \delta(\mathbf{x} - \mathbf{x}_m)$. Inserting this point mass into (6), we obtain a Gaussian density for $\widetilde{p}^1(\mathbf{x})$. For each $i, j \in \{-M, M\}$ on the spatial grid, we set $\hat{p}^1(x_1^i, x_2^j; \theta) = \widetilde{p}^1(x_1^i, x_2^j; \theta)$. Now that we have $\hat{p}^1$, we use (8) repeatedly to compute $\hat{p}^2$, $\hat{p}^3$, and so on until we reach $\hat{p}^F$. The object $\hat{p}^F$ is then a spatially discrete approximation of the transition density from time t_m to time $t_m + Fh = t_{m+1}$. For this last density, instead of evaluating it on the spatial grid used by the trapezoidal rule, we evaluate the density at the data $\mathbf{x}_{m+1}$. This avoids interpolation. In this way, we compute a numerical approximation of $\widetilde{p}(\mathbf{x}_{m+1} \,|\, \mathbf{x}_m, \theta)$, as required for the likelihood function.

2.2 Metropolis Algorithm

Here we embed the DTQ method's likelihood computation into a Metropolis algorithm to sample from the posterior. In the Metropolis algorithm, we construct an auxiliary Markov chain $\{\hat{\boldsymbol{\theta}}_N\}_{N \geq 0}$ which is designed to have an invariant distribution given by the posterior $p(\theta \,|\, \mathbf{x})$. This Markov chain is constructed as $\hat{\boldsymbol{\theta}}_{N+1} = \hat{\boldsymbol{\theta}}_N + Z_{N+1}$, where Z_{N+1} is a random vector with dimension equal to that of the parameter vector θ. In this paper, we choose all components of Z_{N+1} to be independent normal random variables with known means and variances.

The Metropolis algorithm is as follows:

- Choose value q_0 for $\hat{\boldsymbol{\theta}}_0$.
- Once the values $q_0, \ldots, q_N$ of $\hat{\boldsymbol{\theta}}_0, \ldots, \hat{\boldsymbol{\theta}}_N$ have been found:

- Generate a proposal from the auxiliary Markov chain: $q^*_{N+1} = q_N + Z_{N+1}$.
- Calculate the ratio $\rho = \frac{p(q^*_{N+1}\,|\,\mathbf{x})}{p(q_N\,|\,\mathbf{x})}$, where $p(q^*_{N+1}\,|\,\mathbf{x}) \approx \widetilde{p}(\mathbf{x}\,|\,q^*_{N+1})p(q^*_{N+1}) = p(q^*_{N+1})\prod_{m=0}^{M-1}\widetilde{p}(\mathbf{x}_{m+1}\,|\,\mathbf{x}_m, q^*_{N+1})$. Now each term $\widetilde{p}(\mathbf{x}_{m+1}\,|\,\mathbf{x}_m, q^*_{N+1})$ can be computed using the DTQ method discussed in Sect. 2.1.
- Sample $u_N \sim \mathscr{U}(0,1)$. If $\rho > u_N$ set $\hat{\boldsymbol{\theta}}_{N+1} = q^*_{N+1}$; in this case, the proposal is accepted. Else set $\hat{\boldsymbol{\theta}}_{N+1} = q_N$ and the proposal is rejected.

Once we have obtained all the samples $q_0, q_1, \ldots, q_N$ from the Metropolis algorithm, we discard a sufficient number of initial samples to ensure the Markov chain has converged to its invariant distribution.

3 Numerical Tests

We implement the Metropolis algorithm in R. Inside the Metropolis algorithm, we evaluate the likelihood function using the DTQ method, which is implemented in C++ as an R package. To test the method, we first consider the nonlinear SDE

$$\mathrm{d}X_{1,t} = \theta_1\, X_{2,t}\,\mathrm{d}\,t + (0.1 + \theta_4^2\, e^{-X_{1,t}^2})\,\mathrm{d}\,W_{1,t}, \tag{9a}$$

$$\mathrm{d}X_{2,t} = (-\theta_2\, X_{1,t} + \theta_3 X_{2,t}\,(1 - X_{1,t}^2))\,\mathrm{d}\,t + (0.1 + \theta_5^2\, e^{-X_{2,t}^2})\,\mathrm{d}\,W_{2,t}. \tag{9b}$$

This system describes a noisy van der Pol oscillator. The presence of $X_{1,t}$ and $X_{2,t}$ in the diffusion function ensures that the transition density is not Gaussian. To generate simulated data, we start with known values of the parameters: $\theta_1 = 1, \theta_2 = 1, \theta_3 = 4$ and the noise parameters $\theta_4 = \theta_5 = 0.5$. Using a fixed initial condition $(X_{1,0}, X_{2,0})$, we then use the Euler–Maruyama method to step (9) forward in time until a final time $T > 0$. When we carry out this time-stepping, we use a step size of 0.001 and simulate up to $T = 20$. We then retain every 100-th element to yield a data set consisting of 201 observations of X_1 and X_2 with spacing $\Delta t = 0.1$. In this way, we simulate large inter-observation times for a process that in reality operates at a finer time scale.

Using the samples $\{\mathbf{x}_m\}_{m=0}^{M}$ thus constructed, we run the Metropolis algorithm. We infer only the parameters in the drift function, i.e., θ_1, θ_2 and θ_3, keeping other parameters fixed at their known values. We initialize θ at $(0.1, 0.1, 0.1)$, far from the true θ values. We use a diffuse Gaussian prior with mean 0 and standard deviation 100. For the proposal distribution Z_{N+1} in the auxiliary Markov chain, we choose i.i.d. Gaussians with mean 0 and standard deviation 0.05.

Our goal here is to test the performance of the algorithm using simulated data and compare it against Bayesian particle filtering/inference method implemented in the R package "pomp" [6]. This method gives us an alternative, sampling-based approach to approximate the likelihood function. Note that we also compare DTQ against a purely Eulerian approximation of the transition density, i.e., a method where $\tilde{p}(\mathbf{x}_{m+1}|\mathbf{x}_m, \theta)$ is approximated by a Gaussian pdf; this is equivalent to the DTQ method with zero internal quadrature steps, i.e., $h = \Delta t = 0.1$.

When we run the Metropolis algorithm, we discard the first 1000 samples and retain the next 20000 samples. We have settled on 20000 samples because, in this case, using the first 10000 post-burn-in samples does not yield significantly different results than what we obtained, i.e., we see no reason to continue sampling. We record the inferred parameter values, acceptance rate of the method (AR), and mean absolute percentage error (MAPE) for varying values of h for the 3 methods, Euler, DTQ and Pomp.

Parameters	θ_1	θ_2	θ_3	AR	MAPE	Method
$\Delta t = 0.1; h = 0.1/1$	0.747	0.906	3.070	0.296	0.193	Euler
$\Delta t = 0.1; h = 0.1/2$	0.866	1.300	4.260	0.285	0.168	
$\Delta t = 0.1; h = 0.1/4$	0.892	1.160	4.430	0.254	0.124	DTQ
$\Delta t = 0.1; h = 0.1/8$	0.980	1.170	4.210	0.239	0.081	
$\Delta t = 0.1; h = 0.1/2$	1.250	0.257	4.340	0.0003	0.361	
$\Delta t = 0.1; h = 0.1/4$	1.110	0.647	4.060	0.001	0.158	Pomp
$\Delta t = 0.1; h = 0.1/8$	1.040	0.752	3.940	0.0004	0.102	

The first four rows of the table show that using the DTQ method to compute the likelihood yields more accurate posteriors than using a purely Gaussian likelihood (Eulerian method). In comparison to Pomp, our method does slightly better in terms of the means of the posteriors. If we look at the Metropolis samples generated by the two methods, the DTQ method has radically higher acceptance rates than Pomp. The non-adaptive version of the Metropolis sampling for Pomp does not explore the posterior adequately, rejecting thousands of consecutive proposals. A carefully executed adaptive Metropolis algorithm for Pomp does generate better results than the non-adaptive version:

Parameters	θ_1	θ_2	θ_3	AR	MAPE	Method
$\Delta t = 0.1; h = 0.1/2$	0.960	0.809	4.010	0.110	0.078	
$\Delta t = 0.1; h = 0.1/4$	1.000	0.954	3.990	0.164	0.017	Pomp-adaptive
$\Delta t = 0.1; h = 0.1/8$	1.010	1.010	4.020	0.171	0.009	

One should take care to interpret these results: we have invested a great deal of time to tune parameters in the adaptive MCMC scheme for pomp with full knowledge of the "true" parameter vector θ. Overall, what we have learned from this exercise is that there are two main investments of effort that one can make. In the DTQ method, we have invested effort into making the likelihood calculation more accurate, efficient and stable to initial choice of parameters. This allows us to use the DTQ method with a vanilla Metropolis algorithm and obtain reasonable results. One could instead have chosen to improve the vanilla Metropolis algorithm in various ways: adaptive MCMC, sequential MC, Hamiltonian MC, etc. This is the strategy pursued by Pomp. While both strategies have their merits, it seems that the likelihood computed by Pomp is not accurate enough to enable a vanilla Metropolis method to work well.

To understand this point in more detail, we have computed log likelihood surfaces in (θ_2, θ_3) space using both the Pomp and DTQ methods. If we rescale the log likelihood values from both methods so that they achieve a maximum of 0 and then exponentiate, we find that the DTQ likelihood has more reasonable gradients than the Pomp likelihood, which varies over 3 orders of magnitude. The accept/reject ratio depends on the actual density, i.e., the exponential of the log likelihood plus the log prior. Therefore, the sharp dropoff in the likelihood function at points very close to the maximum—seen in Pomp—will cause thousands of consecutive rejected steps. The more gradual dropoff in the DTQ likelihood function leads to a reasonable fraction of accepted steps in a vanilla Metropolis algorithm.

Next, we test the method using the pursuit SDE (2). We set the runner's trajectory equal to a sinusoidal curve $y = \sin(\pi x)$ from $x = -1$ to $x = 1$. We assume the runner covers this trajectory over the time period $0 \leq t \leq 8$. The chaser's trajectory is simulated using the Euler–Maruyama method to step (2) forward in time from a fixed initial condition $\mathbf{X}_0 = (x_0^c, y_0^c)$. During the generation of the data, we use a step size of 10^{-4}. By downsampling this single time series, we generate time series with spacings $\Delta t = 0.4, 0.2, 0.1$.

We set $\nu_1 = 0.15$, $\nu_2 = 0.1$, $\gamma(t) = \gamma_1 = 0.4$ for $0 \leq t < 4$, and $\gamma(t) = \gamma_2 = 1.0$ for $4 \leq t \leq 8$. Because we want all speeds and diffusion constants to be positive, we take $\gamma_i = e^{\theta_i}$ and $\nu_i = e^{\theta_{i+2}}$ for $i = 1, 2$. The priors for θ_1 and θ_2 are normal with variance one and mean equal to the log of the mean speed of the chaser computed over the chaser's entire trajectory. The priors for θ_3 and θ_4 are normal with mean $\log(0.4)$ and variance 1. We use mean zero Gaussian proposals for all components of θ. We choose the variances of these proposals so that the acceptance rate for all runs is near 30%.

Using the samples $\{\mathbf{x}_m\}_{m=0}^{M}$ thus constructed, we run the Metropolis algorithm with $h = \Delta t / i$ with $i = 1, 2, 3, 4$. For each choice of parameters Δt and h, we compute 10100 samples and discard the first 100. To compute the runner's trajectory at intermediate points, we use linear interpolation between times t_m and t_{m+1}. We tabulate the results below; each value of γ_1 represents the mean of e^{θ_1} over all Metropolis samples of θ_1:

Parameters	γ_1	γ_2	ν_1	ν_2	RMSE
$\Delta t = 0.1; h = 0.1/1$	0.301	0.748	0.124	0.088	0.136
$\Delta t = 0.1; h = 0.1/2$	0.311	0.956	0.124	0.085	0.051
$\Delta t = 0.1; h = 0.1/3$	0.307	1.011	0.117	0.080	0.050
$\Delta t = 0.1; h = 0.1/4$	0.308	1.025	0.120	0.082	0.050
$\Delta t = 0.2; h = 0.2/1$	0.306	0.650	0.142	0.114	0.181
$\Delta t = 0.2; h = 0.2/2$	0.310	0.877	0.137	0.119	0.077
$\Delta t = 0.2; h = 0.2/3$	0.309	1.015	0.112	0.084	0.050
$\Delta t = 0.2; h = 0.2/4$	0.304	1.019	0.111	0.085	0.053
$\Delta t = 0.4; h = 0.4/1$	0.292	0.514	0.188	0.201	0.254
$\Delta t = 0.4; h = 0.4/2$	0.312	0.960	0.177	0.177	0.063
$\Delta t = 0.4; h = 0.4/3$	0.307	0.987	0.124	0.144	0.053
$\Delta t = 0.4; h = 0.4/4$	0.303	1.014	0.145	0.113	0.049

Overall, the results show that our algorithm produces mean posterior estimates that are reasonably close to the ground truth values. When the spacing of the data Δt is large, we see dramatic improvement when we use the DTQ method and more internal time steps. For instance, when $\Delta t = 0.4$, the RMS error improves dramatically from 0.254 to 0.049 as we decrease h, i.e., as we take more internal DTQ steps. Similar trends can be seen for the mean estimates of γ_2, ν_1 and ν_2.

Note that all codes and data used in this work are available online at: https://github.com/hbhat4000/sdeinference.

4 NBA Tracking Data

We now turn to real tracking data taken from the game played between the Golden State Warriors and the Sacramento Kings on October 29, 2014. Reviewing this game, we found a fast break where Stephen Curry (of the Warriors) was the runner and Ramon Sessions (of the Kings) was the chaser. The entire fast break lasts 4.12 s. The spatial tracking data is recorded at intervals of 0.04 s, for a total of 104 observations. The tracking data uses the position on a court of dimension 94×50. We have rescaled the data to lie in a square with center (0, 0) and side length equal to one.

To parameterize the chaser's speed $\gamma(t)$, we have used a piecewise constant approximation with 8 equispaced pieces. Combined with the diffusion constants ν_1 and ν_2, this yields a 10-dimensional parameter vector θ. As in the previous simulated data test, we set the true parameters γ_i and ν_i to be the exponentials of the corresponding elements of the θ vector.

For the Metropolis sampler, the priors and proposals are higher-dimensional versions of those described in the simulated data test above. The main difference is that we now generate only 1000 post-burnin samples.

Using the Metropolis samples, we compute a kernel density estimate of each parameter. We then treat the mode of each computed density as the MAP (maximum a posteriori) estimate of the corresponding parameter. We then use the MAP estimates of the parameters in the pursuit SDE (2). We generate 100 sample paths of this SDE using the Euler–Maruyama method with time step 10^{-4}. As shown in Fig. 1, the mean of these sample paths (plotted in black) agrees very well with the chaser's trajectory (plotted in red). This gives evidence that our stochastic pursuit system is an appropriate model for NBA fast breaks involving one runner and one chaser.

To visualize the insight provided by the model, we plot in Fig. 2 the MAP estimated $\gamma(t)$ function over the time period of the fast break, $0 \leq t \leq 4.12$. The speed $\gamma(t)$ is the piecewise constant function plotted in black, while the mean speed computed directly from the data is given by a red horizontal line. The inferred speed shows that the chaser slows down dramatically approximately 1.5 s into the fast break. If one reviews the video footage of the play, this corresponds to the runner confusing the chaser and evading him.

Given our stochastic pursuit model's success in fitting the real data, in future work, we seek to apply the same methodology to a much larger sample of fast breaks. In

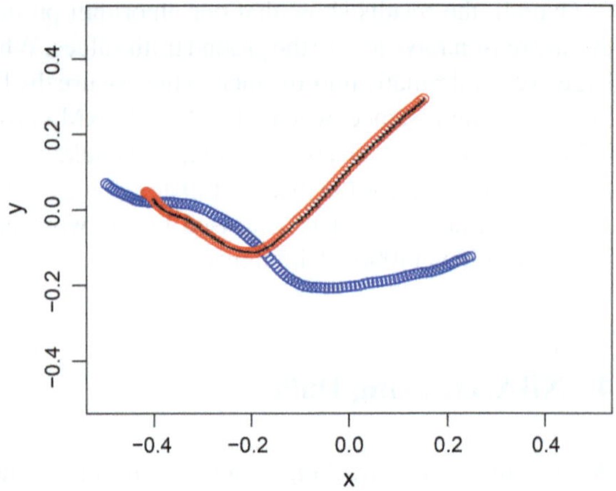

Fig. 1 The agreement between the *black curve* (mean of simulated stochastic pursuit trajectories using MAP estimated parameters) and the *red curve* (chaser's trajectory) shows that the stochastic pursuit model is appropriate. The runner's trajectory is given in *blue*

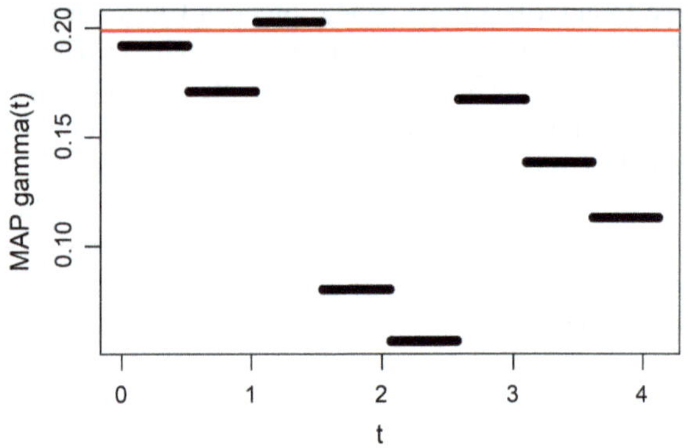

Fig. 2 For the fast break tracking data described in the text, we plot the MAP estimate of the chaser's $\gamma(t)$ in *black*. This gives an estimate of how well the chaser is pursuing the runner at various points in time. The times at which γ dips below 0.1 correspond well to the times where Sessions was not able to keep up with Curry

this way, we can quantify a runner's ability to evade a chaser and/or a chaser's ability to stay near a runner who is actively trying to score.

References

1. Archambeau, C., Cornford, D., Opper, M., Shawe-Taylor, J.: Gaussian process approximations of stochastic differential equations. J. Mach. Learn. Res. Workshop Conf. Proc. **1**, 1–16 (2007)
2. Archambeau, C., Opper, M., Shen, Y., Cornford, D., Shawe-Taylor, J.: Variational inference for diffusion processes. Adv. Neural Inf. Process. Syst. **20**, 17–24 (2007)
3. Bhat, H.S., Madushani, R.W.M.A.: Density tracking by quadrature for stochastic differential equations, arXiv:1610.09572 [stat.CO] (2016)
4. Fuchs, C.: Inference for Diffusion Processes: With Applications in Life Sciences. Springer, Berlin (2013)
5. Iacus, S.M.: Simulation and Inference for Stochastic Differential Equations: With R Examples. Springer Series in Statistics. Springer, New York (2009)

6. King, A.A., Nguyen, D., Ionides, E.L., et al.: Statistical inference for partially observed Markov processes via the R package pomp. J. Stat. Softw. **69**, 1–43 (2016)
7. Picchini, U.: Inference for SDE models via approximate Bayesian computation. J. Comput. Graph. Stat. **23**(4), 1080–1100 (2014)
8. Ruttor, A., Batz, P., Opper, M.: Approximate Gaussian process inference for the drift function in stochastic differential equations. Adv. Neural Inf. Process. Syst. **26**, 2040–2048 (2013)
9. Sørensen, H.: Parametric inference for diffusion processes observed at discrete points in time: a survey. Int. Stat. Rev. **72**(3), 337–354 (2004)
10. Vrettas, M.D., Opper, M., Cornford, D.: Variational mean-field algorithm for efficient inference in large systems of stochastic differential equations. Phys. Rev. E **91**(012148) (2015)

Identification of Patient-Specific Parameters in a Kinetic Model of Fluid and Mass Transfer During Dialysis

Camilla Bianchi, Ettore Lanzarone, Giustina Casagrande and Maria Laura Costantino

Abstract Hemodialysis (HD) is nowadays the most common therapy to treat renal insufficiency. However, despite the improvements made in the last years, HD is still associated with a non-negligible rate of co-morbidities, which could be reduced by means of an appropriate treatment customization. Many differential multi-compartment models have been developed to describe solute kinetics during HD, to optimize treatments, and to prevent intra-dialysis complications; however, they often refer to an average uremic patient. On the contrary, the clinical need for customization requires patient-specific models. In this work, assuming that the customization can be obtained by means of patient-specific model parameters, we propose a Bayesian approach to estimate the patient-specific parameters of a multi-compartment model and to predict the single patient's response to the treatment, in order to prevent intra-dialysis complications. The likelihood function is obtained through a discretized version of a multi-compartment model, where the discretization is in terms of a Runge–Kutta method to guarantee the convergence, and the posterior densities of model parameters are obtained through Markov Chain Monte Carlo simulation.

Keywords Hemodialysis · Patient-specific response · Multi-compartment model · Runge–Kutta discretization · Markov Chain Monte Carlo

C. Bianchi (✉) · G. Casagrande · M.L. Costantino
Department of Chemistry, Materials and Chemical Engineering,
Politecnico di Milano, Milan, Italy
e-mail: camilla.bianchi@polimi.it

G. Casagrande
e-mail: giustina.casagrande@polimi.it

M.L. Costantino
e-mail: marialaura.costantino@polimi.it

E. Lanzarone
Istituto di Matematica Applicata e Tecnologie Informatiche (IMATI),
Consiglio Nazionale delle Ricerche (CNR), Milan, Italy
e-mail: ettore.lanzarone@cnr.it

R. Argiento et al. (eds.), *Bayesian Statistics in Action*, Springer Proceedings in Mathematics & Statistics 194, DOI 10.1007/978-3-319-54084-9_13

1 Introduction

The elective option to treat End Stage Renal Disease (ESRD) is Hemodialysis (HD). However, despite the improvements of the last years, HD is still associated with a non-negligible rate of comorbidities [1]. Indeed, due to therapy discontinuity, HD procedure induces considerable changes in osmotic balances and rapid variations in fluid volumes [2–4] and electrolytic concentrations [5] within the patient's body compartments. Moreover, a need for treatment customization emerged over the years, in order to reduce the associated comorbidities, because the individual tolerance to HD may vary from patient to patient also in the presence of similar treatments [6, 7].

Instruments that simulate and predict the single patient's response to HD treatment in terms of electrolyte and catabolite kinetics are a necessary step toward customization, to identify the most suitable therapy for reducing intra-dialysis complications and the associated long-term dysfunctions. In the literature, many multi-compartment models have been developed to describe solute kinetics during HD, but they often refer to an average uremic patient [8, 9]. On the contrary, the clinical need for customization require patient-specific multi-compartment models [10].

In this work, we refer to the parametric multi-compartment model of Casagrande et al. [10], and we assume that the customization can be obtained by means of patient-specific model parameters. We propose a Bayesian estimation approach to determine the patient-specific parameters of this multi-compartment model and to predict the single patient response to HD treatment. Differently from [10], where a constrained non-linear optimization algorithm was used to get the parameter values, the Bayesian approach allows us to formally include the clinical prior knowledge on the parameters and to directly evaluate the uncertainty associated with the estimates, by means of their posterior probability density functions.

In the literature, Bayesian approaches have been successfully applied to estimate the parameters and the response of differential systems, e.g., we previously applied to estimate the inertance of a hydraulic simulator of the human circulation [11], the thermal conductivity and the temperature profile in a polymer [12], the aortic stiffness from non-invasive measurements [13], and the mortality terms in a stage-structured demographic model [14]. However, to the best of our knowledge, this is the first attempt to apply this framework to HD.

2 Multi-compartment Model

The differential multi-compartment model of Casagrande et al. [10] allows to evaluate the temporal trend of blood concentration for different plasmatic electrolytes and breakdown products (defined as *solutes* in the following), and the blood volume variations for the whole duration of a HD treatment, which is crucial from a clinical viewpoint [15–17]. The model is based on mass and fluid balance equations, and

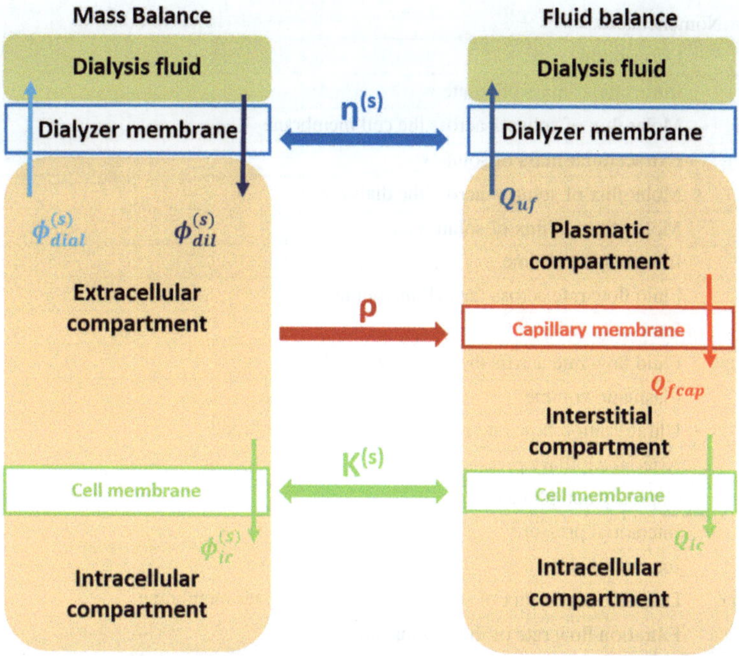

Fig. 1 Body compartments of a HD patient, with fluid and mass exchanges across the biological and artificial membranes (patient-specific parameters ρ, $k^{(s)}$ and $\eta^{(s)}$ highlighted where they act)

both fluid and mass exchanges among patient's body compartments and across the dialyzer membrane are taken into account. The body compartments are represented as two pools (intracellular and extracellular) for the mass exchange of solutes, and as three pools (plasmatic, interstitial and intracellular) for the fluid transfer (Fig. 1). All symbols in Fig. 1 and in the following equations are defined in Table 1.

Mass, fluid and pressure balance equations that account for mass and fluid exchanges among the body compartments are as follows:

$$\frac{dM_{ic}^{(s)}(t)}{dt} = \phi_{ic}^{(s)}(t) \tag{1}$$

$$\frac{dM_{ex}^{(s)}(t)}{dt} = -\phi_{ic}^{(s)}(t) - \phi_{dial}^{(s)}(t) + \phi_{dil}^{(s)} \tag{2}$$

$$\frac{dV_{ic}(t)}{dt} = Q_{ic}(t) \tag{3}$$

$$\frac{dV_{is}(t)}{dt} = -Q_{ic}(t) + Q_{fcap}(t) \tag{4}$$

$$\frac{dV_{pl}(t)}{dt} = -Q_{fcap}(t) - Q_{uf} \tag{5}$$

Table 1 Nomenclature

Name	Description
$M_{ic}^{(s)}(t)$	Intracellular mass of solute s
$\phi_{ic}^{(s)}(t)$	Molar flux of solute s across the cell membrane
$M_{ex}^{(s)}(t)$	Extracellular mass of solute s
$\phi_{dial}^{(s)}(t)$	Molar flux of solute s across the dialyzer
$\phi_{dil}^{(s)}$	Molar dilution flux of solute s
$V_{ic}(t)$	Intracellular volume
$Q_{ic}(t)$	Fluid flow rate across the cell membrane
$V_{is}(t)$	Interstitial volume
$Q_{fcap}(t)$	Fluid flow rate across the capillary membrane
$V_{pl}(t)$	Plasmatic volume
Q_{uf}	Ultrafiltration flow rate across the dialyzer membrane
$P_{ac}(t)$	Arterial capillary pressure
C_c	Capillary compliance
$P_{is}(t)$	Interstitial pressure
E_{is}	Interstitial elastance
$\phi_{diff}^{(s)}(t)$	Diffusive molar flux of solute s across the dialyzer membrane
Q_f	Filtration flow rate of the HD machine
$C_{in}^{(s)}(t)$	Concentration of solute s in plasma water at the dialyzer inlet
L_c	Capillary permeability
$P_n(t)$	Total net filtration pressure (difference between hydraulic and osmotic pressures)
$C_{ic}^{(s)}(t)$	Concentration of solute s in the intracellular compartment
β	Ratio between intracellular and interstitial equilibrium concentrations
$C_{is}^{(s)}(t)$	Concentration of solute s in the interstitial compartment

$$\frac{dP_{ac}(t)}{dt} = \frac{1}{C_c}\frac{dV_{pl}(t)}{dt} \tag{6}$$

$$\frac{dP_{is}(t)}{dt} = E_{is}\frac{dV_{is}(t)}{dt} \tag{7}$$

where (1) and (2) refer to the mass balance for the intracellular and extracellular compartments, respectively, (3)–(5) to the fluid balance in the intracellular, interstitial and plasmatic compartments, respectively, and (6) and (7) to the pressure balance at the arterial capillary side and in the interstitium, respectively. Time is denoted by t, and the index $s = 1, \ldots, 8$ refers to the solutes (sodium, potassium, chlorine, calcium, bicarbonate, magnesium, urea, and creatinine).

Individual HD control in terms of mass and fluid exchange is achieved by identifying the following patient-specific parameters related to the membranes across which fluid and mass transports occur:

- $\eta^{(s)}$: performance of dialyzer membrane in terms of filtration for solute s;

- ρ: relative capillary wall permeability;
- $k^{(s)}$: modulation of mass transfer efficiency across cell membrane for solute s.

Variables $\phi_{dial}^{(s)}(t)$, $Q_{fcap}(t)$ and $\phi_{ic}^{(s)}(t)$ depend on $\eta^{(s)}$, ρ and $k^{(s)}$ as follows:

$$\phi_{dial}^{(s)}(t) = \phi_{diff}^{(s)}(t) + \eta^{(s)} \cdot Q_f \cdot C_{in}^{(s)}(t) \tag{8}$$

$$Q_{fcap}(t) = \rho \cdot L_c \cdot P_n(t) \tag{9}$$

$$\phi_{ic}^{(s)}(t) = -k^{(s)} \left(C_{ic}^{(s)}(t) - \beta C_{is}^{(s)}(t) \right) \tag{10}$$

where (8) is valid for standard hemodialysis. Further details are reported in [10].

3 Bayesian Estimation Approach

The patient-specific parameters to estimate with the Bayesian method are globally denoted as $\Theta = \{\eta^{(1)}, \ldots, \eta^{(8)}, \rho, k^{(1)}, \ldots, k^{(8)}\}$ in the following. The likelihood function is based on a discretized version of the multi-compartment model. Given the structure of the model, the posterior densities of the Θ parameters are numerically obtained with a Markov Chain Monte Carlo (MCMC) algorithm.

3.1 Discretized Model and Likelihood Function

The likelihood function is obtained from the discretized formulation of the ODE system in (1)–(7). To guarantee the convergence, the temporal discretization is performed with the 4th order Runge–Kutta method.

Each patient-specific parameter in Θ is assumed to be a random variable. In this way, each discretized equation is a random process and we can express the density of each state variable $X^i(t_h)$ at discrete time t_h as conditioned to Θ and to all values $X^1(t_{h-1}), \ldots, X^n(t_{h-1})$ at time t_{h-1}:

$$X^i(t_h) \sim \mathscr{L}\left(X^i(t_h) | X^1(t_{h-1}), \ldots, X^n(t_{h-1}), \Theta\right) \quad \forall i, h \tag{11}$$

where $\mathscr{L}$ denotes the conditioned probability law, which does not follow any known form due to the structure of the system, and n is the number of state variables in model (1)–(7). Other variables different from the state variables are also present in the model, but we can express them as function of the state variables at the same time instant t_{h-1}.

We consider that observations X_{obs}^i of X^i $(i = 1, \ldots, n)$ are taken at some time instants t_{h^*}, where the set $\{t_{h^*}\}$ in which the observations are available is a subset of $\{t_h\}$. This holds because the discretization step Δ, chosen according to the differential equations, is usually thicker than the frequency of the observations.

We assume that all observations $X^i_{obs}(t_{h^*})$ are associated with an error (e.g., a measurement error). Thus, we model them as stochastic variables following a *Gamma* distribution with modal value $X^i(t_{h^*})$ and standard deviation $\alpha X^i(t_{h^*})$:

$$X^i_{obs}(t_{h^*}) \sim \mathscr{G}\left(1 + \frac{1+\sqrt{1+4\alpha^2}}{2\alpha^2}, \frac{1+\sqrt{1+4\alpha^2}}{2\alpha^2 X^i(t_{h^*})}\right) \quad \forall i, h^* \tag{12}$$

where $\mathscr{G}$ denotes the Gamma distribution, $1 + \frac{1+\sqrt{1+4\alpha^2}}{2\alpha^2}$ is the shape parameter, and $\frac{1+\sqrt{1+4\alpha^2}}{2\alpha^2 X^i(t_{h^*})}$ the rate parameter.

In particular, we consider three different parameters α based on the state variable, i.e., α_v for V_{is} and V_{pl}, α_{mex} for all $M^{(s)}_{ex}$, and α_{mic} for $M^{(s)}_{ic}$. They are other parameters to estimate; thus, the vector of the parameters to estimate is $\Theta^* = \Theta \cup \{\alpha_v, \alpha_{mex}, \alpha_{mic}\}$.

The combination of (11) and (12) gives $\forall i$ the conditional law of the observations at each instant t_{h^*}. Their product over t_{h^*} and over i gives the likelihood function $f\left(\hat{\mathbf{X}}_{obs}|\Theta^*\right)$ of the observations given the parameter vector Θ^*:

$$f\left(\hat{\mathbf{X}}_{obs}|\Theta^*\right) = \prod_{i,h^*} \mathscr{G}\left(1 + \frac{1+\sqrt{1+4\alpha^2}}{2\alpha^2}, \frac{1+\sqrt{1+4\alpha^2}}{2\alpha^2 X^i(t_{h^*})}\right) \times \widehat{\mathscr{L}}\left(X^i(t_{h^*})|X^1(t_{h^*-1}), \ldots, X^n(t_{h^*-1}), \Theta\right) \tag{13}$$

where $\hat{\mathbf{X}}_{obs}$ denotes the overall set of observations $\forall i, h^*$, and the law $\widehat{\mathscr{L}}$ represents the marginal density of $X^i(t_{h^*})$, which is derived combining all densities $\mathscr{L}$ in (11) between the observation at t_{h^*} and the previous observation at t_{h^*-1}. In the notation, α has to be specified (α_v, α_{mex} or α_{mic}) based on the specific variable X^i.

3.2 Prior Densities and Computation of the Posterior Densities

The following independent prior densities are chosen, according to the HD literature [10, 18] that provides some information about standard values and ranges of parameters in Θ^*:

- $\eta^{(s)}$: *Uniform* density with minimum value equal to 0 and maximum value equal to 1 $\forall s$;
- ρ: *Gamma* density with mean value equal to 1 and standard deviation equal to 0.1;
- $k^{(s)}$: *Gamma* density with mean value equal to $2.50 \cdot 10^{-3}$ for sodium, $3.30 \cdot 10^{-3}$ for bicarbonate, $1.30 \cdot 10^{-3}$ for urea, $1.30 \cdot 10^{-4}$ for creatinine, $1.67 \cdot 10^{-4}$ for all other solutes, and standard deviation equal to the 10% of the respective mean value $\forall s$;

- $\alpha_v, \alpha_{mex}, \alpha_{mic}$: *Gamma* density with mean value and variance equal to 0.05.

The posterior density of Θ^* is numerically obtained with a Markov Chain Monte Carlo (MCMC) method. In particular, we use STAN [19], which implements the Hamiltonian Monte Carlo algorithm to sample from each marginal posterior density. Estimates are obtained with 1200 iterations, including a warm up of 500 iterations. The discretization time step Δ is equal to 5 s.

4 Computational Validation of the Approach

In this chapter, we show the validation of the proposed approach on a test instance that replicates a HD session, in which the Bayesian method is applied using simulated observations of the state variables V_{pl}, V_{is}, $M_{ex}^{(s)}$, $M_{ic}^{(s)}$.

4.1 Test Instance

To obtain the test instance, the ODE system in (1)–(7) was integrated over a period of 4 h (the typical duration of a HD treatment). Clinical data acquired at the beginning of a HD session were used to initialize the model. Moreover, fixed model parameters were assigned considering standard therapy settings, and patient-specific parameters from average literature values (the latters are reported in the last column of Table 2).

In particular, the patient was 151 cm tall and started the treatment with weight equal to 63.6 kg, and sodium, potassium and urea concentrations equal to 138, 5.2 and 32.5 mmol, respectively. The therapy was hemodiafiltration with pre-dilution (blood and dialysis flow rates equal to 300 and 500 ml/min respectively), conducted using a FX100 Cordiax dialyzer (Fresenius Medical Care, Bad Homburg vor der Höhe, Germany) and a dialysis fluid with sodium and bicarbonate concentrations equal to 137 and 35 mmol, respectively. The ultrafiltration volume prescribed by the physician was equal to 3.5 L.

Clinical data used to initialize the integration were acquired during the DialysIS Project at the Dialysis and Nephrology Unit of the Regional Hospital of Lugano (Lugano, Switzerland).

Integration was performed with the *ode15s* solver in Matlab (The MathWorks Inc., Natick, MA, USA) and a value every minute was stored for each state variable.

4.2 Results

Nice traceplots and autocorrelations are obtained, showing a satisfactory convergence of the chain. Moreover, values $\hat{R}$ of the Gelman-Rubin convergence statistics are always equal or higher than 1 for each estimated parameter in Θ^*, confirming the convergence of the chain.

The computational time has been equal to 15 h on a Server with processor X86-64 AMD Opteron 6328 and 64 GB of dedicated RAM. This time is largely lower than the common interval between two consecutive HD sessions, which is a fundamental requirement to apply the method in the clinical practice, where patient-specific parameters must be estimated with the data of a session and then used to optimize the kinetic response of the following session.

Results are reported in Table 2, in terms of the posterior densities of the estimated model parameters, compared with the *real* parameter values (those used for generating the test dataset). Results show that in most of the cases the posterior 25–75% credible intervals contain the *real* value and that the order of magnitude is almost always correctly estimated.

However, some differences are present. Thus, to validate the approach, we compare in Fig. 2 the simulated (generated with the *real* parameter values) and the estimated trends for three of the state variables: a volume, an extracellular mass and an intracellular mass. As the estimated trends are stochastic, due to the posterior density of the estimated parameters, they are reported in terms of their median value and 25–75% confidence band. Results show a good reproducibility of the test dataset, with low differences between simulated and estimated trends.

Table 2 Estimated posterior densities of model parameters versus their *real* values

Parameter	Min	25%	50%	75%	Max	*Real*
$\eta^{(1)}$	0.50	0.50	0.50	0.50	0.50	0.50
$\eta^{(2)}$	0.49	0.49	0.49	0.49	0.49	0.50
$\eta^{(3)}$	0.50	0.50	0.50	0.50	0.50	0.50
$\eta^{(4)}$	0.50	0.50	0.50	0.50	0.50	0.50
$\eta^{(5)}$	0.40	0.41	0.41	0.42	0.42	0.50
$\eta^{(6)}$	0.49	0.50	0.50	0.51	0.51	0.50
$\eta^{(7)}$	0.44	0.45	0.45	0.45	0.46	0.50
$\eta^{(8)}$	0.46	0.47	0.48	0.48	0.50	0.50
ρ	0.86	0.87	0.87	0.87	0.87	1.00
$k^{(1)}$	$1.69 \cdot 10^{-3}$	$2.04 \cdot 10^{-3}$	$2.14 \cdot 10^{-3}$	$2.23 \cdot 10^{-3}$	$2.55 \cdot 10^{-3}$	$2.50 \cdot 10^{-3}$
$k^{(2)}$	$1.67 \cdot 10^{-4}$	$1.73 \cdot 10^{-4}$	$1.75 \cdot 10^{-4}$	$1.77 \cdot 10^{-4}$	$1.84 \cdot 10^{-4}$	$1.67 \cdot 10^{-4}$
$k^{(3)}$	$8.13 \cdot 10^{-5}$	$9.82 \cdot 10^{-5}$	$1.04 \cdot 10^{-4}$	$1.10 \cdot 10^{-4}$	$1.31 \cdot 10^{-4}$	$1.67 \cdot 10^{-4}$
$k^{(4)}$	$8.70 \cdot 10^{-19}$	$9.96 \cdot 10^{-19}$	$1.03 \cdot 10^{-18}$	$1.07 \cdot 10^{-18}$	$1.21 \cdot 10^{-18}$	$1.67 \cdot 10^{-4}$
$k^{(5)}$	$8.89 \cdot 10^{-3}$	$9.56 \cdot 10^{-3}$	$9.80 \cdot 10^{-3}$	$1.00 \cdot 10^{-2}$	$1.11 \cdot 10^{-2}$	$3.30 \cdot 10^{-3}$
$k^{(6)}$	$1.65 \cdot 10^{-4}$	$1.70 \cdot 10^{-4}$	$1.72 \cdot 10^{-4}$	$1.73 \cdot 10^{-4}$	$1.78 \cdot 10^{-4}$	$1.67 \cdot 10^{-4}$
$k^{(7)}$	$1.32 \cdot 10^{-3}$	$1.34 \cdot 10^{-3}$	$1.34 \cdot 10^{-3}$	$1.35 \cdot 10^{-3}$	$1.36 \cdot 10^{-3}$	$1.30 \cdot 10^{-3}$
$k^{(8)}$	$1.28 \cdot 10^{-4}$	$1.30 \cdot 10^{-4}$	$1.30 \cdot 10^{-4}$	$1.31 \cdot 10^{-4}$	$1.33 \cdot 10^{-4}$	$1.30 \cdot 10^{-4}$

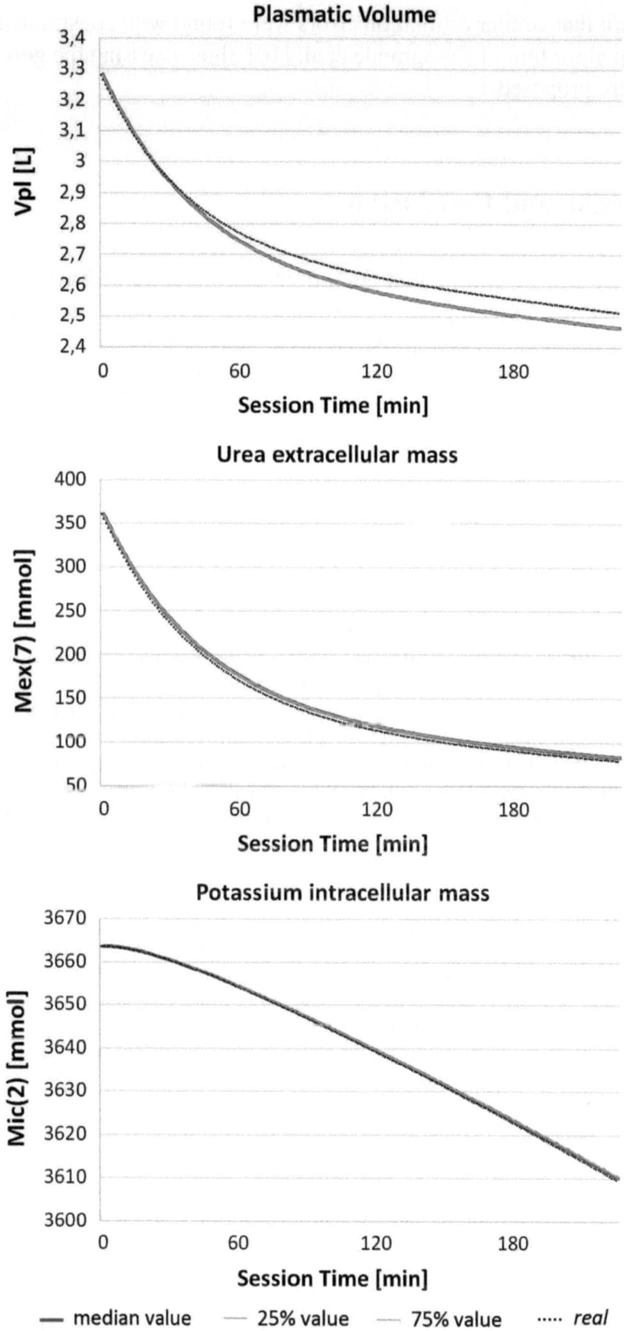

Fig. 2 Comparison between simulated and estimated trends (plasmatic volume, extracellular mass of urea, intracellular mass of potassium)

We remark that similar estimation errors were found with constrained non-linear optimization algorithm of Casagrande et al. [10], thus assessing the goodness of the approach here proposed.

5 Discussion and Conclusion

In this chapter we present the first attempt in the literature of Bayesian estimation applied to HD, and in particular we estimate the patient-specific parameters of a multi-compartment model to improve HD treatment customization. Results from the validation show good performance of the approach, i.e., the capability to detect the *real* values of the parameters and to draw good trajectories. Future work will be conducted to apply the approach to real patient data, and then to use the outcomes of the estimation to optimize HD treatments.

References

1. Palmer, H., Biff, F., William, L.: Recent advances in the prevention and management of intradialytic hypotension. J. Am. Soc. Nephrol. **19**(1), 8–11 (2008)
2. Prakash, S., Reddan, D., Heidenheim, P., et al.: Central, pheripheral, and other blood volume changes during hemodialysis. ASAIO J. **48**(4), 379–382 (2002)
3. Tapolyai, M.B., Faludi, M., Fülöp, T., et al.: Which fluid space is affected by ultrafiltration during hemodiafiltration? Hemodial. Int. **18**(2), 384–390 (2014)
4. Daugirdas, J.T.: Dialysis hypotension: a hemodynamic analysis. Kidney Int. **39**(2), 233–246 (1991)
5. Santos, S.F., Peixoto, A.J.: Sodium balance in maintenance hemodialysis. Semin. Dial. **23**(6), 549–555 (2010)
6. Clark, W.R., Leypoldt, J.K., Henderson, L.W., et al.: Quantifying the effect of changes in the hemodialysis prescription on effective solute removal with a mathematical model. J. Am. Soc. Nephrol. **10**(3), 601–609 (1999)
7. Leypoldt, J.K., Kablitz, C., Gregory, M.C., et al.: Prescribing hemodialysis using a weekly urea mass balance model. Blood Purif. **9**(5), 271–284 (1991)
8. Gotch, F.A.: Kinetic modeling in hemodialysis. In: Nissenson, A.R., Fine, R.N., Gentile, D.E. (eds.) Clinical Dialysis, pp. 156–188. Appelton and Lange, Norwalk (1995)
9. Di Filippo, S., Corti, M., Andrulli, S., et al.: Optimization of sodium removal in paired filtration dialysis by single pool sodium and conductivity kinetic models. Blood Purif. **15**(1), 34–44 (1997)
10. Casagrande, G., Bianchi, C., Vito, D., et al.: Patient-specific modeling of multicompartmental fluid and mass exchange during dialysis. Int. J. Artif. Organs **39**(5), 220–227 (2016)
11. Lanzarone, E., Ruggeri., F.: Inertance estimation in a lumped-parameter hydraulic simulator of human circulation. J. Biomech. Eng. - T. ASME **135**(6), 061012 (2013)
12. Lanzarone, E., Pasquali, S., Mussi, V., Ruggeri., F.: Bayesian estimation of thermal conductivity and temperature profile in a homogeneous mass. Numer. Heat Transf. Part B - Fundam. **66**(5), 397–421 (2014)
13. Auricchio, F., Conti, M., Ferrara, A., Lanzarone, E.: A clinically-applicable stochastic approach for non-invasive estimation of aortic stiffness using computed tomography data. IEEE Trans. Bio-Med. Eng. **62**(1), 176–187 (2015)

14. Lanzarone, E., Pasquali, S., Gilioli, G., Marchesini, E.: A Bayesian estimation approach for the mortality in a stage-structured demographic model. J. Math. Biol. (2017). doi:10.1007/s00285-017-1099-4
15. Andrulli, S., Colzani, S., Mascia, F., et al.: The role of blood volume reduction in the genesis of intradialytic hypotension. Am. J. Kidney Dis. **40**(6), 1244–1254 (2002)
16. Ishibe, S., Peixoto, A.: Methods of assessment of volume status and intercompartmental fluid shifts in hemodialysis patients: implications in clinical practice. Semin. Dial. **17**(1), 37–43 (2004)
17. Dasselaar, J., Lub-de Hooge, M.N., Pruim, J., et al.: Relative blood volume changes underestimate total blood volume changes during hemodialysis. Clin. J. Am. Soc. Nephrol. **2**, 669–674 (2007)
18. Ursino, M., Colí, L., Brighenti, C., et al.: Prediction of solute kinetics, acid-base status, and blood volume changes during profiled hemodialysis. Ann. Biomed. Eng. **28**(2), 204–216 (2000)
19. Stan Development Team: Stan modeling language users guide and reference manual, version 2.9.0 (2015)

A Bayesian Nonparametric Approach to Ecological Risk Assessment

Guillaume Kon Kam King, Julyan Arbel and Igor Prünster

Abstract We revisit a classical method for ecological risk assessment, the Species Sensitivity Distribution (SSD) approach, in a Bayesian nonparametric framework. SSD is a mandatory diagnostic required by environmental regulatory bodies from the European Union, the United States, Australia, China etc. Yet, it is subject to much scientific criticism, notably concerning a historically debated parametric assumption for modelling species variability. Tackling the problem using nonparametric mixture models, it is possible to shed this parametric assumption and build a statistically sounder basis for SSD. We use Normalized Random Measures with Independent Increments (NRMI) as the mixing measure because they offer a greater flexibility than the Dirichlet process. Indeed, NRMI can induce a prior on the number of components in the mixture model that is less informative than the Dirichlet process. This feature is consistent with the fact that SSD practitioners do not usually have a strong prior belief on the number of components. In this short paper, we illustrate the advantage of the nonparametric SSD over the classical normal SSD and a kernel density estimate SSD on several real datasets. We summarise the results of the complete study in [18], where the method is generalised to censored data and a systematic comparison on simulated data is also presented, along with a study of the clustering induced by the mixture model to examine patterns in species sensitivity.

Keywords Bayesian Nonparametrics · Ecotoxicology · HC_5 · Mixture models · Normalized random measures · Species Sensitivity Distribution

G. Kon Kam King (✉)
University of Torino, Italy and Collegio Carlo Alberto, Via Real Collegio, 30, 10024 Moncalieri, TO, Italy
e-mail: guillaume.konkamking.work@gmail.com

J. Arbel
Inria Grenoble - Rhône-Alpes & Université Grenoble Alpes, 655 Avenue de l'Europe, 38330 Montbonnot-Saint-Martin, France
e-mail: julyan.arbel@inria.fr

I. Prünster
Department of Decision Sciences, BIDSA and IGIER, Bocconi University, Via Röntgen 1, 20136 Milan, Italy
e-mail: igor@unibocconi.it

R. Argiento et al. (eds.), *Bayesian Statistics in Action*, Springer Proceedings in Mathematics & Statistics 194, DOI 10.1007/978-3-319-54084-9_14

1 Introduction

Assessing the response of a community of species to an environmental stress is critical for ecological risk assessment. Methods for this purpose vary in levels of complexity and realism. Species Sensitivity Distribution (SSD) represents an intermediate tier, more refined than rudimentary assessment factors [22] but practical enough for routine use by environmental managers and regulators in most developed countries (Australia, Canada, China, EU, South Africa, USA …). The SSD approach is intended to provide, for a given contaminant, a description of the tolerance of all species possibly exposed using information collected on a sample of those species. This information consists of Critical Effect Concentrations (CECs), a concentration specific to a species which marks a limit over which the species suffers a critical level of effect. This is for instance the concentration at which 50% of the tested organisms died (Lethal Concentration 50% (LC_{50})), or the concentration which inhibited growth or reproduction by 50% compared to the control experiment (Effect Concentration 50% (EC_{50})). Each CEC is the summary of long and costly bioassay experiments for a single species, so they are rarely available in large number. Typical sample sizes range from 10 to 15 [7].

To describe the tolerance of all species to be protected, the distribution of the CECs is then estimated from the sample. In practice, a parametric distributional assumption is often adopted [8]: the CECs are assumed to follow a log-normal [1], log-logistic [19], triangular [28, 34] or BurrIII [25] distribution.

Once the response of the community is characterised by the distribution, the goal of risk assessment is to define a safe concentration protecting all or most of the species. In the case of distributions without a lower threshold strictly above 0, a cut-off value is often chosen as the safe concentration. Typically, this is the Hazardous Concentration for 5% of the Species (HC_5), which is the 5th percentile of the distribution. Reasonings behind this choice include: that the lowest bound of the confidence interval around the 5th percentile will be used instead of the estimate, that a safety factor will be subsequently applied to that value and that ecosystems have a certain resilience to perturbations.

The lack of justification for the choice of any given parametric distribution has sparked several research directions. Some authors [10, 12, 28, 31, 32, 34] have sought to find the best parametric distribution by model comparison using goodness-of-fit measures. The general understanding is that no single distribution seems to provide a superior fit and that the answer is dataset dependent [8]. Therefore, the log-normal distribution has become the customary choice, notably because it readily provides confidence intervals on the HC_5, and because model comparison and goodness of fit tests have relatively low power on small datasets, precluding the emergence of a definite answer to the question. Another research direction consisted in seeking to avoid any reference to a distribution, using so-called nonparametric or distribution-free approaches. Those efforts included using the empirical distribution function [14, 26], methods based on ranks [4, 27], bootstrap resampling [12, 29] or nonparametric kernel density estimation [30]. All these approaches have in common that they require large sample sizes to be effectively applicable. Finally, authors have considered the possibility that the distribution of the CECs might not be a single distribution but rather a mixture of distributions [33], datasets being an assemblage of several log-normally distributed subgroups [5, 16]. This is more realistic from an ecological point of view because several factors influence the tolerance of a species to a contaminant such as the taxonomic group or the mode of action, and contaminant such as pesticides might even target specific species groups. Therefore, there is strong evidence in favour of the presence of groups of CECs, although the CECs within a group might remain log-normally distributed.

Ignorance of the group structure is a strong motivation for a nonparametric approach. However, the method must remain applicable to small datasets, which suggests trying to improve on the existing frequentist nonparametric methods. Bayesian nonparametric mixture models offer an interesting solution for both large and small datasets, because the complexity of the mixture model adapts to the size of the dataset. It offers a good compromise between a simplistic one-component parametric model and a kernel density method which in a certain sense lacks flexibility and might cause overfitting. Moreover, the low amount of information available in small datasets to estimate the groups parameters can be complemented via the prior, as some a priori degree of information is generally available from other species or contaminants [2, 5, 6]. This paper summarises the results of the complete study in [18].

The rest of the article is organised as follows. In Sect. 2 we present the Bayesian nonparametric (BNP) model and existing frequentist models for SSD and explain how to obtain a density estimate. Then in Sect. 3 we compare the different methods on a real dataset, illustrating the benefits of the BNP SSD. We conclude with a final discussion in Sect. 4.

2 Models for SSD

Given that concentrations vary on a wide range, it is common practice to work on log-transformed concentrations. Consider a sample of n log-concentrations denoted by $\boldsymbol{X} = (X_1, \ldots, X_n)$. We propose to carry out density estimation for the SSD based on sample $\boldsymbol{X}$ by use of nonparametric mixtures. Bayesian nonparametric mixtures were introduced in [21] with Dirichlet process mixtures (DPM). Generalizations of the DPM correspond to allowing the mixing distribution to be any discrete nonparametric prior. A large class of such prior distributions is obtained by normalizing increasing additive processes [24]. The normalization step, under suitable conditions, gives rise to so-called normalized measures with independent increments (NRMI) as defined by Regazzini et al. [23], see also [3] for a recent review. An NRMI mixture model is defined hierarchically as:

$$
\begin{aligned}
X_i|\mu_i,\sigma &\overset{\text{ind}}{\sim} k(\cdot|\mu_i,\sigma), \quad \mu_i|\tilde{P} \overset{\text{i.i.d.}}{\sim} \tilde{P}, \quad i=1,\ldots,n, \\
\tilde{P} &\sim \text{NRMI}, \quad \sigma \sim \text{Ga}(a_\sigma, b_\sigma).
\end{aligned} \tag{1}
$$

where k is a kernel, which we assume parametrized by some $\theta = (\mu, \sigma) \in \mathbb{R} \times \mathbb{R}_+$, and $\tilde{P}$ is a random probability on $\mathbb{R}$ whose distribution is an NRMI. In our model, all clusters have a common variance. This is easier to fit on a small dataset, because information about the variance is pooled across clusters. Similar mixture SSD models described in [5] also assume common variance. As described in the Introduction, concentrations are commonly fitted with a log-normal distribution. Our aim is to move from this parametric model to the nonparametric one in (1). In order to allow comparisons to be made, we stick to the normal specification for k on the log-concentrations $\boldsymbol{X}$ by letting: $k(x|\mu,\sigma) = \mathcal{N}(x|\mu,\sigma)$. Under this framework, density estimation is carried out by evaluating the posterior predictive density along the lines of [3]:

$$
\hat{f}(x|\tilde{P}, \boldsymbol{X}) = \iint k(x|\mu,\sigma)\mathrm{d}\pi(\sigma)\mathrm{d}\tilde{P}(\mu) \tag{2}
$$

for any x in $\mathbb{R}$, where π denotes the posterior distribution of σ.

To specify the prior, we choose as mixing random measure the normalized stable process [17] with:

(i) a stability parameter $\gamma = 0.4$, which controls the flatness of the prior on the number of clusters. The parameter γ can take values in (0, 1). Taking the limit $\gamma \to 0$ reduces the model to a Dirichlet process, larger values of γ lead to less informative priors on the number of clusters. The parameter γ was chosen as a good compromise between model flexibility and numerical stability. The total mass parameter is, without loss of generality, set equal to 1.

(ii) a base measure (which corresponds to the mean of the random probability measure) $P_0(\cdot) = \mathcal{N}(\cdot\,|\,\varphi_1, \varphi_2)$ with mean φ_1 and standard deviation φ_2, hyperparameters fixed a priori to specify a certain knowledge in the degree of smoothness.

(iii) a common variance for all the clusters with a vaguely informative prior distribution $Ga(0.5, 0.5)$.

Recent years have witnessed the appearance of a wealth of softwares dedicated to implement Bayesian nonparametric models and sample from their posterior. To cite a few, the R package DPpackage [13], is a rather comprehensive bundle of functions for Bayesian nonparametric models, while Bayesian Regression [15] is a software for Bayesian nonparametric regression. For posterior sampling, we use the R package BNPdensity and the function MixNRMI1 which implements BNP density models under a general specification of normalized random measures based on the generalised gamma processes (see [3]). The package is available from the Comprehensive R Archive Network (CRAN).

To illustrate the interest of the Bayesian nonparametric SSD, we compare our proposed BNP model to two commonly used frequentist models: the normal distribution [1] and the nonparametric Kernel Density Estimate (KDE) recently proposed by Wang at al. [30]. For both frequentist approaches, the data is assumed to be iid. Density estimates take on respectively the following form ($\hat{\mu}$ and $\hat{\sigma}$ are MLE)

$$\hat{f}_{\mathcal{N}}(x) = \mathcal{N}(x \mid \hat{\mu}, \hat{\sigma}) \quad \text{and} \quad \hat{f}_{KDE}(x) = \frac{1}{n}\sum_{i=1}^{n} \mathcal{N}(x \mid X_i, 1.06\hat{\sigma}n^{-\frac{1}{5}}). \tag{3}$$

2.1 *Model Comparison and Cross-Validation*

For the purpose of comparing the predictive performance of the model, we resort to Leave-One-Out (LOO) cross-validation. We compute the LOOs for each of the methods as $\forall i$, $\text{LOO}_i = \hat{f}(X_i \mid \boldsymbol{X}_{-i})$ where $\hat{f}(x \mid \boldsymbol{X}_{-i})$ is the density for one of the three methods estimated from $\boldsymbol{X}$ with X_i left out. The LOOs for the BNP model correspond to the conditional predictive ordinates (CPOs) statistics which are commonly used in applications, see [9]. A CPO statistic is defined for each log-concentration X_i as follows:

$$\text{CPO}_i = \hat{f}(X_i|\boldsymbol{X}_{-i}) = \int k(X_i|\theta)\mathrm{d}\pi(\theta|\boldsymbol{X}_{-i}) \tag{4}$$

where $\boldsymbol{X}_{-i}$ denotes the whole sample $\boldsymbol{X}$ but X_i, $\mathrm{d}\pi(\theta|\boldsymbol{X}_{-i})$ is the posterior distribution associated to $\boldsymbol{X}_{-i}$ and $\hat{f}$ is the posterior predictive distribution of Eq. (2). As shown by [3], CPOs can be easily approximated by Monte Carlo as

$$\widehat{\text{CPO}_i} = \left(\frac{1}{T}\sum_{t=1}^{T}\frac{1}{k(X_i|\theta^{(t)})}\right)^{-1} \tag{5}$$

where $\{\theta^{(t)}, t = 1, 2, \ldots, T\}$ is an MCMC sample from the posterior distribution.

2.2 *Quantile Estimation and HC_5*

The quantity of interest for ecological risk assessment is the HC_5, which corresponds to the 5th percentile of the SSD distribution. We choose as an estimator the median of the posterior distribution of the 5th percentile, while the 95% credible bands are formed by the 2.5 and 97.5% quantiles of the posterior distribution of the 5th percentile. The 5th percentile of the KDE is obtained by numerical inversion of the cumulative distribution function, and the confidence intervals using nonparametric bootstrap. The 5th percentile of the normal SSD and its confidence intervals are obtained following the classical method of [1].

3 Application to Real Data

We applied this model to a selection of contaminants extracted from a large database collected by the National Institute for Public Health and the Environment (RIVM). This database was prepared, studied and published by Hickey et al. [11]. We only considered non censored data. Left or right censored data were discarded, while interval censored data were replaced by the centre of the interval. Kon Kam King et al. [18] will describe how the method can be adapted to include censored data. Using a continuous distribution for the CECs implies that the model does not support ties (or, in other words, observing ties has zero probability). However, ties may appear in the dataset due to the rounding of concentrations. Hence, we used a small jittering of the data.

We selected three example datasets which feature three typical sample sizes: a relatively large `carbaryl` dataset (CAS: 63-25-2, insecticide, 55 species), a medium-sized `temephos` dataset (CAS: 3383-96-8, mosquito larvicide, 21 species), and a small `captan` dataset (CAS: 133-06-2, fungicide, 13 species). Datasets for new contaminants are always small, the minimum requirement set by the European Chemical Agency being 10 species. The datasets can be visualised on the histograms of Fig. 1 (left panel).

These datasets illustrate different features of the three approaches: when there is a clear multimodality in the data, the BNP SSD is more flexible than the fixed bandwidth KDE SSD (Fig. 1, `carbaryl` and `captan`). When the data do not exhibit strong multimodality, as for `temephos`, the BNP reduces to the normal SSD model, whereas the KDE remains by construction a mixture of many normal components.

One might think to increase the flexibility of the KDE by simply decreasing the bandwidth. However, that would also decrease the robustness of the method. On the middle panel of Fig. 1, the LOOs give an indication of the robustness to over-fitting of the three methods. For `carbaryl` and `captan`, they show that the superior flexibility of the BNP SSD compared to the KDE SSD does not come at the expense of robustness, because the median CPO of the BNP SSD is higher than the other two. In the case of `temephos`, the median LOO likelihood estimate of the normal model is very similar to the median CPO for the BNP SSD, sign that there is little over-fitting. This generally illustrates the fact that model complexity in a BNP model scales with the amount and structure of the data. On the right panel of Fig. 1, the credible intervals of the HC_5 for the BNP SSD are generally larger than the confidence interval of the normal SSD, which is coherent with the model uncertainty of the nonparametric approach.

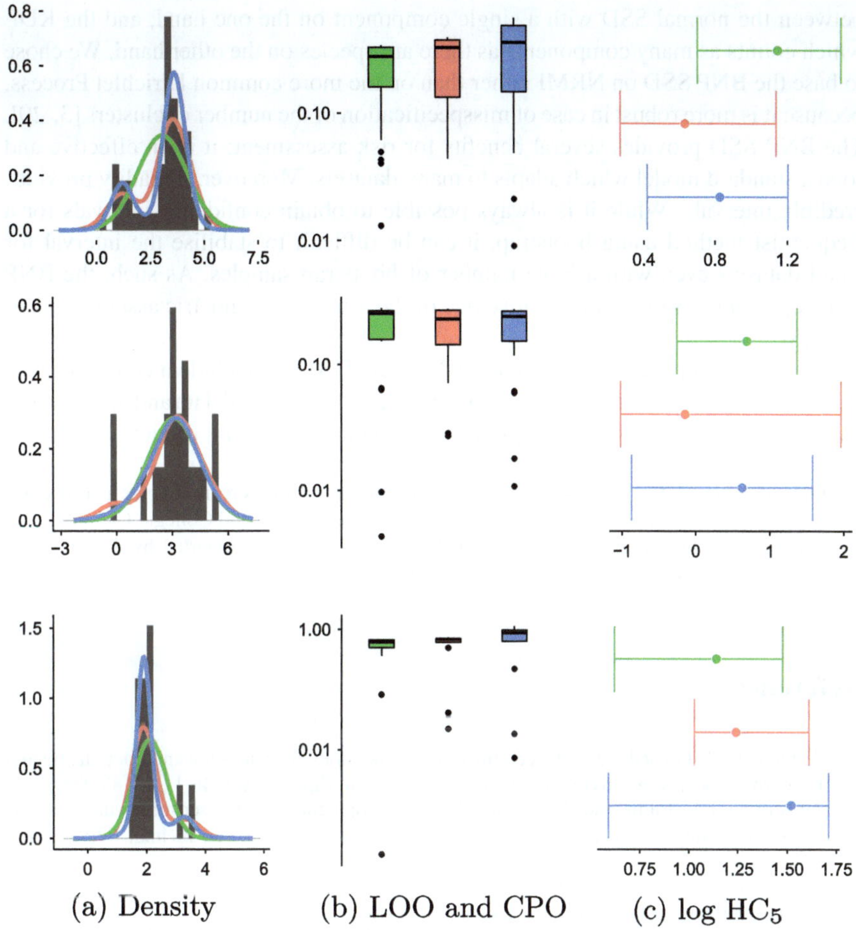

Fig. 1 The top panel represents the large-size carbaryl dataset, the middle panel represents the medium-sized temephos dataset, the bottom panel represents the small-sized captan dataset. The Normal is in green, The KDE in *red* and the BNP in blue. Concentrations are log transformed. *Left*: Histogram and density estimates. *Centre*: Boxplot for the LOOs (for Normal and KDE) and the CPO (for BNP) on logarithmic scale. The horizontal line corresponds to the median. The box hinges extend to the inner quartiles. The whiskers extend to cover points up to one and a half times the inter-quartile distance away from the hinges. For both frequentist methods, the n LOOs are obtained by fitting the model n times, while an analytical expression is available for the BNP method (Eq. 5). *Right*: log HC_5 and associated confidence/credible intervals (for Normal, KDE and BNP)

4 Discussion

The BNP SSD seems to perform well when the dataset deviates from a normal distribution. Its great flexibility is an asset to describe the variability of the data, while it does not seem prone to over-fitting. It can be thought of as an intermediate model between the normal SSD with a single component on the one hand, and the KDE which counts as many components as there are species on the other hand. We chose to base the BNP SSD on NRMI rather than on the more common Dirichlet Process, because it is more robust in case of misspecification of the number of clusters [3, 20]. The BNP SSD provides several benefits for risk assessment: it is an effective and robust standard model which adapts to many datasets. Moreover, it readily provides credible intervals. While it is always possible to obtain confidence intervals for a frequentist method using bootstrap, it can be difficult to stabilise the interval for small datasets even with a large number of bootstrap samples. As such, the BNP SSD represents a safe tool to remove one of the arbitrary parametric assumptions of SSD [8].

The extended paper supporting the BNP SSD [18] will include a comparison of methods on simulated data, an extension to the case of censored data and an emphasis on the potential benefits of the approach from a biological point of view.

Acknowledgements The authors would like to thank two anonymous reviewers for helpful comments and suggestions. J. Arbel was a postdoc at Bocconi University and Collegio Carlo Alberto, Italy, when this article was submitted. J. Arbel and I. Prünster are supported by the European Research Council (ERC) through StG "N-BNP" 306406.

References

1. Aldenberg, T., Jaworska, J.S.: Uncertainty of the hazardous concentration and fraction affected for normal species sensitivity distributions. Ecotoxicol. Environ. Saf. **46**(1), 1–18 (2000)
2. Awkerman, J.A., Raimondo, S., Barron, M.G.: Development of species sensitivity distributions for wildlife using interspecies toxicity correlation models. Environ. Sci. Technol. **42**(9), 3447–3452 (2008)
3. Barrios, E., Lijoi, A., Nieto-Barajas, L.E., Prünster, I.: Modeling with normalized random measure mixture models. Stat. Sci. **28**(3), 313–334 (2013)
4. Chen, L.: A conservative, nonparametric estimator for the 5th percentile of the species sensitivity distributions. J. Stat. Plann. Inference **123**(2), 243–258 (2004)
5. Craig, P.S.: Exploring novel ways of using species sensitivity distributions to establish PNECs for industrial chemicals: final report to Project Steering Group. Technical report (2013)
6. Craig, P.S., Hickey, G.L., Luttik, R., Hart, A.: Species non-exchangeability in probabilistic ecotoxicological risk assessment. J. R. Stat. Soc. Ser. A Stat. Soc. **175**(1), 243–262 (2012)
7. ECHA.: Characterisation of dose concentration-response for environment. In: Guidance on Information Requirements and Chemical Safety Assessment, Number May, chapter R.10. European Chemicals Agency, Helsinki (2008)
8. Forbes, V.E., Calow, P.: Species sensitivity distributions revisited: a critical appraisal. Hum. Ecol. Risk Assess. **8**(3), 473–492 (2002)
9. Gelfand, A.E.: Model determination using sampling-based methods. In: Markov Chain Monte Carlo in Practice, pp. 145–161 (1996)
10. He, W., Qin, N., Kong, X., Liu, W., Wu, W., He, Q., Yang, C., Jiang, Y., Wang, Q., Yang, B., Xu, F.: Ecological risk assessment and priority setting for typical toxic pollutants in the water from Beijing-Tianjin-Bohai area using Bayesian matbugs calculator (BMC). Ecol. Indic. **45**, 209–218 (2014)
11. Hickey, G.L., Craig, P.S., Luttik, R., de Zwart, D.: On the quantification of intertest variability in ecotoxicity data with application to species sensitivity distributions. Environ. Toxicol. Chem. **31**(8), 1903–1910 (2012)

12. Jagoe, R.H., Newman, M.C.: Bootstrap estimation of community NOEC values. Ecotoxicology **6**(5), 293–306 (1997)
13. Jara, A., Hanson, T.E., Quintana, F.A., Müller, P., Rosner, G.L.: DPpackage: Bayesian semi- and nonparametric modeling in R. J. Stat. Softw. **40**(5), 1 (2011)
14. Jones, D.S., Barnthouse, L.W., Suter II, G.W., Efroymson, R.A., Field, J.M., Beauchamp, J.J.: Ecological risk assessment in a large river-reservoir: 3. Benth. Invertebr. Environ. Toxicol. Chem. **18**(4), 599–609 (1999)
15. Karabatsos, G.: A menu-driven software package of Bayesian nonparametric (and parametric) mixed models for regression analysis and density estimation. In: Behavior Research Methods, pp. 1–28 (2016)
16. Kefford, B.J., Hickey, G.L., Gasith, A., Ben-David, E., Dunlop, J.E., Palmer, C.G., Allan, K., Choy, S.C., Piscart, C.: Global scale variation in the salinity sensitivity of riverine macroinvertebrates: Eastern Australia, France. Isr. South Afr. PLoS ONE **7**(5), e35224 (2012)
17. Kingman, J.: Random discrete distributions. J. R. Stat. Soc. Ser. B **37**(1), 1–22 (1975)
18. Kon Kam King, G., Arbel, J., Prünster, I.: Species Sensitivity Distribution Revisited: A Bayesian Nonparametric Approach (In preparation) (2016)
19. Kooijman, S.: A safety factor for LC 50 values allowing for differences in sensitivity among species. Water Res. **21**(3), 269–276 (1987)
20. Lijoi, A., Mena, R.H., Prünster, I.: Controlling the reinforcement in Bayesian non-parametric mixture models. J. R. Stat. Soc. Ser. B Stat. Methodol. **69**(4), 715–740 (2007)
21. Lo, A.Y.: On a class of Bayesian nonparametric estimates: I. Density Estim. Ann. Stat. **12**(1), 351–357 (1984)
22. Posthuma, L., Suter II, G.W., Trass, P.T.: Species Sensitivity Distributions in Ecotoxicology. CRC Press, Boca Raton (2002)
23. Regazzini, E., Lijoi, A., Prünster, I.: Distributional results for means of normalized random measures with independent increments. Ann. Stat. **31**(2), 560–585 (2003)
24. Sato, K.: Lévy Processes and Infinitely Divisible Distributions. Cambridge Studies in Advanced Mathematics, vol. 68. Cambridge University Press, Cambridge (1999)
25. Shao, Q.: Estimation for hazardous concentrations based on NOEC toxicity data: an alternative approach. Environmetrics **11**(5), 583–595 (2000)
26. Suter II, G.W., Barnthouse, L.W., Efroymson, R.A., Jager, H.: Ecological risk assessment in a large river-reservoir: 2. Fish Community Environ. Toxicol. Chem. **18**(4), 589–598 (1999)
27. Van Der Hoeven, N.: Estimating the 5-percentile of the species sensitivity distributions without any assumptions about the distribution. Ecotoxicology **10**(1), 25–34 (2001)
28. Van Straalen, N.M.: Threshold models for species sensitivity distributions applied to aquatic risk assessment for zinc. Environ. Toxicol. Pharmacol. **11**(3–4), 167–172 (2002)
29. Wang, B., Yu, G., Huang, J., Hu, H.: Development of species sensitivity distributions and estimation of HC(5) of organochlorine pesticides with five statistical approaches. Ecotoxicology **17**(8), 716–724 (2008)
30. Wang, Y., Wu, F., Giesy, J.P., Feng, C., Liu, Y., Qin, N., Zhao, Y.: Non-parametric kernel density estimation of species sensitivity distributions in developing water quality criteria of metals. Environ. Sci. Pollut. Res. **22**(18), 13980–13989 (2015)
31. Xing, L., Liu, H., Zhang, X., Hecker, M., Giesy, J.P., Yu, H.: A comparison of statistical methods for deriving freshwater quality criteria for the protection of aquatic organisms. Environ. Sci. Pollut. Res. **21**(1), 159–167 (2014)
32. Xu, F.-L., Li, Y.-L., Wang, Y., He, W., Kong, X.-Z., Qin, N., Liu, W.-X., Wu, W.-J., Jorgensen, S.E.: Key issues for the development and application of the species sensitivity distribution (SSD) model for ecological risk assessment. Ecol. Indic. **54**, 227–237 (2015)
33. Zajdlik, B.A., Dixon, D.G., Stephenson, G.: Estimating water quality guidelines for environmental contaminants using multimodal species sensitivity distributions: a case study with Atrazine. Hum. Ecol. Risk Assess. **15**(3), 554–564 (2009)
34. Zhao, J., Chen, B.: Species sensitivity distribution for chlorpyrifos to aquatic organisms: model choice and sample size. Ecotoxicol. Environ. Saf. **125**, 161–169 (2016)

Approximate Bayesian Computation Methods in the Identification of Atmospheric Contamination Sources for DAPPLE Experiment

Piotr Kopka, Anna Wawrzynczak and Mieczyslaw Borysiewicz

Abstract Sudden releases of harmful material into a densely-populated area pose a significant risk to human health. The apparent problem of determining the source of an emission in urban and industrialized areas from the limited information provided by a set of released substance concentration measurements is an ill-posed inverse problem. When the only information available is a set of measurements of the released substance concentration in urban and industrial areas, it is difficult to determine the source of emission. The Bayesian probability framework provides a connection between model, observational and additional information about the contamination source. The posterior distribution of the source parameters was sampled using an Approximate Bayesian Computation (ABC) algorithm. The stochastic source determination method was validated against the real data set acquired in a highly disturbed flow field in an urban environment. The datasets used to validate the proposed methodology include the dispersion of contaminant plumes in a full-scale field experiment performed within the project 'Dispersion of Air Pollutants and their Penetration into the Local Environment in London (DAPPLE)'.

Keywords Atmospheric contamination sources · Approximate Bayesian computation

1 Stochastic Event Reconstruction Procedure

In emergency response management it is important to know the extent of the area that might become contaminated following the release of dangerous material in cities and the subsequent movement of polluted air. The lack of pertinent experimental information means there is a gap in the understanding of short-range dispersion behavior in highly urbanized areas. Given a gas source and wind field, we can apply an appropriate atmospheric dispersion model to calculate the expected gas concentration for any location. Conversely, given concentration measurements and knowledge of the arrangement of buildings, wind field, and other atmospheric air parameters, identify-

P. Kopka (✉) · A. Wawrzynczak · M. Borysiewicz
National Centre for Nuclear Research, Otwock, Poland
e-mail: piotr.kopka@ncbj.gov.pl

R. Argiento et al. (eds.), *Bayesian Statistics in Action*, Springer Proceedings in Mathematics & Statistics 194, DOI 10.1007/978-3-319-54084-9_15

ing the actual location of the release source and its parameters is difficult. This problem has no unique solution and can be analyzed using probabilistic frameworks. In the framework of Bayesian approach, all quantities included in the mathematical model are modeled as random variables with joint probability distributions. This randomness can be interpreted as a lack of knowledge of parameter values, and is reflected in the uncertainty of the true values. Bayesian methods formulate this problem into searching for a posterior distribution based on efficient sampling of an ensemble of simulations using priori knowledge and observed data. Previously [1, 2], we have tested the methodology by combining Bayesian inference with Markov chain Monte Carlo (MCMC) and Sequential Monte Carlo (SMC) methods and applied these to the problem of dynamic, data-driven contaminant source localization, based on data from the synthetic experiment.

We propose algorithms to locate the source of contamination based on the data from the central London DAPPLE experiment that was performed in May and June 2007 (see Sect. 2) [3]. We used the fast running QUIC-URB model for computing mean flow fields around buildings and QUIC-PLUME [4] as the forward model to predict the concentrations at the sensor locations (Sect. 3). As a sampling approach in the event reconstruction procedure we used a modern algorithm from the class of likelihood-free Bayesian methods [5] with some extension, described in Sect. 4.

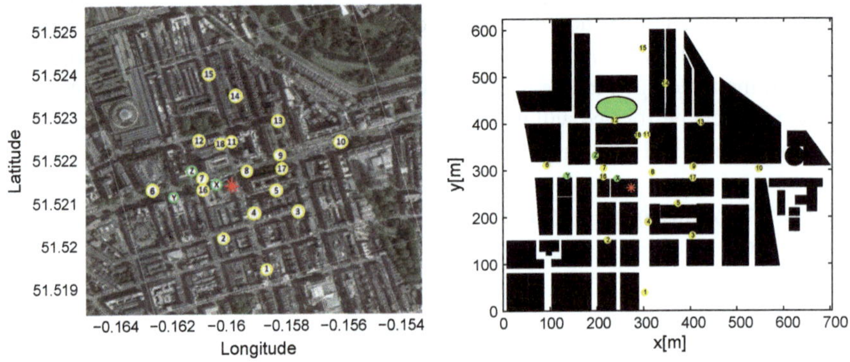

Fig. 1 *Left panel* - The map shows the DAPPLE area of central London and is centered at the focal intersection, that of Marylebone Road and Gloucester Place (at 51.5218N 0.1597W). The sampling receptors are numbered 1–18 (*yellow dots*). Three fixed-point tracer sources (*green dots* X, Y and Z); red star - Westminster City Council (WCC). The white rectangle shows the computational domain. *Right panel* - The rotated DAPPLE area, with the selected buildings (*black rectangles*) and greenery (*green* ellipses)

2 DAPPLE Experiment

The DAPPLE experiment took place in central London (see Fig. 1). The two major roads in the vicinity are Marylebone Road, which runs from west to east, and Gloucester Place, which intersects perpendicularly with Marylebone Road near the Westminster City Council building (the red star in Fig. 1) [3]. A release was carried out on the fourth day of the experiment, 28th June 2007, in which a sequence of ten samples was taken over a 30 min sampling period at each of the 18 receptor positions. The sampling process included the collection of ten 150 s samples at each of the 18 sites, each sample separated from the next by 30 s. The source locations (green X point) and monitoring sites (numbered yellow points) are shown on the map included in Fig. 1. The total mass emitted from the point-source release was 323 mg of *perfluoromethyl-cyclohexane* ($PMCH$, $C7F14$), in accordance with experimental requirements. Two sets of long-term reference measurements were taken to generate the wind data sets: the rooftop Westminster City Council (WCC) (18 m) and tower top (190 m) winds.

In Fig. 1 the rectangle area was selected as a computing domain (white line). The positions of all the objects (sensors, source, buildings, wind direction, etc.) have been rotated by a 17° angle, to bring the main streets parallel to the edges of the domain. The $latitude - longitude$ geographic coordinate system was changed to the metric system with a reference point (0, 0). This reference point denotes the lower left corner of the white rectangle, both for the convenience of creating a domain and the presentation of results.

3 Dispersion Model

The Quick Urban Industrial Complex (QUIC) Dispersion Modeling System is intended for applications where the dispersion of air pollutants released near buildings must be computed very quickly [4]. The QUIC system comprises a wind model - QUIC-URB, a dispersion model QUIC-PLUME, and a graphical user interface. The modeling strategy adopted in QUIC-URB was originally developed by Rockle [6] and uses a 3D mass-consistent wind model to combine properly resolved time-averaged wind fields around buildings [7]. The mass-consistent technique is based on a 3D complex terrain diagnostic wind model. The basic methodology involves first generating an initial wind field that includes various empirical parameterizations to account for the physics of flow around buildings. Next, this velocity field is forced to be divergence free, subject to the weak constraint that the variance of the difference between the initial velocity field and mass consistent final velocity field is minimized. The QUIC-PLUME is Lagrangian particle models which describe gas dispersion by simulating the releases of particles and moving them with an instantaneous wind composed of a mean and turbulent wind. The Eq. (1) describe the positions of particles in x, y, z domain:

$$\mathrm{x} = \mathrm{x}_p + U\Delta t + \frac{u'_p + u'}{2}\Delta t, \mathrm{y} = \mathrm{y}_p + V\Delta t + \frac{v'_p + v'}{2}\Delta t,$$
$$\mathrm{z} = \mathrm{z}_p + W\Delta t + \frac{w'_p + w'}{2}\Delta t \quad (1)$$

where x, y, and z are the current position of the particle in the domain and the subscript p invoke to the previous positions. The U, V, W are the mean winds, while u', v', w' are the fluctuating components of the instantaneous wind. The Δt factor is the time step in dispersion model. The fluctuating constituents of the instantaneous winds are calculated from:

$$u' = u'_p + du, \quad v' = v'_p + dv, \quad w' = w'_p + dw \quad (2)$$

The equations for du, dv, and dw are quite complicated and described in detail in [4]. The average concentrations C^{ijk}, normalized to unit release, are estimated by summing over all particles Q that are found within the sampling box i, j, k during the concentration averaging time t_{ave}:

$$C^{ijk} = \sum \frac{Q\Delta t_c}{N_{total} d\mathrm{x}_{box} d\mathrm{y}_{box} d\mathrm{z}_{box} t_{ave}} \quad (3)$$

where N_{total} is the total number of particles released during the gas simulation, $d_{\mathrm{x}_{box}}$ is the sampling box size in direction x, $d_{\mathrm{y}_{box}}$ is the sampling box size in y direction, $d_{\mathrm{z}_{box}}$ is the sampling box size in the direction z, Δt_c is particle time step. All others unique procedures related to turbulence associated with walls and rooftops, reflection by walls and non-local mixing are presented in [4].

4 Approximate Bayesian Computation Algorithm

In $ABC - SMC$ methods, the set of samples with weights, called particles, sampled from the population with the prior distribution $\pi(\theta)$, are propagated through a sequence of intermediate posterior distributions $\pi(\theta|\rho(x^t, x^t_{obs}) < \varepsilon_t), t = 1, \ldots, T$, until it represents a sample from the target distribution $\pi(\theta|\rho(x^T, x^T_{obs}) < \varepsilon_T)$. In [5] the authors propose strategies called $ABCSMC$ with Adaptive Weights ($ABC - SMC - AW$). Algorithm 1 shows the description of $ABC - SMC - AW$.

In $ABC - SMC - AW$ procedure after initialization of the threshold schedule, first N samples are simulated based on the predefined a priori distribution $\pi(\theta)$ and the corresponding acceptance condition $\rho(x^1, x^1_{obs}) < \varepsilon_1$. In time step $t = 2$ simple uniform weights are changed based on additional kernel $K_{x,t}(x^t_{obs}|x^{t-1}_i)$ proposed in [5]. Samples, denoted by a tilde are drawn from the previous generation with probabilities v^{t-1}_j. Using perturbation kernel $K_{\theta,t}(\theta^t_i|\tilde{\theta}_i)$ new "fresh" samples θ^t_i are obtained, with the veracity of the condition $\rho(x^t, x^t_{obs}) < \varepsilon_t$. The weights are calculated according to the formula in step (11); in step (12) the weights are normalized and the time

Algorithm 1 ABC-SMC-AW

1. Initialize threshold schedule $\varepsilon_1 > \varepsilon_2 > \cdots > \varepsilon_T$
2. Set $t = 1$

for $i = 1$ to N **do**
  3. Simulate $\theta_i^t \sim \pi(\theta)$ and $x^t \sim \pi(x^t|\theta_i^t)$
  4. Until $\rho(x^t, x_{obs}^t) < \varepsilon_t$
  5. Set $w_i^t = \frac{1}{N}$

end for

for $t = 2$ to T **do**
  6. Compute new weights $v_i^{t-1} \propto w_i^{t-1} K_{x,t}(x_{obs}^t|x_i^{t-1})$ for $i = 1, \ldots, N$
  7. Normalize weights v_i^{t-1} for $i = 1, \ldots, N$
  for $i = 1$ to N **do**
    8. Pick $\tilde{\theta}_i$ from the set $\{\theta_j^{t-1}\}_{1 \leq j \leq N}$ with probabilities $\{v_j^{t-1}\}_{1 \leq j \leq N}$
    9. Draw $\theta_i^t \sim K_{\theta,t}(\theta_i^t|\tilde{\theta}_i)$ and $x^t \sim \pi(x^t|\theta_i^t)$
    10. Until $\rho(x^t, x_{obs}^t) < \varepsilon_t$
    11. Compute new weights as
    $w_i^t \propto \frac{\pi(\theta_i^t)}{\sum_j v_j^{t-1} K_{\theta,t}(\theta_i^t|\theta_j^{(t-1)})}$
    12. Normalize weights w_i^t for $i = 1, \ldots, N$
  end for

end for

step is increased - $t = t + 1$. The procedure is repeated until $t \leq T$. In the following sections the details are discussed, along with the motivation for choosing specific components of the Algorithm 1 for the problem of stochastic event reconstruction.

4.1 Data and Model

To compute the $\rho(x^t, x_{obs}^t)$ value we use data from the sensor network which measures gas concentration $\hat{C}_i^{Sj}$ where i corresponds to the time step and Sj is the sensor identifier in the following way $x_{obs}^t \equiv |\hat{C}_1^{Sj}, \hat{C}_2^{Sj}, \ldots, \hat{C}_t^{Sj}|$ for $j = 1, 2, \ldots, 18$, because in this test case we have 18 sensors $(S1, S2, \ldots, S18)$, whose positions are given in Fig. 1 as yellow dots. We assume that the substance concentrations registered by the sensors arrive subsequently at time intervals, hereafter referred to as 'time steps'. It is important to know that for time step t only data $\hat{C}_1^{Sj} \hat{C}_2^{Sj} \ldots \hat{C}_t^{Sj}$ are available and finally we have ten time steps $(t = 10)$. The reconstruction algorithm starts to search a source location (x, y) and release rate (q) just after the first 6 min $(t = 2)$. To get the predicted concentration a QUIC-PLUME forward model is running and it refers to the procedure $x^t \sim \pi(x^t|\theta_i^t)$ in Algorithm 1. To run a dispersion model and obtain data x^t we use source parameter vector θ_i^t and the information obtained from the QUIC-URB subsystem. The simulated data also have a form of concentration value $x^t \equiv |C_1^{Sj}, C_2^{Sj}, \ldots, C_t^{Sj}|$ for $j = 1, 2, \ldots, 18$ where Sj corresponds to the known locations of j sensor.

4.2 Distance Measure

The choice of distance measure or summary statistics is a crucial step in ABC. Since distance measures are not sufficient in many cases, this choice involves a trade-off between loss of information and reduction of dimensionality. In those cases we chose to normalize approximation error between all the data obtained to the current time step t which is also called Fractional Bias (FB) [8]. The FB is used to indicate a bias towards underprediction or overprediction of concentration data by the model. Due to the data type for all sensors in time step t the $\rho(x^t, x^t_{obs})$ measure is as follows:

$$\rho(x^t, x^t_{obs}) = \frac{1}{18}\sum_{j=1}^{18}\left(\frac{1}{t}\sum_{i=1}^{t}\frac{|C_i^{Sj} - \hat{C}_i^{Sj}|}{C_i^{Sj} + \hat{C}_i^{Sj}}\right), \tag{4}$$

under additional definition, that $\frac{|C_i^{Sj} - \hat{C}_i^{Sj}|}{C_i^{Sj} + \hat{C}_i^{Sj}} = 0$ when $C_i^{Sj} = 0$ and $\hat{C}_i^{Sj} = 0$. Given that the concentration $C_i^{Sj} \geq 0$, the value of $\rho(x^t, x^t_{obs})$ is always between 0 and 1. Let us notice that $\rho(x^t, x^t_{obs}) = 0$ is the situation when our prediction is perfect. In the opposite case, when $\rho(x^t, x^t_{obs}) = 1$ the prediction is wrong. In finding source parameters one of the most important areas is the detection time window, when there is a measurement in the current sensor. The measure (1) supports this approach, because when we have non-zero concentration in some time steps but our model shows that there should be 0 concentration value, the penalty value for this step will be 1. The situation is the same, if the observed value is equal to 0 and the model shows a positive value of the concentration. On the other hand, if $C_i^{Sj} > 0$ and $\hat{C}_i^{Sj} > 0$ then the absolute difference also has an impact on the value of $\rho(x^t, x^t_{obs})$ measure. Additionally, each sensor has an equal contribution to the $\rho(x^t, x^t_{obs})$ measure, regardless of the level of concentration, which is of course smaller in sensors located further from the source.

4.3 Weights and Threshold Schedule

The most commonly used adaptive scheme for threshold choice is based on the quantile of the empirical distribution of the distances between the simulated data and observations from the previous population, (see [9, 10]). The method determines ε_t at the beginning of the t time-iteration by sorting the measure $\rho(x_i^{t-1}, x_{obs}^{t-1})_{1<i\leq N}$ and setting ε_t such that α_t percent of the simulated data $\rho(x_i^{t-1}, x_{obs}^{t-1})_{1<i\leq N}$ are below it, for some predetermined α_t. In [11] the authors show a new strategy based on an acceptance rate curve but also discuss a cumulative number of simulation versus different threshold schedules. In this, and many other cases, quantile-based methods seem to be an easy and appropriate solution of estimating ε_t. Based on our own preprocessing experience we set quantile $\alpha_2 = 0.7$ in the second time step, that subse-

quently decreases to $\alpha_{10} = 0.3$ for $t = 10$ [11]. The additional kernel $K_{x,t}(x^t_{obs}|x_i^{t-1})$ in Algorithm 1 state 6, which is used in calculating the weights, depends on observed and simulated data. Since weights are normalized in step (7), in Algorithm 1 we can simply use the $\rho(x^t, x^t_{obs})$ measure as the proposed kernel. Due to the restriction $0 \leq \rho(x^t, x^t_{obs}) \leq 1$ we can define $K_{x,t}(x^t_{obs}|x_i^{t-1}) \equiv 1 - \rho(x_i^{t-1}, x^t_{obs})$, because the greater weight should correspond to a better solution.

4.4 *Transition Kernel*

We chose transition kernel $K_{\theta,t}(\cdot|\cdot)$ to be a Gaussian kernel. Unfortunately in this type of inverse problems the parameters are often highly correlated and multimodality is very common. Especially when the (x, y) domain contains a lot of prohibited regions, like buildings. Samples may tend to split in a disjointed group by filling out different street canyons. In such cases it is interesting to consider the use of a local mean and covariance matrix. Instead of computing the covariance matrix based on all the samples from $(t-1)$, a better idea is to use only limited information about the local correlation. In [12] one of the proposed methods is to use the multivariate normal kernel based on the M neighbours. The authors in [12] pay attention to the disadvantages of choosing this perturbation kernel. First, the parameter M typically has to be fixed before any of the information about the posterior are known (too small a value of M may lead to a lack of exploration of parameter space, while too large would offer little or no advantage compared to the standard multivariate normal kernel). In our case the number of samples allocated to one time step is $N = 1000$ samples for each time step. Based on pre-processed experiments we determined the number of neighbors $M = 70$ and Mahalanobis distance.

5 Results and Conclusions

Figure 2 shows the locations of the buildings in the DAPPLE London area, together with all the samples generated in subsequent time steps $t = 2, 3, \ldots, 10$ which are decomposed directly to $6, 9, \ldots, 30$ experimental minutes. As we can see, samples after the $4th$ time step converge from all possible (x, y) space to the vicinity of the actual source location. Using these samples, we construct the marginal probability distributions for the source location and release rate, as shown in Fig. 3 for all time intervals. As time goes on, the mass of probability distribution is concentrating in the vicinity of the proper values of x and y. This looks quite different for emissions amounts, where posterior distribution for the parameter q looks like a bimodal distribution.

A color pattern reflected in Fig. 4 was used to show empirical 2D probability distribution of all parameters combinations. The colored contour lines are enveloping higher probability at the joint posterior distributions. The diagonal plots are

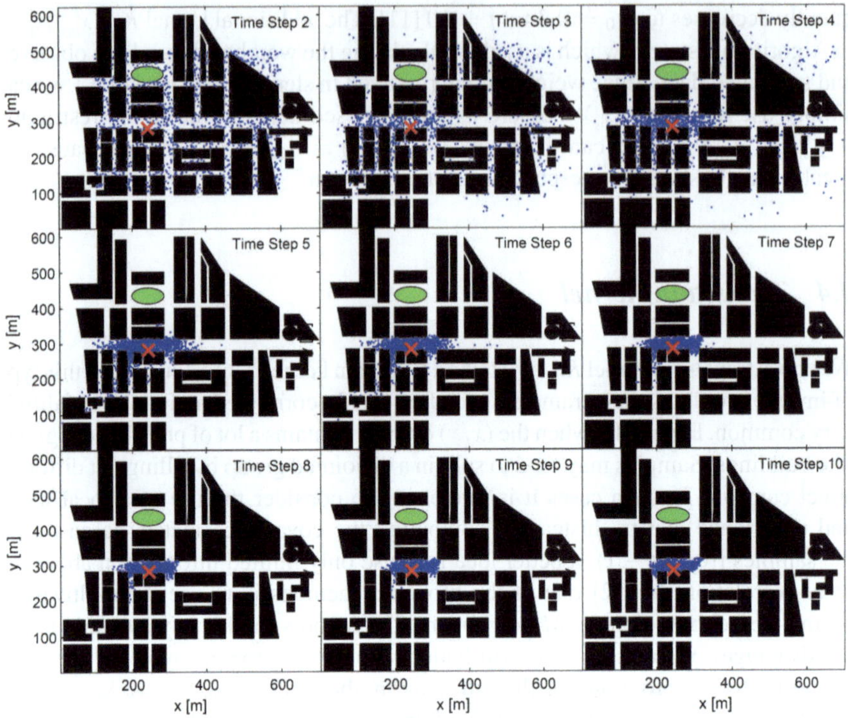

Fig. 2 A scatter plot of all samples generated in the subsequent time steps $t = 2, 3, \ldots, 10$ in (x, y) space of source location. The *red* cross marks the true source position

marginal empirical posterior distributions of the forward model parameters. The real parameter values from the field experiment are highlighted with red vertical lines in diagonal plots and black cross markers on the other subplot, which are successfully captured by the high posterior probability region. The target contamination source parameters obtained after the transformation of the relative domain are $x = 243$ m, $y = 282$ m and $q = 323$ mg, where the estimated most probable parameters values are $P(x = 223.0 \pm 7.6\,\text{m}) = 0.0632$, $P(y = 291.4 \pm 6.7\,\text{m}) = 0.1990$ and $P(q = 144.9 \pm 5.3\,\text{mg}) = 0.0218$.

Posterior probability distributions of model parameters were used to build prior distribution when new concentration data became available. Although the ABC framework is, generally a comprehensive operational event reconstruction tool it needs to address various release scenarios. The present study focused on steady point source releases in a highly urbanized area. However, possible release scenarios may include moving sources. Furthermore, the scale of the event may range from local sites to areas of greater size. Future work will concentrate on adding new possible hazardous scenarios to the presented stochastic event reconstruction tool, not necessarily the release of gasses into the atmosphere.

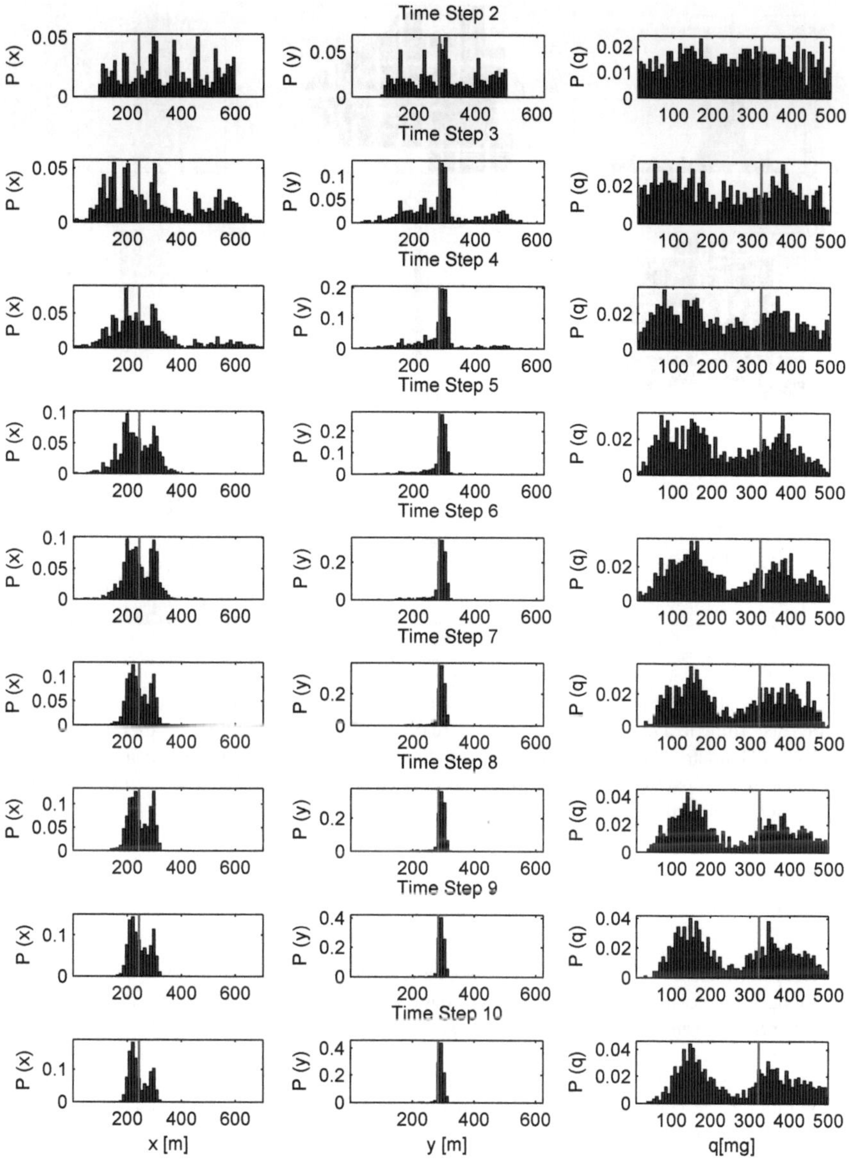

Fig. 3 Evolution of the marginal posteriori probability distribution for x, y and q parameters for time steps $t = 2, 3, \ldots, 10$. The *red vertical line* represents target value of parameters

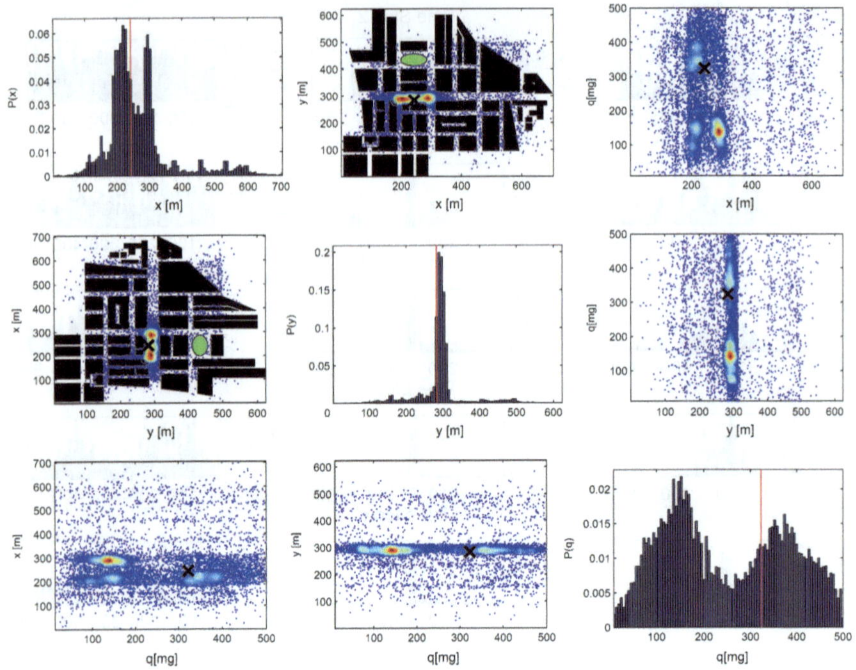

Fig. 4 Bivariate and marginal posterior distributions for all parameters $\theta \equiv (x, y, q)$. The plot is colored according to a probability density, where the most probable regions are colored the deepest *red* (i.e., a heatmap). The *red vertical lines* in diagonal plots (*black* cross in bivariate) show the real value of each parameter. The distributions are built based on all the samples generated in the reconstruction procedure

References

1. Borysiewicz, M., Wawrzynczak, A., Kopka, P.: Stochastic algorithm for estimation of the model's unknown parameters via Bayesian inference, in 2012 Federated Conference on Computer Science and Information Systems (FedCSIS), IEEE, pp. 501–508 (2012)
2. Wawrzynczak, A., Kopka, P., Borysiewicz, M.: Sequential Monte Carlo in Bayesian assessment of contaminant source localization based on the sensors concentration measurements. In: Parallel Processing and Applied Mathematics, pp. 407–417. Springer, Berlin (2014)
3. Wood, C.R., Barlow, J.F., Belcher, S.E., Dobre, A., Arnold, S.J., Balogun, A.A., Lingard, J.J., Smalley, R.J., Tate, J.E., Tomlin, A.S., et al.: Dispersion experiments in central London: the 2007 DAPPLE project. Bull. Am. Meteorol. Soc. **90**(7), 955–969 (2009)
4. Williams, M.D., Brown, M.J., Singh, B., Boswell, D.: QUIC-PLUME theory guide. Los Alamos National Laboratory (2004)
5. Bonassi, F.V., West, M., et al.: Sequential Monte Carlo with adaptive weights for approximate Bayesian computation. Bayesian Anal. **10**(1), 171–187 (2015)
6. Rockle, R.: Bestimmung der Stromungsverhaltnisse im Bereich komplexer Bebauungsstrukturen (1990)
7. Sherman, C.A.: A mass-consistent model for wind fields over complex terrain. J. Appl. Meteorol. **17**(3), 312–319 (1978)

8. Cox, W.M.: Protocol for determining the best performing model. Technical report, Environmental Protection Agency, Research Triangle Park, NC (United States). Technical Support Division (1992)
9. Beaumont, M.A., Zhang, W., Balding, D.J.: Approximate Bayesian computation in population genetics. Genetics **162**(4), 2025–2035 (2002)
10. Del Moral, P., Doucet, A., Jasra, A.: An adaptive sequential Monte Carlo method for approximate Bayesian computation. Stat. Comput. **22**(5), 1009–1020 (2012)
11. Silk, D., Filippi, S., Stumpf, M.P.: Optimizing threshold-schedules for approximate Bayesian computation sequential Monte Carlo samplers: applications to molecular systems (2012). arXiv:1210.3296
12. Filippi, S., Barnes, C.P., Cornebise, J., Stumpf, M.P.: On optimality of kernels for approximate Bayesian computation using sequential Monte Carlo. Stat. Appl. Genet. Mol. Biol. **12**(1), 87–107 (2013)

Bayesian Survival Analysis to Model Plant Resistance and Tolerance to Virus Diseases

Elena Lázaro, Carmen Armero and Luis Rubio

Abstract Viruses constitute a major threat to large-scale production of crops worldwide producing important economical losses and undermining sustainability. We evaluated a new plant variety for resistance and tolerance to a specific virus through a comparison with other well-known varieties. The study is based on two independent Bayesian accelerated failure time models which assess resistance and tolerance survival times. Information concerning plant genotype and virus biotype were considered as baseline covariates and error terms were assumed to follow a modified standard Gumbel distribution. Frequentist approach to these models was also considered in order to compare the results of the study from both statistical methodologies.

Keywords Accelerated failure regression model · Interval-censoring · Plant breeding

1 Introduction

Viruses constitute a major threat to large-scale production of crops worldwide thus producing important economical losses and undermining sustainability [2]. Introgression of genes conferring resistance and/or tolerance by plant breeding is the most efficient and simplest strategy for disease control [7]. Most breeding programs are aimed to find and implement resistance based on the absence of systemic infection. But, new proposals suggest that considering degrees of resistance (reduction

E. Lázaro (✉) · C. Armero
Universitat de València, Calle Dr. Moliner, 50, 46100 Burjassot, Spain
e-mail: elena.lazaro@uv.es

C. Armero
e-mail: carmen.armero@uv.es

L. Rubio
Instituto Valenciano de Investigaciones Agrarias, CV-315,
Km 10.7, 46113 Moncada, Spain
e-mail: lrubio@ivia.es

R. Argiento et al. (eds.), *Bayesian Statistics in Action*, Springer Proceedings in Mathematics & Statistics 194, DOI 10.1007/978-3-319-54084-9_16

of virus infectivity and/or multiplication), and/or tolerance (reduction of symptom severity) may be useful to rescue valuable phenotypes. This requires developing new analytical tools to assess the different levels of resistance and tolerance [9].

In this context, Bayesian survival analysis can provide a suitable framework to assess resistance and tolerance patterns among different plant valuable varieties as well as to capture the uncertainties associated to these estimations. This approach can deal with censoring issues and small sample sizes avoiding the usual frequentist requirement of implementing asymptotic calculations to make exact inferences in complex models. Furthermore, thanks to recent advances in computing, software development such as BUGS [8], and practical methods for prior elicitation, Bayesian survival analysis has become feasible for both practitioners and researchers [4].

The main scientific question addressed in this study was to evaluate a new plant variety, characterised by its genotype, for resistance and tolerance to a specific virus through a comparison with other well-known varieties. The comparison was based on two independent Bayesian accelerated failure time models which assess resistance and tolerance survival times, respectively. The same baseline covariates were used in both models to facilitate the posterior comparison between them. Variables related to plant genotypes and virus biotype crosses were considered as baseline covariates in the analysis and error terms in the regression model were assumed to follow a modified standard Gumbel error. Models were also estimated under the frequentist approach in order to compare the outcomes from both statistical methodologies.

2 Data Description

Three genotype characterizations (G1 for susceptible plants, G2 for resistant, and G3 for plants to evaluate) and two different virus biotype (V1 with capacity to only infect susceptible plants (G1) and V2 with a resistance-breaking capacity to infect also resistant plants (G2)) were considered. A total of 180 plants belonging to genotypes G1, G2 and G3 were inoculated with virus biotype V1 and V2 according to a balanced two-factor factorial design which generated six groups with 30 plants each. All plants were evaluated in terms of resistance and tolerance at monitoring times 7, 14, 21, and 28 days post inoculation (*dpi*).

Resistance was defined as the time, in days, from virus inoculation to virus infection detection, and measured through a reliable molecular detection test. Tolerance was the time, in days, from virus inoculation to the appearance of severe symptoms. It was assessed by expert visual observation. Time zero was in both survival processes synchronised with the time at virus inoculation. Survival times were interval-censored when the event of interest occurred between two consecutive monitoring times or right-censored when it was not observed at the end of the study (28 *dpi*).

Table 1 shows the observed resistance and tolerance frequency for the plants of each of the six groups considered. Groups G2V1 and G3V1 contain a great number of individuals right censored for both events. This is not the case of the G1V1 group

Table 1 Frequency of resistance and tolerance survival times regarding to plant genotype and virus biotype

Genotype	Virus	Resistance (*dpi*)					Tolerance (*dpi*)				
		(0, 7]	(7, 14]	(14, 21]	(21, 28]	28<	(0, 7]	(7, 14]	(14, 21]	(21, 28]	28<
G1	V1	8	14	7	1	0	0	2	23	5	0
	V2	21	9	0	0	0	1	3	26	0	0
G2	V1	0	0	0	0	30	0	0	0	0	30
	V2	2	12	3	6	7	0	4	11	15	0
G3	V1	1	2	1	3	23	0	0	0	0	30
	V2	2	12	3	6	7	0	0	0	0	30

where all plants experienced both events before the end of the study. Remarkably, for virus V2 the number of right censored plants was at most 7 in nearly all groups. However, in the G3V2 group neither of the plants developed severe symptoms at the end of the study.

3 Accelerated Failure Time Regression Models

We considered an *Accelerated Failure Time* (AFT) regression model [6]. Time-to-event, T_i, for individual i, $i = 1, \ldots, n$ is expressed as

$$\log(T_i) = \boldsymbol{x}_i'\boldsymbol{\beta} + \sigma\epsilon_i, \quad i.i.d.\ \epsilon_i \sim S_\epsilon(\cdot), \tag{1}$$

where $\boldsymbol{x}_i'$ is a p-dimensional vector of covariates, $\boldsymbol{\beta}$ the subsequent vector of regression coefficients, σ a scale parameter, and $S_\epsilon(\cdot)$ a known baseline survival function which is assumed to follow a modified standard Gumbel distribution. This AFT model implies a conditional (on $\boldsymbol{\beta}$ and σ) Weibull survival model for T_i with shape $1/\sigma$ and scale $(\log(2)\, \mathrm{e}^{-\boldsymbol{x}_i'\boldsymbol{\beta}/\sigma})$ parameters [1].

This model was independently estimated for resistance and tolerance survival times. In both cases the baselines covariates were indicator variables for identifying the relevant plant genotype (G1, G2 and G3) and virus biotype (V1 and V2) crosses in the study; G1 plants inoculated with biotype V1 (G1V1) was always the reference category. Both Bayesian models were completed with the specification of a prior distribution for the subsequent parameters. We assumed a prior independent default scenario. The marginal prior distribution for each regression coefficient $\beta_{\mathrm{G}_j\mathrm{V}_k}$, $j = 1, 2, 3$, $k = 1, 2$, was elicited as a normal distribution centered at zero and a wide variance, $\pi(\beta_{\mathrm{G}_j\mathrm{V}_k}) = \mathrm{N}(0, 1000)$. A uniform distribution Un(0, 100) was selected as the marginal prior distribution for σ.

The likelihood function of $(\boldsymbol{\beta}, \sigma)$ for the observed data was the product of the likelihood function for each individual. Individual time-to-event data was right or interval censored. A right censored data corresponds to individuals that have not experienced the event of interest at the end of the period of the study, 28 *dpi*. Its contribution to the likelihood is $P(T_i > 28 \mid \boldsymbol{x}_i, \boldsymbol{\beta}, \sigma)$, its survival function at 28 *dpi*. Interval censored data for individual i arises when the event of interest occurred between the current monitoring time (t_{iu}) and the previous one (t_{il}) and its contribution to the likelihood is $S_i(t_{il} \mid \boldsymbol{x}_i, \boldsymbol{\beta}, \sigma) - S_i(t_{iu} \mid \boldsymbol{x}_i, \boldsymbol{\beta}, \sigma)$. Consequently

$$\begin{aligned} L(\boldsymbol{\beta}, \sigma) &= \prod_{i=1}^{n} L_i(\boldsymbol{\beta}, \sigma) \\ &= \prod_{i \in \mathcal{R}} S_i(28 \mid \boldsymbol{x}_i, \boldsymbol{\beta}, \sigma) \prod_{i \in \mathcal{I}} [S_i(t_{il} \mid \boldsymbol{x}_i, \boldsymbol{\beta}, \sigma) - S_i(t_{iu} \mid \boldsymbol{x}_i, \boldsymbol{\beta}, \sigma)], \end{aligned}$$

where $\mathcal{R}$ ($\mathcal{I}$) is the set of right (interval) censored data, and $S_i(t \mid \boldsymbol{x}_i, \boldsymbol{\beta}, \sigma)$ the survival function for individual i

$$S_i(t \mid \boldsymbol{x}_i, \boldsymbol{\beta}, \sigma) = \exp\{-\log(2)\, t^{(1/\sigma)}\, \mathrm{e}^{-\boldsymbol{x}_i'\boldsymbol{\beta}/\sigma}\}.$$

4 Results

The posterior distribution of the parameters for each model was approximated by means of Markov Monte Carlo methods (MCMC) through WinBUGS software [8]. Specifically, simulation was run considering three Markov chains with 100 000 iterations and a burn-in period with 10 000. In addition, the chains were thinned by storing every 10th iteration in order to reduce autocorrelation in the saved sample and avoid space computer problems. Trace plots of the simulated values of the chains appear overlapping one another, indicating stabilization. Convergence of the chains to the posterior distribution was assessed using the potential scale reduction factor, $\hat{R}$, and the effective number of independent simulation draws, neff. In all cases, the $\hat{R}$ values were equal or close to 1 and neff > 100, thus indicating that the distribution of the simulated values between and within the three chains was practically identical, and that sufficient MCMC samples had been obtained, respectively [3].

Both models were also estimated under the frequentist approach in order to compare Bayesian and frequentist results. Frequentist estimation was performed through the function `survreg()` of the `survival` R package [10, 11]. Note that in this implementation, the error term ϵ of (1) is assumed to follow a standard Gumbel distribution instead of a modified Gumbel distribution.

Results were arranged in two parts for tolerance and resistance separately. However, as both survival times were studied with the same type of model the outcomes are presented under the same scheme to detect similarities and differences between them. We focused on the effect of covariates on the estimated probabilities of remaining free of infection and free of the appearance of severe symptoms. A small subsection for comparing Bayesian and frequentist results is also included.

4.1 Resistance

Posterior summaries of the approximated posterior distribution for the regression coefficients and the error scale parameter are shown in Table 2. Genotype plants G1 have the shortest resistance times among the plants inoculated with V1. Posterior probabilities $P(\beta_{G2V1} > 0 \mid \mathcal{D}) = 1$ and $P(\beta_{G3V1} > 0 \mid \mathcal{D}) = 1$, where $\mathcal{D}$ represent the data, provide strong evidence that G2 and G3 plants show a better resistance behaviour with regard to G1 under V1 infection. In addition, genotype G2 is the most resistant variety with $P(\beta_{G2V1} > \beta_{G3V1} \mid \mathcal{D}) = 1$ despite of the wide variability of its estimated coefficient. Under biotype infection V2, resistance is worse for all genotypes although G3 genotype improves resistance in relation to G2 ($P(\beta_{G3V2} > \beta_{G2V2} \mid \mathcal{D}) = 0.99$).

Figure 1 shows the posterior mean of the probability of remaining free of infection red over time (from 0 to 28 dpi) for each genotype plant under virus infection V1 and V2. For both virus, G1 plants show the lowest probability values in all the monitoring

Table 2 Summary of the MCMC approximate posterior distribution for the resistance model: mean, standard deviation, 95% credible interval, and posterior probability that the subsequent parameter is positive. Group G1V1 is the reference category

Parameter	Mean	Sd	$CI_{95\%}$	$P(\cdot > 0)$
β_{G1V1}	2.27	0.12	[2.02, 2.51]	1.00
β_{G2V1}	4.97	1.24	[2.62, 7.00]	1.00
β_{G3V1}	1.64	0.26	[1.15, 2.24]	1.00
β_{G1V2}	−0.66	0.15	[−0.96, −0.36]	0.00
β_{G2V2}	0.22	0.16	[−0.08, 0.55]	0.93
β_{G3V2}	0.65	0.16	[0.34, 0.98]	1.00
σ	0.55	0.06	[0.46, 0.67]	

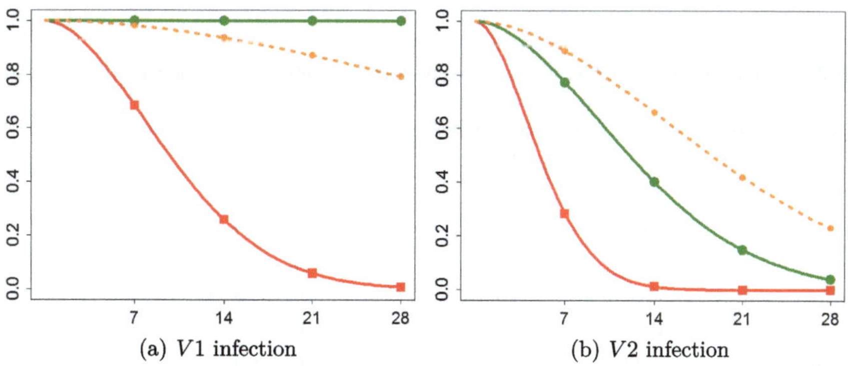

Fig. 1 Posterior mean of the probability of remaining free of infection over time (from 0 to 28 *dpi*) for G1 (in *solid red line*), G2 (in *solid green line*) and G3 (in *dotted orange line*) genotypes under infection V1 and V2. Monitoring times 7, 14, 21 and 28 *dpi* are highlighted with dots

times (7, 14, 21 and 28 *dpi*). Plants G2 exhibit the highest probability values under V1 infection and G3 under V2 infection. Remarkably, the pattern of the differences between genotypes G2 and G3 under virus V1 and V2 is very different. Under V2 infection, differences among posterior probabilities (in favour of no infection for G3) are enough stable from 14 dpi and for any time they exceed the value of 0.27. In the case of V1, there is an increasing difference over time in favour of no infection for G2 with a maximum distance of 0.21 at 28 *dpi*. Posterior mean of the probability of remaining free of infection decreases with time for all genotypes under infection V2 bringing to light V2 resistance-breaking capacity. At 14 *dpi* (the midpoint of the monitoring times), the estimated mean of that probability is 0.26, 1, and 0.93 for groups G1V1, G2V1 and G3V1, and 0.02, 0.40, and 0.65 for G1V2, G2V2 and G3V2, respectively.

4.2 Tolerance

Table 3 shows a summary of the posterior distribution for the regression coefficients and the error scale parameter in the AFT model for tolerance times. Estimation in terms of the sign of the posterior outcomes are quite similar to the subsequent results of the resistance model, but we can also appreciate some noticeable differences. It is worth mentioning the similar effect of biotype V1 on G2 and G3 plants and the overwhelming estimated effect related to G3 genotype under V2 infection. Plants G3 show a similar tolerance pattern for both virus biotypes.

The posterior mean of the probability of remaining free of the appearance of severe symptoms during time study (from 0 to 28 *dpi*) for biotype and virus groups is displayed in Fig. 2. Under V1 infection, plants G2 and G3 exhibit similar probabilities values very close to one. They are higher than G1 values, which shows a decreasing trend with a strong slope between 14 and 21 *dpi*'s. Plants G1 and G3 behave analogously under V1 and V2 infection. However, probabilities for G2 are very different

Table 3 Summary of the MCMC approximate posterior distribution for the tolerance model: mean, standard deviation, 95% credible interval, and posterior probability that the subsequent parameter is positive. Group G1V1 is the reference category

Parameter	Mean	Sd	$CI_{95\%}$	$P(\cdot > 0)$
β_{G1V1}	2.91	0.04	[2.84, 2.98]	1.00
β_{G2V1}	3.90	1.77	[1.08, 6.95]	1.00
β_{G3V1}	4.09	1.74	[1.16, 6.93]	1.00
β_{G1V2}	−0.12	0.05	[−0.23, −0.03]	0.00
β_{G2V2}	0.12	0.05	[0.02, 0.21]	1.00
β_{G3V2}	4.00	1.81	[1.07, 6.89]	1.00
σ	0.15	0.02	[0.12, 0.19]	

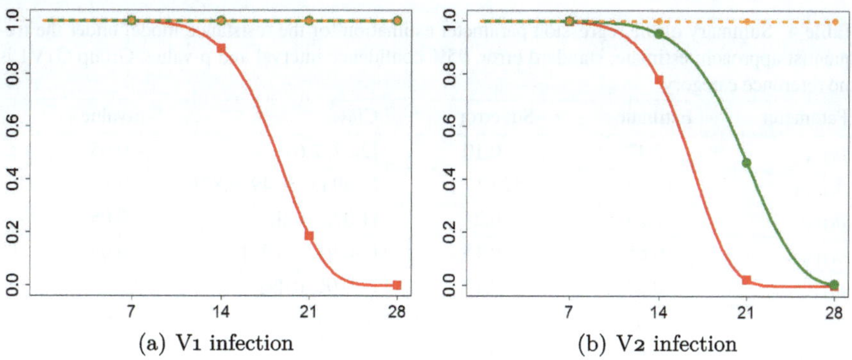

Fig. 2 Posterior mean of the probability of remaining free of the appearance of severe symptoms over time (from 0 to 28 *dpi*) for G1 (in *solid red line*), G2 (in *solid green line*) and G3 (in *dotted orange line*) genotypes under infection V1 and V2. Monitoring times 7, 14, 21 and 28 *dpi* are highlighted with dots

for both virus: G2 is similar to G3 for infection V1 but its behaviour changes under V2 infection. In particular, G2 shows a decreasing probability of remaining free of infection from 14 *dpi* on, which at the end of the monitoring time is equal to the value of variety G1. At 14 *dpi* (the midpoint of the monitoring times), the posterior mean of the probability of remaining free of the appearance of severe symptoms is 0.89, 1, and 1 for G1V1, G2V1 and G3V1 crosses, and 0.77, 0.95, and 1 for G1V2, G2V2 and G3V2, respectively.

4.3 *Tolerance and Resistance: Frequestist and Bayesian Models*

Results in this subsection are focused on the frequentist approach to the resistance (Table 4) and the tolerance (Table 5) model. Both tables try to reproduce the structure of Table 2 (Bayesian resistance model) and Table 3 (Bayesian tolerance model) with regard to the frequentist concepts (estimate, standard error, 95% confidence interval, and p-value) which could be considered as somehow *parallel* to Bayesian posterior mean, standard deviation, 95% credible interval, and posterior probability for a positive regression coefficient.

At first glance, most of the numerical (but not conceptual) results provided by the two approaches seem not to be very different. But a more leisurely observation of them shows relevant differences in the punctual and interval estimation of the regression coefficients, particularly in those groups in which all the observations were right censored. This is the case of the G2V1 group for the resistance model and groups G2V1, G3V1 and G3V2 for tolerance. In the case of the resistance model for group G2V1, the punctual frequentist and Bayesian estimates are 11.54 and 4.97,

Table 4 Summary of the regression parameter estimation for the resistance model under the frequentist approach: estimate, standard error, 95% confidence interval and p-value. Group G1V1 is the reference category

Parameter	Estimate	Sd. error	$CI_{95\%}$	p-value
β_{G1V1}	2.47	0.10	[2.27, 2.67]	<0.05
β_{G2V1}	11.54	2523.17	[−4933.79, 4956.87]	1.00
β_{G3V1}	1.55	0.24	[1.09, 2.02]	<0.05
β_{G1V2}	−0.65	0.15	[−0.94, −0.35]	<0.05
β_{G2V2}	0.22	0.14	[−0.06, 0.49]	0.13
β_{G3V2}	0.63	0.15	[0.34, 0.93]	<0.05
$log(\sigma)$	−0.65	0.10		<0.05

Table 5 Summary of the regression parameter estimation for the tolerance model under the frequentist approach: estimate, standard error, 95% confidence interval and p-value. Group G1V1 is the reference category

Parameter	Estimate	Sd. error	$CI_{95\%}$	p-value
β_{G1V1}	2.97	0.03	[2.90, 3.03]	<0.05
β_{G2V1}	3.60	1710	[−3340.72, 3347.92]	1.00
β_{G3V1}	3.60	1710	[−3340.72, 3347.92]	1.00
β_{G1V2}	−0.12	0.05	[−0.22, −0.02]	<0.05
β_{G2V2}	0.12	0.05	[0.02, 0.21]	<0.05
β_{G3V2}	3.60	1710	[−3340.72, 3347.92]	1.00
$log(\sigma)$	−1.92	0.11		<0.05

respectively. But the more relevant differences between them are in their variability, which gives enormous confidence intervals and p-values close to 1. This is also the case of the frequentist results for tolerance in groups G2V1, G3V1 and G3V2 which all have the same enormous 95% confidence interval.

The inferences achieved indicate that the AFT frequentist model has difficulties in the estimation corresponding to groups with data with very little signal. This is not the case of the Bayesian analyses that accommodate very well for the particular data of the study. This situation agrees with the general comment in [4] about the advantages of the Bayesian methodology over the frequestist one in survival analysis with regard to the fitting problems in the presence of complex censoring data. Moreover, the Bayesian results are more compatible with the agronomic expectations based on preliminary studies.

5 Conclusions

Agronomical conclusions indicated that genotype G3 did not suppose an improvement in terms of resistance with respect to G2. However, they showed a very high tolerance to V2 virus. Future research will probably focus on data from experiences in open field with commercial plants in which the source of tolerance is incorporated. This process is not easy because it is necessary to identify the sources of tolerance and subsequently select the appropriate procedures to be included in the study.

Bayesian survival regression models provide a useful tool for quantifying differences among the different genotype $\times$ virus biotype groups as well as to assess the degree of resistance and of tolerance. They also make possible the incorporation of censoring and truncation mechanisms that are frequent in this type of studies with good inference results. Frailty models and bivariate modeling based on right truncated data [5] are in progress in order to approach a more suitable model that can better capture all the uncertainties of the real problem.

Acknowledgements This research has been partially supported by research grants FPU/02042/2013 from the Spanish Ministry of Education, Culture and Sports, MTM2016-77501-P from the Spanish Ministry of Economy and Competitiveness, and RTA2013-00047-C02 from the INIA. We wish to acknowledge two anonymous referees for their valuable comments and suggestions that have improved very much the original version of the paper.

References

1. Christensen, R., Johnson, W.O., Branscum, A.J., Hanson, T.E.: Bayesian Ideas and Data Analysis: An Introduction to Scientists and Statisticians. Chapman and Hall/CRC, Boca Raton (2011)
2. Gallitelli, D.: The ecology of Cucumber mosaic virus and sustainable agriculture. Virus Res. **71**(1), 9–21 (2000)
3. Gelman, A., Rubin, D.B.: Inference from iterative simulation using multiple sequences. Stat. Sci. **7**, 457–511 (1992)
4. Ibrahim, J.G., Chen, M.H., Sinha, D.: Bayesian Survival Analysis. Springer, New York (2005)
5. Kalbfleisch, J.B., Lawless, J.F.: Regression models for right truncated data with applications to AIDS incubation times and reporting lags. Stat. Sin. **1**, 29–32 (1991)
6. Klein, J.P., Moeschberger, M.L.: Survival Analysis: Techniques for Censored and Truncated Data. Springer, New York (2003)
7. Lecoq, H., Moury, B., Desbiez, C., Palloix, A., Pitrat, M.: Durable virus resistance in plants through conventional approaches: a challenge. Virus Res. **100**, 31–39 (2004)
8. Lunn, D.J., Thomas, A., Best, N., Spiegelhalter, D.: WinBUGS - a Bayesian modelling framework: concepts, structure, and extensibility. Stat. Comput. **10**, 325–337 (2000)
9. Soler, S., Debreczeni, D.E., Vidal, E., Aramburu, J., López, C., Galipienso, L., Rubio, L.: A new Capsicum baccatum accession shows tolerance to wildtype and resistance-breaking isolates of Tomato spotted wilt virus. Ann. Appl. Biol. **167**(3), 343–353 (2015)
10. Therneau, T.M., Grambsch, P.M.: Modeling Survival Data: Extending the Cox Model. Springer, New York (2000)
11. Therneau, T.: A package for survival analysis in S. version 2.38 (2015). http://CRAN.R-project.org/package=survival (2015)

Randomization Inference and Bayesian Inference in Regression Discontinuity Designs: An Application to Italian University Grants

Federica Licari

Abstract Motivated by the evaluation of Italian University grants, we will address the problem of multiplicities in (fuzzy) Regression Discontinuity (RD) settings. Following Li, Mattei and Mealli [1], we adopt a probabilistic formulation of the assignment mechanism underling RD designs and we select suitable subpopulations around the cutoff point on the basis of observed covariates using both randomization tests and a Bayesian model-based approach both accounting for the problem of multiple testing. We then conduct our analysis studying the effect of university grants on two binary outcomes, dropout and a variable equal to one for students who realize at least one University Credit (CFU), using both the Fisher-exact P-value approach and a model-based Bayesian approach. In both cases we account for the multivariate nature of the outcome by (a) proposing a multiple testing approach, and (b) defining estimands on the joint outcome.

Keywords Regression discontinuity designs · Multiple outcome · Causal effect · Fisher exact p-value · MCMC

1 Introduction

The entrance in the university world results in a radical change in students' life. It requires financial resources that may obstacle the students' University career, prompting students from low income families to dropout. In order to give equal opportunity to achieve higher education to motivated students, every year Italian universities offer financial aids to a limited number of eligible students who apply for a grant. The assignment rules underlying the Italian university grants define a Fuzzy

F. Licari (✉)
Department of Statistica, Informatica, Applicazioni, University of Florence, viale Morgagni 59, 50134 Florence, Italy
e-mail: f.licari@disia.unifi.it

R. Argiento et al. (eds.), *Bayesian Statistics in Action*, Springer Proceedings in Mathematics & Statistics 194, DOI 10.1007/978-3-319-54084-9_17

Regression Discontinuity (FRD) design: students need to be eligible and apply for a grant to receive a grant. In this work we focus on first year students, where the eligibility status is only based on an indicator of the economic family situation, i.e., the ISEE indicator. Under the framework proposed by Li, Mattei, Mealli [1] we adopt a probabilistic formulation of the assignment mechanism underling RD designs and we select suitable subpopulations around the cutoff point on the basis of observed covariates using both randomization tests and a Bayesian model-based approach both accounting for the problem of multiple testing. We then evaluate causal effects of university grants on two outcomes: dropout, a binary variable equal to one for students who dropout by the end of the first academic year, and zero otherwise; and CFU, a binary variable equal to one for students who realize at least one university credit during the academic year, and zero otherwise. For inference we use both the Fisher-exact P-value approach and a model-based Bayesian approach accounting for the multivariate nature of the outcome by proposing multiple testing methods and defining new estimands for the joint outcome.

2 Causal Estimands in FRD Design with Multiple Outcomes

We first introduce the notation in the context of our application. The forcing variable plays an important role in RD designs: it determines the assignment to treatment or to control. Formally, consider a sample of N units (students) indexed by $i = 1 \dots N$. Let S_i be the forcing variable for student i, which is a combined economic measurement of the student's family income and assets (the ISEE indicator). Let Z_i be the eligibility status, which is a deterministic function of S_i. In our study we have $Z_i = \mathbf{1}_{\{S_i \le s_0\}}$, where $\mathbf{1}(\cdot)$ is the indicator function and s_0 is the threshold, which is equal to 15000 euros. In our study there are students who do not comply with the eligibility status. Therefore, Z plays the role of an "instrument" or an "encouragement" variable as in randomized experiments with non-compliance. Formally, let $W_i(z)$ be an indicator for grant receipt status given eligibility status z, and let $W_i^{obs} = W_i(Z_i)$ be the actual treatment received, equal to 1 if student i receives a grant and 0 otherwise. In our study $W_i(0) = 0$ by design, because ineligible students cannot receive any grant. Therefore, based on the grant status W, subjects can be classified into two (latent) principal strata: never-takers (NT), students who would never take the grant irrespective of their eligibility status ($W_i(0) = W_i(1) = 0$); and compliers (C), students who would always comply with their eligibility status ($W_i(z) = z$ for $z = 0, 1$). Let G_i be the indicator of the compliance status for student i: $G_i = g$, for $g \in \{C, NT\}$ Let $Y_{drop_i}(z)$ and $Y_{CFU_i}(z)$ be the potential outcomes for Dropout and CFU, respectively, given eligibility status z, and let $Y_{drop_i}^{obs} = Y_{drop_i}(Z_i)$ and $Y_{CFU_i}^{obs} = Y_{CFU_i}^{obs}(Z_i)$ be the corresponding observed outcomes. Finally, let $\mathbf{Y}_i(z) = (Y_{drop_i}(z), Y_{CFU_i}(z))$ and $\mathbf{Y}_i^{obs} = (Y_{drop_i}^{obs}, Y_{CFU_i}^{obs})$, for $i = 1 \dots N$.

Following Li, Mattei and Mealli (2016), we adopt a probabilistic formulation of the assignment mechanism underling RD designs viewing the forcing variable as a random variable with a probability distribution. This framework is based on three key assumptions:

Assumption 1 (*Local overlap*) Let $\mathcal{U}$ be the random sample (or population) of units in the study. There exists a subset of units, $\mathcal{U}_{s_0}$, such that for each $i \in \mathcal{U}_{s_0}$, $\mathbb{P}(S_i \leq s_0) > \varepsilon$ and $\mathbb{P}(S_i > s_0) > \varepsilon$ for some sufficiently large $\varepsilon > 0$.

Assumption 2 (*Local RD-SUTVA*) For each units $i \in \mathcal{U}_{s_0}$, consider two eligibility statuses $Z_i^{'} = \mathbf{1}_{\{S_i^{'} \leq s_0\}}$ and $Z_i^{''} = \mathbf{1}_{\{S_i^{''} \leq s_0\}}$, with possibly $S_i^{'} \neq S_i^{''}$. If $Z_i^{'} = Z_i^{''}$, this means, if either $S_i^{'} \leq s_0$ and $S_i^{''} \leq s_0$, or $S_i^{'} > s_0$ and $S_i^{''} > s_0$, then $W_i(Z^{'}) = W_i(Z^{''})$, and $\mathbf{Y}_i(Z^{'}) = \mathbf{Y}_i(Z^{''})$.

Assumption 3 (*Local Randomization*) For each $i \in \mathcal{U}_{s_0}$,

$$\mathbb{P}(S_i \mid Y_i(1), Y_i(0), W_i(1), W_i(0), \mathbf{X_i}) = \mathbb{P}(S_i),$$

where $\mathbf{X_i}$ is a vector of covariates.

Assumption 1 assumes that there exists at least one subpopulation of units, $\mathcal{U}_{s_0}$, such that for units belonging to such a population the probability of having a value of the forcing variable lying in both sides of the threshold is sufficiently faraway from both zero and one.

Assumption 2 makes the SUTVA for this subpopulation. Finally the last assumption formalizes the concept of RD design as a local randomized experiment.
An important issue in practice is the selection of the subpopulation, $\mathcal{U}_{s_0}$. In principle $\mathcal{U}_{s_0}$ can have any shape. For convenience we focus on subpopulations that comprise units with a realized value of the forcing variable S in a symmetric intervals around the threshold. Specifically we make the following assumption:

Assumption 4 (*Shape of the Subpopulation*) $\exists h > 0$: $\forall \varepsilon > 0$, $\mathbb{P}(s_0 - h \leq S_i \leq s_0 + h) > 1 - \varepsilon$, $\forall i \in \mathcal{U}_{s_0}$.

Under local randomization, Assumption 3, we expect that in the subpopulation $\mathcal{U}_{s_0}$, pre-treatment variables are well balanced in the two subsamples defined by assignment. We use this balanced property to select potential subpopulations. We adopt both a Bayesian approach and a Fisher exact P-value approach to multiple testing to find subpopulations of units where our RD assumptions hold. As we can see in Tables 1 and 2, the two approaches lead to similar results suggesting that suitable subpopulations are defined using the following bandwidths: $h = 500, 1000$ and 1500.

Given the target population we can define local causal estimands within it. We focus on causal effects for compliers within $\mathcal{U}_{s_0}$ on Dropout and CFU, separately, and on the joint outcome, $\mathbf{Y}$, defined previously. Specifically, for the joint outcome, we are interested in assessing if the receipt of the grant reduces the proportion of students who dropout without taking any exam, therefore we focus on the event $(1, 0)$. Formally, the causal estimands of interest are:

$$CACE_{drop_{\mathcal{U}_{s0}}} = \mathbb{P}(Y_{drop_i}(1) = 1 \mid G_i = C, i \in \mathcal{U}_{s0}) - \mathbb{P}(Y_{drop_i}(0) = 1 \mid G_i = C, i \in \mathcal{U}_{s0}) \quad (1)$$

$$CACE_{CFU_{\mathcal{U}_{s0}}} = \mathbb{P}(Y_{CFU_i}(1) = 1 \mid G_i = C, i \in \mathcal{U}_{s0}) - \mathbb{P}(Y_{CFU_i}(0) = 1 \mid G_i = C, i \in \mathcal{U}_{s0}) \quad (2)$$

$$CACE_{10_{\mathcal{U}_{s0}}} = \mathbb{P}(\mathbf{Y}_i(1) = (1,0) \mid G_i = C, i \in \mathcal{U}_{s0}) - \mathbb{P}(\mathbf{Y}_i(0) = (1,0) \mid G_i = C, i \in \mathcal{U}_{s0}) \quad (3)$$

3 Randomization Based Inference and Bayesian Inference in FRD Designs with Multiple Outcomes

We conduct randomization-based inference and Bayesian inference under the RD assumptions both accounting for the fuzzy nature of the design and the multivariate nature of the outcome variable. Specifically, we conduct two different types of analysis. First we focus on causal effects of the treatment on dropout and CFU, separately, testing the sharp null hypothesis of no treatment effect adjusting inferences for multiple comparisons. Then, we focus on the joint outcome, $\mathbf{Y}$, aiming at assessing causal effects of the treatment on Dropout and CFU jointly. This approach allows us to explicitly take into account the association structure between the two outcomes.

We conduct the analyses under the following additional assumption:

Table 1 Posterior probabilities that pre-treatment variables are well balanced in the two subsamples defined by assignment

Bayesian approach				
bandwidth (h)	500 Euro	1000 Euro	1500 Euro	2000 Euro
Variable	(n=1042)	(n=2108)	(n=3166)	(n=4197)
Sex	0.95	0.96	0.98	0.97
HS type (other)				
Humanity	0.95	0.96	0.98	0.97
Science	0.91	0.93	0.95	0.89
Tech	0.81	0.81	0.82	0.62
HS grade	0.96	0.98	0.97	0.98
Year (2004)				
2005	0.96	0.93	0.97	0.98
2006	0.92	0.91	0.96	0.93
University	0.92	0.98	0.69	0.10
(PI versus FI)				
Major in University (other)				
Humanity	0.90	0.80	0.80	0.93
Science	0.86	0.75	0.78	0.90
Social science	0.82	0.71	0.76	0.86
BioMed	0.78	0.68	0.74	0.84
Tech	0.63	0.62	0.70	0.79

s_0 =15000, first year students, Florence and Pisa, 2004–2006

Table 2 Adjusted p-values testing that pre-treatment variables are well balanced in the two subsamples defined by assignment

Randomization approach				
bandwidth (h)	500 Euro	1000 Euro	1500 Euro	2000 Euro
Variable	(n=1042)	(n=2108)	(n=3166)	(n=4197)
Sex	1	1	1	1
HS type				
Humanity	1	0.99	1	1
Science	1	0.99	1	0.89
Tech	1	1	1	1
Other	0.44	0.79	0.35	0.30
HS grade	0.97	1	1	1
Year of enrollement				
2004	1	0.99	0.99	1
2005	1	0.91	1	1
2006	1	0.99	1	1
University (PI versus FI)	0.99	1	0.10	0.01
Major in University				
Humanity	0.92	0.32	0.45	0.89
Science	1	1	0.99	0.99
Social science	1	1	1	1
BioMed	1	0.85	1	1
Tech	0.92	0.96	1	1
Other	0.98	1	1	1

s_0 =15000, first year students, Florence and Pisa, 2004–2006

Assumption 5 (*Exclusion Restriction for Never-Takers*)

$$Y_{ij}(0) = Y_{ij}(1) \text{ for each } i : G_i = NT,\ j = drop, CFU,$$

which rules out direct effects of eligibility on dropout and CFU for never-takers.

Note that we do not use information on the covariates in the outcome analyses; covariate are only used for the selection of the subpopulations. Indeed, once a subpopulation has been selected, under the RD assumptions, adjusting for the covariates is not required, although they can help improving inferences.

Randomization Based Inference

We conduct randomization-based inference in the context of our FRD design appropriately handling issues of non-compliance by deriving Bayesian posterior predictive $p-$values for complier causal effects [2, 3].

Table 3 P-values adjusted for multiple comparisons across the three bandwidths

	$h = 500$		$h = 1000$		$h = 1500$	
	Estimates	p-value	Estimates	p-value	Estimates	p-value
$CACE_{drop}$	−0.16	0.367	−0.15	0.240	−0.25	0.007
$CACE_{CFU}$	0.12	0.444	0.09	0.456	0.15	0.157
$CACE_{10}$	−0.18	0.176	−0.08	0.413	−0.19	0.072

In the first type of analysis we appropriately adjust for multiple tests, using the procedure proposed by Lee et al. [3] Specifically, we obtain familywise tests using the following procedure. We first derive nominal posterior predictive $p-$values for complier causal effects imputing the missing compliance statuses, drawing from their posterior predictive distribution according to a compliance model that assumes the null hypothesis. Then we calculate adjusted posterior predictive $p-$values using the joint randomization distribution of the nominal p-values. In the second type of analysis we derive Bayesian posterior predictive $p-$values for complier causal effects on the joint outcome, specifying a multinomial distribution for $\mathbf{Y}$. Table 3 shows the results obtained by the two types of analyses and the estimates of the statistics test used, $\widehat{CACE}$. We can observe that the p-values are all quite high for the subpopulations defined by bandwidths $h = 500$ and $h = 1000$, showing no evidence against the null hypothesis of no effect. For the subpopulation defined by the bandwidth $h = 1500$ we do not find any significant effect of university grants on CFU either, but we find statistical evidence that university grants reduce students' dropout: in fact the p-value based on $CACE_{drop}$ is very small (0.007). Similarly, the p-value for $CACE_{10}$ is quite small (0.072), suggesting that there exist statistical evidence that university grants positively affect students' performances. These results for the subpopulation defined by the bandwidth $h = 1500$ may be at least partially due to the fact that this subpopulation is larger and includes students with values of the forcing variable more heterogeneous.

Bayesian Inference

In the Bayesian analysis we focus on the joint outcome, $\mathbf{Y}$, which has a multinomial distribution defined by the four nominal category: $(0, 0)$, $(0, 1)$ $(1, 0)$ and $(1, 1)$. Our attention is prompt on the category $(1, 0)$ aiming at evaluating causal effect on students who dropout and realized zero credits.

We assume a multinomial logistic regression model for the potential outcomes. Formally:

$$\mathbb{P}\left(\mathbf{Y}_i(z) = j | G_i = g, \mathbf{X}_i, i \in \mathscr{U}_{s_0}; \boldsymbol{\theta}\right) = \frac{\exp\{\beta_{0,j}^{(zg)} + \beta_{1,j}^{(zg)} S_i^* + \mathbf{X}_i \boldsymbol{\beta}_j\}}{1 + \sum_{j \in \{(0,1),(1,0),(1,1)\}} \exp\{\beta_{0,j}^{(zg)} + \beta_{1,j}^{(zg)} S_i^* + \mathbf{X}_i \boldsymbol{\beta}_j\}} \equiv \pi_{jz_i}^{(g)}$$

for $z = 0, 1$; $g = C, NT$; $j \in \{(0,0), (0,1), (1,0), (1,1)\}$;
with $\beta_{0,(0,0)}^{(zg)} = 0$, $\beta_{1,(0,0)}^{(zg)} = 0$, and $\boldsymbol{\beta}_{(0,0)} = \mathbf{0}$. Under exclusion restriction for Never Takers, Assumption 5, we assume equality of the coefficients under $z = 0, 1$, (e.g. $\beta_{0j}^{(1NT)} = \beta_{1j}^{(0NT)}$; $\beta_{1j}^{(0NT)} = \beta_{1j}^{(1NT)}$), so we have $\pi_{j0_i}^{(NT)} = \pi_{j1_i}^{(NT)} \equiv \pi_{j_i}^{(NT)}$.

Another model involved in our analysis is the model for the compliance status, which is dichotomous. We can assume for it a logistic model. Formally:

$$\mathbb{P}(G_i = c | \mathbf{X}_i, i \in \mathscr{U}_{s_0}; \boldsymbol{\theta}) = \frac{\exp\{\alpha_0 + \alpha_1 S_i^* + \mathbf{X}_i \boldsymbol{\alpha}\}}{1 + \exp\{\alpha_0 + \alpha_1 S_i^* + \mathbf{X}_i \boldsymbol{\alpha}\}} \equiv \pi_{C_i}$$

where $S_i^* = (S_i - s_0)/1000$.

We use as priors for the parameters independent Cauchy distributions with location parameter zero and scale parameter 2.5. We denote with $\boldsymbol{\theta} = \{\beta_{0,j}^{(zg)}, \beta_{1,j}^{(zg)}, \boldsymbol{\beta}_j, \alpha_0, \alpha_1, \boldsymbol{\alpha}\}$ the full parameters vector involved in our models and with $p(\boldsymbol{\theta})$ its prior probability. Let f_{i,C_z} be the outcome probability for $\mathbf{y}$ for students who are Compliers by eligibility status:

$$f_{i,C_z} \equiv \pi_{00z_i}^{(C)\mathbf{1}\{\mathbf{y}_i=(0,0)\}} \cdot \pi_{01z_i}^{(C)\mathbf{1}\{\mathbf{y}_i=(0,1)\}} \cdot \pi_{10z_i}^{(C)\mathbf{1}\{\mathbf{y}_i=(1,0)\}} \cdot \pi_{11z_i}^{(C)\mathbf{1}\{\mathbf{y}_i=(1,1)\}}, \text{ by z = 0,1.}$$

For Never Takers the outcome probability is:

$$f_{i,NT} \equiv \pi_{00_i}^{(NT)\mathbf{1}\{\mathbf{y}_i=(0,0)\}} \cdot \pi_{01_i}^{(NT)\mathbf{1}\{\mathbf{y}_i=(0,1)\}} \cdot \pi_{10_i}^{(NT)\mathbf{1}\{\mathbf{y}_i=(1,0)\}} \cdot \pi_{11_i}^{(NT)\mathbf{1}\{\mathbf{y}_i=(1,1)\}}$$

Then the distribution of $\boldsymbol{\theta}$ is:

$$\begin{aligned} p(\boldsymbol{\theta} \mid \mathbf{Y}, (X), Z, W) \propto p(\boldsymbol{\theta}) \times [&\prod_{i \in O(0,0)} [\pi_{C_i} \cdot f_{i,C_0} + (1 - \pi_{C_i}) f_{i,NT}] \\ \times &\prod_{i \in O(1,0)} (1 - \pi_{C_i}) \cdot f_{i,NT} \\ \times &\prod_{i \in O(1,1)} \pi_{C_i} \cdot f_{i,C_1}], \end{aligned} \tag{4}$$

where $O(0,0)$, $O(1,0)$ and $O(1,1)$ are the three observed groups defined by the combination of Z_i^{obs} and W_i^{obs}: ($Z^{obs} = 0$, $W^{obs} = 0$), ($Z^{obs} = 1$, $W^{obs} = 0$), ($Z^{obs} = 1$, $W^{obs} = 1$). By design students have no way to obtain a grant if not eligible, so the realization of the fourth combination ($Z^{obs} = 0$, $W^{obs} = 1$) is not possible.

Table 4 Posterior distributions of finite-population causal estimands: summary statistics

	$h = 500$			$h = 1000$			$h = 1500$		
	50%	2.5%	97.5%	50%	2.5%	97.5%	50%	2.5%	97.5%
$CACE_{drop}$	−0.17	−0.42	0.09	−0.22	−0.40	−0.02	−0.28	−0.41	−0.12
$CACE_{CFU}$	0.18	−0.03	0.43	0.13	0.00	0.29	0.17	0.03	−0.35
$CACE_{10}$	−0.17	−0.42	0.01	−0.12	−0.28	0.00	−0.16	−0.33	−0.04

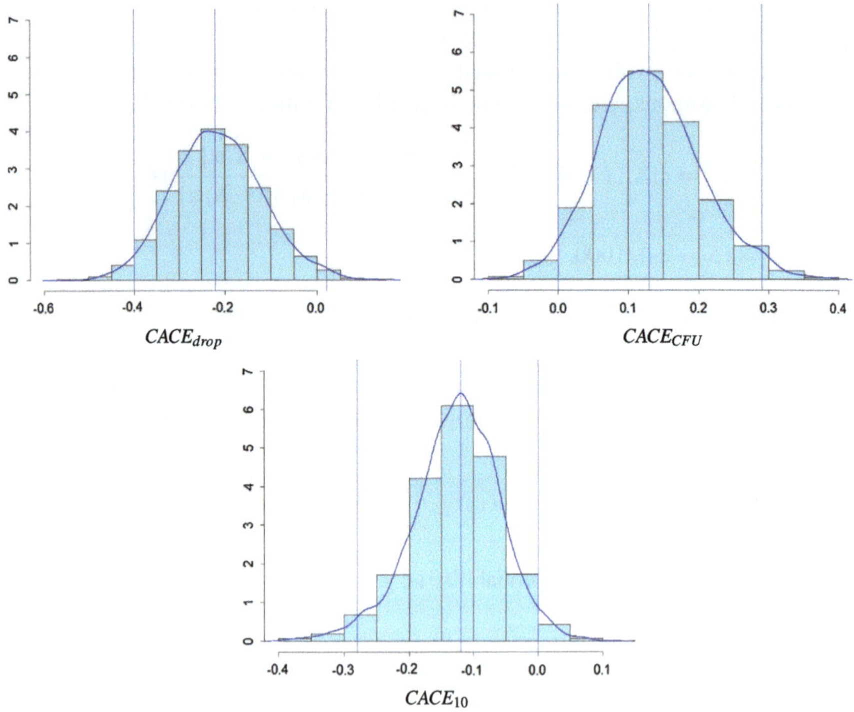

Fig. 1 Plot of the posterior density of $CACE_{drop}$, $CACE_{CFU}$ and $CACE_{10}$ in the subpopulation defined by bandwidth $h = 1000$

We use MCMC methods to derive the posterior distribution of the parameters, $\boldsymbol{\theta}$, and we run 62500 iterations, burning 12500 and saving each 10.

Our interest is on the Complier Average Causal Effect (CACE) relatively to the category $\mathbf{Y} = (1, 0)$, that is a measure of the effect of the grants on the university careers (Eq. 3). Nevertheless, we can obtain the CACE for dropout and CFU, respectively, by marginalization.

Table 4 summarizes the quantiles of the posterior distribution of the three causal estimands, across the three subpopulations defined by the bandwidths $h = 500$, 1000, 1500. Figure 1 shows the posterior density of the three CACE for the subpopulation defined by the bandwidth $h = 1000$.

Comparing the results obtained by the Randomization approach and by the Bayesian approach we can observe that they do not differ so much. Our data provide some evidence of positive effect of the university grants on the accademic career of the students: the university grant reduces dropout and prompts students to successfully take exams.

References

1. Li, F., Mattei, A., Mealli, F.: Evaluating the effect of university grants on student dropout: evidence from a regression discontinuity design using bayesian principal stratifications analysis. Ann. Appl. Stat. **9**(4), 1906–1931 (2015)
2. Rubin, D.B.: More powerful randomization-based p-values in double-blind trials with non-compliance. Stat. Med. **17**, 371385 (1998)
3. Lee, J.J., Miratrix, L., Pillai, S.N.: More Powerful Multiple Testing in Randomized Experiments with Non-compliance (Unpublished manuscript) (2016)
4. Berry, S.M., Berry, D.A.: Accounting for multiplicities in assessing drug safety: a three level hierarchical mixture model. Biometrics **60**(2), 418–426 (2004)

Bayesian Methods for Microsimulation Models

Consuelo R. Nava, Cinzia Carota and Ugo Colombino

Abstract This article proposes Bayesian methods for microsimulation models and for policy evaluations. In particular, the Bayesian Multinomial Logit and the Bayesian Multinomial Mixed Logit models are presented. They are applied to labour-market choices by single females and single males, enriched with EUROMOD microsimulated information, to evaluate fiscal policy effects. Estimates using the two Bayesian models are reported and compared to the results stemming from a standard approach to the analysis of the phenomenon under consideration. Improvements in model performances, when Bayesian methods are introduced and when random effects are included, are outlined. Finally, ongoing work, based on nonparametric model extensions and on analysis of work choices by couples is briefly described.

Keywords Bayesian Inference · Microsimulation · Multinomial Logit Model · Multinomial Mixed Logit Model · Random Utility Models

1 Introduction

In recent decades, describing and learning from individual choice behaviours have become increasingly important in social sciences, especially in microeconomics and marketing. In the presence of mutually exclusive discrete alternative choices, well-established random utility models (RUM) [35] are employed and referred to as "discrete choice models". Indeed, they provide an interesting extension of the classic theory of utility maximization among multiple discrete alternatives, with challenging theoretical and empirical statistical implications. For an extensive and

C.R. Nava (✉) · C. Carota · U. Colombino
Dipartimento di Economia e Statistica Cognetti De Martiis, Università degli studi di Torino, Lungo Dora Siena 110, 10153 Torino, Italy
e-mail: consuelorubina.nava@unito.it

C. Carota
e-mail: cinzia.carota@unito.it

U. Colombino
e-mail: ugo.colombino@unito.it

R. Argiento et al. (eds.), *Bayesian Statistics in Action*, Springer Proceedings in Mathematics & Statistics 194, DOI 10.1007/978-3-319-54084-9_18

general review of discrete choice models applied to economic fields see [6, 43, 49]. Moreover, the economic literature takes advantage of RUMs also for policy evaluations [1, 2, 8, 9, 50] which, in general, are performed by comparing events before and after the policy's implementation. However, recently, public policy implications have also been explored by means of microsimulation models, which are often used to anticipate and estimate the effects of socio-economic interventions. They are, in fact, powerful tools with which to predict effects of tax and benefit reform interventions [3, 11, 15, 16, 18, 32, 46]. Such microsimulation models combine results from observational studies, real data and expert opinions to simulate decision-maker choices statically or dynamically. Here we make use of the static EUROMOD model for tax and benefits, briefly introduced in Sect. 3 and based on the Italian Survey on Household Income and Wealth (SHIW) data. As well known, microsimulation matched with discrete choice models has high potentialities [9] especially in the evaluation of public programs and in the estimation of labour supply reactions.[1] Moreover, the nature of the data and the research question suggest that it is necessary to capture the heterogeneity of decision-makers, as well as to achieve a high level of accuracy in the covariate effect estimation.

Within this framework, we propose more flexible RUMs obtained by adopting a Bayesian approach, in order to increase flexibility in the use, interpretation, and estimation of microsimulation models. In general, Bayesian procedures on the one hand do not require any function maximization, and on the other, achieve consistency and efficiency under assumptions milder than those of classic procedures. The interest in Bayesian methods for social sciences, therefore, is rapidly growing in order to deal with both theoretical, see e.g. [7, 25, 29, 31, 41], and empirical, see e.g. [13, 24, 26], problems. In particular, considerable research attention is paid to parametric [4] and non parametric [12, 19, 39] prior elicitation for multinomial discrete choice models.

Here we focus on a Bayesian hierarchical interpretation of a Multinomial Logit model (MLM)[2] for categorical variables [37]. We then enrich this model by introducing random effects. The resulting Multinomial Mixed Logit model (MMLM) is endowed with suitable priors in Sect. 2. Recent studies, in fact, have shown that a Bayesian approach guarantees a higher level of tractability and accuracy, as well as an easier implementation of discrete choice models [39, 49]. This applies to the entire class of Generalized Linear Mixed models [36], of which the Multinomial Mixed Logit model is a special case.

Hence, the new applied contributions of this article are: (i) Bayesian methods integrated into microeconometric models addressed also to the analysis of (partially) simulated data; (ii) Bayesian estimates more interpretable and accurate than those obtained under the usual frequentist assumptions [15, 16]; (iii) introduction of a general framework enabling potential advances in (static) microeconomic simulation

[1]For evaluations of local reforms see, for instance [2, 15, 16] for Italy, [3] for Italy, Norway and Sweden, [32] for Spain, [5] for France and [10] for Germany.

[2]For the sake of simplicity, we identify the Multinomial Logit model (MLM) with a general model in which the underlying utilities depend on individual characteristics, choice attributes and/or variables combining individuals and choices. Sometimes, in econometrics, the latter is considered to be a generalization of the MLM combined with a Conditional Logit model.

studies; (iv) flexible Bayesian methods for (fiscal) policy evaluations based on more parsimonious models which, in turn, are based on assumptions milder than those of classic studies [2, 9, 15, 16, 32, 46]. To the best of our knowledge, in fact, this is one of the few applications of Bayesian procedures to microsimulation models (excluding calibration issues [44]), as well as to policy evaluations [17, 33].

The article is organized as follows. Section 2 presents the Bayesian multinomial models applied and comparatively discussed in Sect. 3, where a brief description of the data is also provided. Some final comments and an outline of ongoing work conclude the article.

2 Models

The microeconomic problem addressed in what follows is based on the jth ($j = 1, \ldots, J$) decision-maker, which selects the ith choice among a finite set $C = \{1, \ldots, I\}$ of mutually exclusive and exhaustive alternatives. This can be described by a binary random variable Y_{ji}, driven by a random utility maximization, so that

$$\pi_{ji} = \mathsf{Pr}(Y_{ji} = 1|C) = \mathsf{Pr}\left(\mathsf{U}_{ji} = \max_{h=1,\ldots,I} \mathsf{U}_{jh}\middle|C\right) \quad \text{with } \mathsf{U}_{ji} = \mathbf{x}_{ji}^{'}\boldsymbol{\beta} + \varepsilon_{ji}. \quad (1)$$

In the linear utility function U_{ji}, $\mathbf{x}_{ji}$ represents the $r \times 1$ vector of observed explanatory variables (for individual j and choice i), $\boldsymbol{\beta}$ is a $r \times 1$ vector of fixed effects and ε_{ji} is an error component. Both $\mathbf{x}_{ji}$ and ε_{ji} can be individual-specific for the jth decision-maker and/or choice-specific according to i, characterizing U_{ji} in (1). In all cases $\mathbf{x}'_{ji}\boldsymbol{\beta}$ represents the systematic part of the utility function, i.e. the representative utility, while ε_{ji} is the stochastic one.

The selection of an error distribution leads to different econometric models. Here we are interested in i.i.d. standard Gumbel (or Extreme Value Type I [30]) errors ε_{ji} that lead to the MLM [37], i.e. to a special Generalized Linear model. Interestingly, according to the extreme value theory, the asymptotic distribution of the maximum of $\varepsilon_{j1}, \ldots \varepsilon_{jI}$ i.i.d standard normal random variables converges to a standard Gumbel distribution (see the Fisher-Tippett-Gnedenko theorem [23, 27]). Therefore, its use is close to assuming independent normal errors, except that the heavy tails allow more robust analyses taking into account a "slightly more aberrant behavior than the normal" (p. 39 [49]), crucial for RUMs. Moreover, the advantages of this error component specification are that the difference between two Extreme Value Type I random variables is a Logistic[3] random variable and that the Extreme Value Type I

[3] It can be proved that the difference between two standard i.i.d. Gumbel random variables is a Logistic by means of the characteristic function (c.f.). In general, if $\varepsilon \sim \mathsf{Gumbel}(0, 1)$, its characteristic function is $\phi_\varepsilon(t) = \mathsf{E}(e^{it\varepsilon}) = \Gamma(1 - it)$. Thus, $\phi_{\varepsilon_i - \varepsilon_j}(t) = \phi_{\varepsilon_i}(t)\phi_{\varepsilon_j}(-t) = \Gamma(1 - it)\Gamma(1 + it) = \Gamma(1 - it)\Gamma(it)it$. By Euler's reflection formula $\Gamma(z)\Gamma(1 - z) = \frac{\pi}{\sin(\pi z)}$ and by property $i^2 = -1$, hence $\phi_{\varepsilon_i - \varepsilon_j}(t) = \frac{\pi i t}{\sin(\pi i t)} = \frac{\pi t}{-i \sin(\pi i t)}$ which is the c.f. for the Logistic distribution.

is closed under maximization. Therefore, the choice probability (1) becomes[4]:

$$\pi_{ji} = \frac{\exp\{\mathbf{x}_{ji}^{'}\boldsymbol{\beta}\}}{\sum_{h=1}^{I}\exp\{\mathbf{x}_{jh}^{'}\boldsymbol{\beta}\}}. \tag{2}$$

As said, we propose to embed result (2) in the Bayesian hierarchical MLM described below and implemented in Sect. 3:

$$Y_{ji} \sim \mathsf{Bern}(\pi_{ji}) \ \forall\, j = 1, \dots, J \ \text{ and } \ i = 1 \dots, I \tag{3}$$

$$\mathsf{logit}(\pi_{ij}) = \mathbf{x}_{ji}^{'}\boldsymbol{\beta} \tag{4}$$

$$\boldsymbol{\beta} \sim \mathsf{N}(\mu_{\beta}, V_{\beta}). \tag{5}$$

Successively, we weaken the assumption (4) by adding a random component[5] according to individual-specific and/or choice-specific features:

$$\mathsf{logit}(\pi_{ji}) = \mathbf{x}_{ji}^{'}\boldsymbol{\beta} + \mathbf{w}_{ji}\mathbf{b} \tag{6}$$

$$\mathbf{b} \sim \mathsf{N}(0, V_b); \quad V_b \sim \mathsf{IW}(\Psi, \nu) \tag{7}$$

[4]The form of the choice probability under a MLM model follows from the representation of $\mathsf{Pr}(Y_{ji}|C)$ as $\mathsf{Pr}(\mathsf{U}_{ji} > \mathsf{U}_{jh}\ \forall\, h{=}1, \dots, I\ \ h \neq i)$, which, for $h \neq i$, reduces to

$$\begin{aligned}\pi_{ji} &= \mathsf{Pr}(\mathbf{x}_{ji}^{'}\boldsymbol{\beta}+\varepsilon_{ji} > \mathbf{x}_{jh}^{'}\boldsymbol{\beta}+\varepsilon_{jh}\ \forall\, h{=}1, \dots, I) = \mathsf{Pr}(\varepsilon_{ji} - \varepsilon_{jh} > \mathbf{x}_{jh}^{'}\boldsymbol{\beta} - \mathbf{x}_{ji}^{'}\boldsymbol{\beta}\ \forall\, h = 1, \dots, I)\\ &= \int_{\varepsilon} \mathsf{I}(\varepsilon_{ji} - \varepsilon_{jh} > \mathbf{x}_{jh}^{'}\boldsymbol{\beta} - \mathbf{x}_{ji}^{'}\boldsymbol{\beta}\ \forall\, h = 1, \dots, I) f(\boldsymbol{\varepsilon}_j) d\boldsymbol{\varepsilon}_j\end{aligned}$$

Assuming that the errors are i.i.d. Gumbel distributed and resorting to the substitution $t = \exp(-\varepsilon_{ji})$

$$\begin{aligned}\pi_{ji} &= \mathsf{Pr}(Y_{ji} = 1|C, \varepsilon_{ji})\mathsf{Pr}(\varepsilon_{ji}) = \int_{-\infty}^{\infty}\left(\prod_{h\neq i} e^{-e^{-(\varepsilon_{ji}+\mathbf{x}_{ji}^{'}\beta-\mathbf{x}_{jh}^{'}\beta)}}\right) e^{-\varepsilon_{ji}} e^{-e^{-\varepsilon_{ji}}} d\varepsilon_{ji}\\ &= \int_{-\infty}^{\infty}\left(\prod_{h} e^{-e^{-(\varepsilon_{ji}+\mathbf{x}_{ji}^{'}\beta-\mathbf{x}_{jh}^{'}\beta)}}\right) e^{-\varepsilon_{ji}} d\varepsilon_{ji} = \int_{-\infty}^{\infty} \exp\left\{\sum_{h} e^{-(\varepsilon_{ji}+\mathbf{x}_{ji}^{'}\beta-\mathbf{x}_{jh}^{'}\beta)}\right\} e^{-\varepsilon_{ji}} d\varepsilon_{ji}\\ &= \int_{-\infty}^{\infty} \exp\left\{-e^{-\varepsilon_{ji}}\sum_{h} e^{-(\mathbf{x}_{ji}^{'}\beta-\mathbf{x}_{jh}^{'}\beta)}\right\} e^{-\varepsilon_{ji}} d\varepsilon_{ji} = \int_{0}^{\infty} \exp\left\{-t\sum_{h} e^{-(\mathbf{x}_{ji}^{'}\beta-\mathbf{x}_{jh}^{'}\beta)}\right\} dt\\ &= \left.\frac{\exp(-t\sum_{h} e^{-(\mathbf{x}_{ji}^{'}\beta-\mathbf{x}_{jh}^{'}\beta)})}{-\sum_{h} e^{-(\mathbf{x}_{ji}^{'}\beta-\mathbf{x}_{jh}^{'}\beta)}}\right|_{0}^{\infty} = \frac{1}{\sum_{h} e^{-(\mathbf{x}_{ji}^{'}\beta-\mathbf{x}_{jh}^{'}\beta)}} = \frac{\exp\{\mathbf{x}_{ji}^{'}\boldsymbol{\beta}\}}{\sum_{h=1}^{I}\exp\{\mathbf{x}_{jh}^{'}\boldsymbol{\beta}\}}\end{aligned}$$

as proved in [20] p. 63 and in [49], pp. 78–79.

[5]Alternatively, as suggested in [49] p. 143, random coefficients can be simply considered as part of the utility error component, inducing correlations among alternative utilities.

where $\mathbf{w}_{ji}$ is a $s \times 1$ known design vector, while $\mathbf{b}$ is a vector of s random effects. We also assign an Inverse Wishart (IW) prior distribution[6] to the random effect (co)variance matrix V_b.[7] According to the design of $\mathbf{w}_{ji}$, which can contain dummy variables, covariates or both, the MMLM may exhibit, respectively, random intercepts, random slopes, or both [45].

This extension is known as the MMLM (see for instance [38, 49]) and it overcomes some Logit model limitations given that it allows for random taste variation as well as substitution patterns.

The main econometric advantages of the MMLM are, on the one hand, the absence of the independence assumption from irrelevant alternatives (see [49] pp. 49–50), and on the other, the possibility to account for individual heterogeneity.[8] Moreover, in the presence of the mixing distribution characterizing the MMLM, a Bayesian approach guarantees significant computational advantages [39, 43, 49].

The next section describes the application of a Bayesian MLM and a Bayesian MMLM to EUROMOD simulations based on SHIW data.

3 Data and Results

For model and method comparisons, we used EUROMOD, a static tax-benefit microsimulation model covering member states of the European Union (here we consider the Italian case), for fiscal reform effect analysis.[9] The simulations were based on individual micro-data representative of the national population of interest, given the European Union Statistics on Income and Living Conditions (EU-SILC) data [48]. Furthermore, this model makes it possible to store simulated and real data, respectively from EUROMOD and SHIW, at a micro level, which enabled the statistical analyses presented in Sect. 2. To this end, we used the illustrated micro-econometric model of household labour supply (1).

We restricted the study to households with only one decision-maker, i.e. singles, ignoring couple data. Agents were aged between 20 and 55, neither retired nor students. In particular, we observed 657 singles, 291 women and 366 men, who were given the possibility to choose among 10 different types of jobs (i.e. $i = 1, \ldots, 10$) and non labour-market participation (indexed as job 0). The inclusion of self-employment, moreover, required wage rates independent from hours of work. Variables used as predictors were: weekly hours of work, gross wages, age

[6]Some other priors less informative than the Inverse-Wishart can also be considered following [28].

[7]This is one way among others to overcome possible limitations [47] due to the restrictive i.i.d. assumption about the random component of the utility function in the MLM. Hence, MMLM allows a proper representation of choice behaviour.

[8]This is a crucial feature if microsimulation models have to evaluate policy impacts properly [9].

[9]The fiscal policies considered were the ones measured in the EUROMOD model, applied to real data provided by the Bank of Italy from the SHIW-1998. Thus, data were partially microsimulated by EUROMOD, according to the 1998 Italian fiscal policy.

class, taxes and benefits. The last two variables were simulated with EUROMOD and the age class was considered as the grouping variable for random (job) intercepts in the MMLM. Other details on data description can be found in [15, 16].

A Chi square test was performed to confirm differences in the 10+1($= I$) job type frequencies in the two sub-populations (single females and single males). P-value $< 2.2 \cdot 10^{-16}$ reinforce and justified our decision to make use of Multinomial models.

Bayesian estimates were obtained by Markov Chain Monte Carlo (MCMC) methods: a Gibbs sampling algorithm (a special case of the Metropolis-Hastings algorithm) was used to update most of model parameters. Hence, models were run for 50000 iterations with a burn-in phase of 15000 and a thinning interval equal to 5. Standard diagnostic tools confirmed the convergence of runs. Hyperprior parameters were set to be as follows: $\mu_\beta = 0$, $V_\beta = \mathbf{I}_4 \cdot 10^{10}$, with $\mathbf{I}_4$ denoting a 4×4

Table 1 Classic (first two columns) and Bayesian (last four columns) point estimates for single female and single men, without and with random intercepts (grouped by age classes). In the Bayesian case, they are denoted by "*" and "." respectively when the corresponding 95% and 90% HPD intervals are bounded away from zero

	MLM		Bayesian MLM		Bayesian MMLM	
Variable names	Female	Men	Female	Men	Female	Men
Intercept job 1	−3.39307 *	−1.76703 *	−2.52738 *	−1.61559 *	−2.70898 *	−1.12120 *
Intercept job 2	−1.50327	−1.59548	−2.34357 *	−1.59101 *	−2.83937 *	−1.74396 *
Intercept job 3	0.60008	0.15216	−2.25951 *	−1.66466 *	−2.83933 *	−1.41215 *
Intercept job 4	0.99819	0.54394	−2.36011 *	−1.56762 *	−2.62890 *	−1.33762 *
Intercept job 5	3.50702	3.42909	−2.10496 *	−1.41377 *	−2.68308 *	−0.59950 *
Intercept job 6	2.38043	2.43712	−2.41557 *	−1.63019 *	−2.89699 *	−1.33678 *
Intercept job 7	1.39848	2.67504	−2.20461 *	−1.67959 *	−2.71274 *	−1.41510 *
Intercept job 8	3.05579	2.49211	−2.32966 *	−1.63187 *	−2.91137 *	−1.06941 *
Intercept job 9	2.78303	1.92209	−2.31842 *	−1.70870 *	−2.65600 *	−1.31812 *
Intercept job 10	3.27083	2.16576	−2.16333 *	−1.79336 *	−2.70004 *	−1.07083 *
Hours	−0.07634	−0.05002	0.06969 *	0.04348 *	0.07353 *	0.03383 *
Wage	0.06270 *	0.03776 *	0.06693 *	0.04275 *	0.06225 *	0.04295 *
Taxes	−0.00082 *	−0.00052 .	−0.00118 *	−0.00072 *	−0.00103 *	−0.00060 *
Benefits	0.00229 *	0.00254 .	0.00079 .	−0.00094	0.00052	0.00143 .

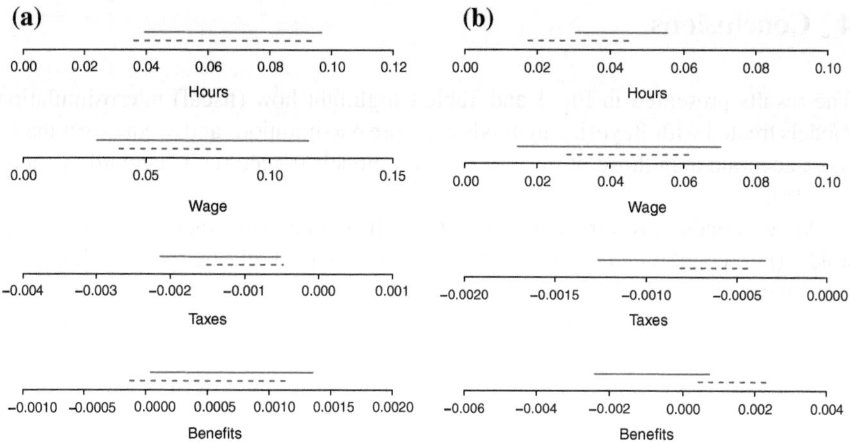

Fig. 1 Single female (**a**) and single male (**b**) 95% HPD intervals given the Bayesian Multinomial Logit model (*red - continuous line*) and the Bayesian Multinomial Mixed Logit model (*blue - dashed line*) for hours, wages, taxes and benefits variables

identity matrix. The residual covariance matrix was $\frac{1}{I} \cdot (\mathbf{I}_{I-1} + \mathbf{U})$, where $\mathbf{I}_{I-1}$ is a $(I-1) \times (I-1)$ identity matrix and $\mathbf{U}$ is a $(I-1) \times (I-1)$ unit matrix, as suggested in [28], given I as the number of possible choices. Finally, for the inverse-Wishart prior, V_b was set to be equal to $\mathbf{I}_{I-1}$.

Models did not include a global intercept, so that the first 10 estimated coefficients represented actual job type specific intercepts. Moreover, to enhance differences among the employment behaviours of single men and single women, we estimated model parameters separately for these two sub-populations. Point estimates are set out in Table 1 under the two Bayesian models (last four columns) and compared with the classic estimates of the MLM (first two columns), using the R package `MCMCglmm` [28].

These estimates are even more encouraging when compared to those of the standard models usually applied to such partially microsimulated data (see [16]). Our Bayesian models, in fact, were more parsimonious with respect to [16] and produced more reasonable results: for instance, significant 95% highest posterior density (HPD) intervals for the majority of job type intercepts and for the amount of working hours (see Table 1). Moreover, Fig. 1 shows that MMLM results in tighter HPD intervals [13, 31] for both single female and single male estimates. Therefore the introduction of random effects into the Bayesian model yields more accurate estimates while still preserving the significant 95% HPD intervals. This guarantees higher precision when comparing different tax and benefit reform scenarios, as well as when computing labour supply elasticity.

4 Conclusions

The results presented in Fig. 1 and Table 1 highlight how (fiscal) microsimulation models treated with Bayesian methods are more parsimonious and produce estimates more accurate than those of microsimulation models treated with standard methods [15, 16].

We are conducting further research to evaluate decisions undertaken by households (i.e. couples) and to compare labour demand elasticity with Bayesian and frequentist methods. A Bayesian approach may also have advantages in terms of model convergence, given the higher number of levels (11×11) in the dependent variable. Moreover, to enhance population heterogeneity, the proposed methods and suitably adjusted models are applied to a sample including also the inactive population (pensioners, students, etc.) besides the active one considered in the present article [32].

Finally, we are currently considering model extensions obtained by assigning Bayesian nonparametric priors (like, for instance, the widely-used Dirichlet process [12–14, 19, 43] or other more sophisticated process priors, see e.g. [22, 34, 40, 42]) to random effects, so as to induce, among other things, more accurate estimates as suggested in [13, 26, 31].

References

1. Aaberge, R., Colombino, U., Strøm, S.: Labor supply responses and welfare effects of tax reforms. Scand. J. Econ. **97**(4), 635–659 (1995)
2. Aaberge, R., Colombino, U., Strøm, S.: Labor supply in Italy an empirical analysis of joint household decisions, with taxes and quantity constraints. J. Appl. Econ. **14**, 403–422 (1999)
3. Aaberge, R., Colombino, U., Strøm, S.: Welfare effects of proportional taxation: empirical evidence from Italy, Norway and Sweden. J. Popul. Econ. **13**, 595–621 (2000)
4. Athey, S., Imbens, G.: Discrete Choice Models with Multiple Unobserved Choice Characteristics. Working paper, Harvard University (2007)
5. Bargain, O.: On Modelling Household Labour Supply with Taxation. IZA DP No. 1455 (2005)
6. Ben-Akiva, M., Lerman, S.R.: Discrete Choice Analysis. MIT Press, Cambridge (1985)
7. Bernardo, J., Smith, A.F.M.: Bayesian Theory. Wiley, New York (1994)
8. Blundell, R., MaCurdy, T.: Labour supply: a review of alternative approaches. In: Ascenfelter, O., Card, D. (eds.) Handbook of Labour Economics, pp. 1559–1695. North Holland, Amsterdam (1999)
9. Blundell, R., Duncan, A., McCrae, J., Meghir, C.: The labour market impact of the working families - tax credit. Fisc. Stud. **21**, 75–104 (2000)
10. Bonin, H., Kempe, W., Schneider, H.: Household Labour Supply Effects of Low-Wage Subsidies in Germany. IZA DP, No. 637 (2002)
11. Bourguignon, F., Spadaro, A.: Microsimulation as a tool for evaluating redistribution policies. J. Econ. Inequal. **4**, 77–106 (2006)
12. Burda, M., Harding, M., Hausman, J.: A Bayesian mixed logit-probit model for multinomial choice. Technical Report, University of Toronto, Department of economics (2008)
13. Carota, C., Nava, C.R., Soldani, I., Ghiglione, C., Schiaparelli, S.: Statistical models for species richness in the Ross Sea. In: Proceedings of the GRASPA2015 Conference, Bari, pp. 49–52 (2015)

14. Carota, C., Filippone, M., Leombruni, R., Polettini, S.: Bayesian nonparametric disclosure risk estimation via mixed effects log-linear models. Ann. Appl. Stat. **9**, 525–546 (2015)
15. Colombino, U.: A new equilibrium simulation procedure with discrete choice models. Int. J. Microsimul. **6**(3), 25–49 (2013)
16. Colombino, U.: Five crossroads on the way to basic income. An Italian tour. Ital. Econ. J. **1**, 353–389 (2015)
17. Cooper, N.J., Sutton, A.J., Abrams, K.R.: Decision analytical economic modelling within a Bayesian framework: application to prophylactic antibiotics use for caesarean section. Stat. Methods Med. Res. **11**, 491–512 (2002)
18. Creedy, J., Duncan, A.S.: Behavioural microsimulation with labour supply responses. J. Econ. Surv. **16**(1), 1–38 (2002)
19. De Blasi, P., James, L., Lau, J.: Bayesian nonparametric estimation and consistency of mixed multinomial Logit choice models. Bernoulli **16**, 679–704 (2010)
20. Domencich, T.A., McFadden, D.: Urban Travel Demand: A Behavioral Analysis. North Holland Publishing, Amsterdam (1975)
21. Dufrese, F., Gerber, H.U., Shiu, E.S.W.: Risk theory and the gamma process. ASTIN Bull. **10**, 47–53 (1991)
22. Favaro, S., Lijoi, A., Nava, C.R., Nipoti, B., Prünster, I., Teh, Y.W.: On the stick-breaking representation for homogeneous NRMIs. Bayesian Anal. **11**, 697–724 (2016)
23. Fisher, R.A., Tippett, L.H.C.: Limiting forms of the frequency distribution of the largest or smallest member of a sample. Proc. Camb. Philos. Soc. **24**, 180–290 (1928)
24. Gelman, A., Carlin, J.B., Stern, H.S., Rubin, D.B.: Bayesian Data Analysis, 2nd edn. Chapman & Hall/CRC, Boca Raton (2003)
25. Gill, J.: Bayesian Methods: A Social Behavioral Sciences Approach, Third edn. Chapman and Hall/CRC, Boca Raton (2014)
26. Gill, J., Casella, G.: Nonparametric priors for Ordinal Bayesians Social Science Models: Specification and Estimation. Journal of the American Statistical Association **104**, 453–464 (2009)
27. Gnedenko, B.V.: Sur la distribution limite du terme maximum d'une série aléatoire. Ann. Math. **44**, 423–453 (1943)
28. Hadfield, J.D.: MCMC methods for multi-response generalized linear mixed models. The MCMCglmm R package. J. Stat. Softw. **33**(2), 1–22 (2009)
29. Jeffreys, S.H.: Theory of Probability, 3rd edn. Clarendon Press, Oxford (1998)
30. Johnson, N., Kotz, S.: Distributions in Statistics: Continuous Univariate Distributions. Wiley, New York (1970)
31. Kyung, M., Gill, J., Casella, G.: Estimation in Dirichlet random effects models. Ann. Stat. **28**, 979–1009 (2010)
32. Labeaga, J., Oliver, X., Spadaro, A.: Discrete choice models of labour supply, behavioural microsimulation and the Spanish tax reforms. J. Econ. Inequal. **6**(3), 247–273 (2008)
33. Li, G., Haining, R., Richardson, S., Best, N.: Evaluating Neighbourhood Policing using Bayesian Hierarchical Models: No Cold Calling in Peterborough, England, London: Imperial College. http://eprints.ncrm.ac.uk/1824/ (2011). Accessed 12 Aug 2016
34. Lijoi, A., Prünster, I.: Models beyond the Dirichlet process. In: Bayesian Nonparametrics. Cambridge Series in Statistical and Probabilistic Mathematics, pp. 80–136. Cambridge University Press, Cambridge (2010)
35. Marschak, J.: Binary choice constraints on random utility indications. In: Arrow, K. (ed.) Stanford Symposium on Mathematical Methods in the Social Sciences, pp. 312–329. Stanford University Press, Stanford (1960)
36. McCullagh, P., Nelder, J.A.: Generalized Linear Models, second edn. CRC Press, Boca Raton (1989)
37. McFadden, D.: Conditional logit analysis of qualitative choice behavior. In: Zarembka, P. (ed.) Frontiers in Econometrics, pp. 105–142. Academic Press, New York (1974)
38. McFadden, D., Train, K.E.: Mixed MNL models of discrete response. J. Appl. Econ. **15**, 447–470 (2000)

39. Nava, C.R.: Some advances in Bayesian modeling and computation. Ph.D thesis (2014)
40. Polettini, S.: A generalized semiparametric Bayesian Fay-Herriot model for small area estimation shrinking both means and variances. Bayesian Analysis (to appear) (2016)
41. Robert, C.: The Bayesian Choice: From Decision Theoretic Foundations to Computational Implementation, Second edn. Springer, Berlin (2007)
42. Roychowdhury, A., Kulis, B.: Gamma Processes, Stick-Breaking, and Variational Inference. arXiv preprint arXiv:1410.1068 (2014)
43. Rossi, P., Allenby, G., McCulloch, R.: Bayesian Statistics and Marketing. Wiley, New York (2005)
44. Rutter, C.M., Miglioretti, D.L., Savarino, J.E.: Bayesian calibration of microsimulation models. J. Am. Stat. Assoc. **104**, 1338–1350 (2009)
45. Skrondal, A., Rabe-Hesketh, S.: Generalized Latent Variable Modeling. Chapman & Hall (CRC), Boca Raton (2004)
46. Spadaro, A.: Micro-simulation and normative policy evaluation: an application to some EU tax benefits systems. J. Public Econ. Theory **7**(4), 593–622 (2005)
47. Stopher, P., Meyburg, A., Brög, W.: Travel behaviour research: a perspective. In: Stopher, P., Meyburg, A., Brög, W. (eds.) New Horizons in Travel Behaviour Research. Lexington Books, Lexington (1981)
48. Sutherland, H., Figari, F.: EUROMOD: the European Union tax-benefit microsimulation model. Int. J. Microsimul. **6**(1), 4–26 (2013)
49. Train, K.E.: Discrete Choice Methods with Simulation. Cambridge University Press, Cambridge (2003)
50. van Soest, A.: Structural models of family labor supply: a discrete choice approach. J. Hum. Res. **30**, 63–88 (1995)

A Bayesian Model for Describing and Predicting the Stochastic Demand of Emergency Calls

Vittorio Nicoletta, Ettore Lanzarone, Alessandra Guglielmi, Valérie Bélanger and Angel Ruiz

Abstract Emergency Medical Service (EMS) systems aim at providing immediate medical care in case of emergency. A careful planning is a major prerequisite for the success of an EMS system, in particular to reduce the response time to emergency calls. Unfortunately, the demand for emergency services is highly variable and uncertainty should not be neglected while planning the activities. Thus, it is of fundamental importance to predict the number of future emergency calls and their interarrival times to support the decision-making process. In this paper, we propose a Bayesian model to predict the number of emergency calls in future time periods. Calls are described by means of a generalized linear mixed model, whose posterior densities of parameters are obtained through Markov Chain Monte Carlo simulation. Moreover, predictions are given in terms of their posterior predictive probabilities. Results from the application to a relevant real case show the applicability of the model in the practice and validate the approach.

Keywords Emergency medical services · Demand prediction · Generalized linear mixed model · Posterior predictive probabilities · Markov chain Monte Carlo

---

V. Nicoletta (✉) · A. Ruiz
Department of Operations and Decision Systems, Université Laval,
Quebec, Canada
e-mail: vittorio.nicoletta.1@ulaval.ca

A. Ruiz
e-mail: angel.ruiz@osd.ulaval.ca

E. Lanzarone
Consiglio Nazionale delle Ricerche (CNR), Istituto di Matematica Applicata e Tecnologie Informatiche (IMATI), Milan, Italy
e-mail: ettore.lanzarone@cnr.it

A. Guglielmi
Dipartimento di Matematica, Politecnico di Milano, Milan, Italy
e-mail: alessandra.guglielmi@polimi.it

V. Bélanger
Department of Logistics and Operations Management, HEC Montréal,
Montreal, Canada
e-mail: valerie.belanger@cirrelt.ca

R. Argiento et al. (eds.), *Bayesian Statistics in Action*, Springer Proceedings in Mathematics & Statistics 194, DOI 10.1007/978-3-319-54084-9_19

1 Introduction

Emergency Medical Service (EMS) consists of pre-hospital medical care and transport to a medical facility. Almost all EMS requests arrive by phone, through calls to an emergency number. The urgency of each request is evaluated and the location is obtained. Then, an ambulance is dispatched to the call site and, if needed, the patient is transported to a medical facility. Demand for such services is constantly increasing throughout the world, according to population growth and aging, while we observe a continuous pressure of governments to reduce health care costs; thus, an efficient use of resources is fundamental to guarantee a good quality of the service while maintaining the economic sustainability.

Several optimization planning models have been developed in the literature for EMS systems (see Bélanger et al. [3] for an extensive review). Unfortunately, the EMS demand is highly variable, and the uncertainty should not be neglected while planning the activities. Hence, it is fundamental to fairly predict the future number of emergency calls and their interarrival times.

The goal of this paper is thus to propose and validate a Bayesian model to predict the number of emergency calls in future time slots. The number of calls is described by means of a generalized linear mixed model, and the inference is based on the posterior density of model parameters, which is obtained through a Markov Chain Monte Carlo simulation scheme. Then, predictions are given in terms of their posterior predictive probabilities.

We demonstrate the applicability of the approach using the information available from the city of Montréal, Québec, Canada. Results show the convergence of the approach, good fitting, and low prediction errors.

The paper is organized as follows. A review of previous works dealing with stochastic modeling of EMS calls is presented in Sect. 2; the general features of an EMS system and the typical structure of the demand dataset are described in Sect. 3. Then, the Bayesian model is proposed in Sect. 4, and its application to the Montréal case is presented in Sect. 5. Conclusions of the work are finally given in Sect. 6.

2 Literature Review

Several studies deal with EMS calls prediction under a frequentist approach. An interesting survey of the works dated before 1982 can be found in Kamenetsky et al. [10]. In addition, the authors also presented regression models to predict EMS demand as a function of population, employment, and other demographic variables. Socio-economic parameters such as median income and percentage of people living below poverty line have been considered by Cadigan and Bugarin [5]. More recently, McConnell and Wilson [12] focused on the increasing impact of age distribution on EMS demand, while Channouf et al. [6] developed ARIMA models.

To the best of our knowledge, Bayesian approaches have not been considered for the EMS demand yet, even though they have been successfully applied in the health care literature. In fact, Bayesian approaches allow combining the available data with prior information within a solid theoretical framework, and results can be used as prior information once new data are available, which are important features in health applications. A good example of application to another health care service (i.e., the home care service) can be found in [1, 2].

3 Problem Description

An EMS system consists of an operations center and a certain number of ambulances, including the related staff. Ambulances are located in predetermined sites, ready to serve EMS requests. Requests arrive at the operations center via telephone, where they are evaluated. If a call requires an intervention, it is assigned to one of the available vehicles. The aim of an EMS is to serve all calls as fast as possible, maximizing the number of calls served within a given threshold that depends on the type of area (urban or rural).

For this purpose, due to the high uncertainty related to EMS calls, the decision maker needs accurate estimates of the demand as input for any optimization model underlying ambulance dispatching.

The typical EMS dataset includes several information about the calls and the provided service. For the aim of developing a prediction model, we focus on the calls. Three types of information are available:

- *Type*: required service and patient characteristics; this information is usually summarized into a priority associated to the call.
- *Arrival time*: day and time of the call.
- *Coordinates*: latitude and longitude of the call, or alternatively the address.

Usually, for managing purposes, the territory is divided into zones; thus, coordinates are translated into the zone z ($z = 1, \ldots, Z$) of the call. Moreover, concerning the arrival times, in this work we group the time into slots. Thus, day i ($i = 1, \ldots, I$) and slot t ($t = 1, \ldots, T$) are associated to the call, and for each day i we register the number of calls $N^i_{z,t}$ arisen in slot t and zone z. In particular, we consider slots of two hours, i.e., $T = 12$.

4 The Bayesian Model

We propose the following generalized linear mixed model for the number of calls $N^i_{z,t}$:

$$N^i_{z,t} | \lambda^i_{z,t} \overset{ind}{\sim} Poisson\left(\lambda^i_{z,t}\right) \tag{1}$$

$$log\left(\lambda^i_{z,t}\right) = \beta_1 p_z + \beta_2 a_z + \sum_{k=1}^{K} \beta_{3,k} \phi_{k,z} + \beta_4 h_i + \gamma_t \tag{2}$$

where: p_z and a_z are the population and the area of zone z, respectively; h_i is a binary covariate equal to 1 if day i is holiday and 0 otherwise; $\Phi_z = [\phi_{k,z}]$ is a dummy vector of dimension K describing the type of zone z.

Zones z are classified into $K + 1$ types (e.g., residential, commercial, industrial); $\phi_{k,z} = 1$ if zone z is of type k (with $k = 1, \ldots, K$) and 0 otherwise, while $\phi_{k,z}$ is always equal to 0 if zone z is of type $K + 1$, to avoid identifiability problems.

Model (1) and (2) is a generalized linear mixed model with four fixed-effects parameters β_1, β_2, $\boldsymbol{\beta_3}$ and β_4 (where $\boldsymbol{\beta_3}$ is K-dimensional), and a random-effects parameter γ_t. The latter takes into account the similarity of the number of calls in different zones during the same time slot t. In this formulation $\lambda^i_{z,t}$ is the parameter responsible for EMS calls: the higher the parameter $\lambda^i_{z,t}$ is, the higher the expected number of calls is.

Finally, independent non-informative priors, i.e., Gaussian distributions with 0 mean and large variance equal to 100, are chosen for β_1, β_2, β_4, γ_t, and for the components of vector $\boldsymbol{\beta_3}$:

$$
\begin{aligned}
\beta_j &\overset{iid}{\sim} \mathcal{N}(0, 100) \qquad && j = 1, 2, 4 \\
\beta_{3,k} &\overset{iid}{\sim} \mathcal{N}(0, 100) \qquad && k = 1, \ldots, K \\
\gamma_t &\overset{iid}{\sim} \mathcal{N}(0, 100) \qquad && \forall t
\end{aligned}
$$

5 Application to the Dataset

Data adopted in this work are those adopted in [4, 7, 11]. They refer to EMS calls arisen in the city of Montréal and the near suburb of Laval, Québec, Canada, i.e., a region with about 2.3 million of inhabitants and a territory of 744 km^2. According to these data, the region is divided into $Z = 595$ demand zones. In addition to the EMS data, information from Municipality of Montréal have been used to define the vector Φ_z for each zone. Eleven different types of zone are present, as described in Table 1; moreover, to avoid collinearity due to the low number of zones belonging to some types, types are regrouped as follows:

- *Residential* ($k = 1$);
- *Workplace*, regrouping commercial, office, industrial and institutional ($k = 2$);
- *Street* ($k = 3$);
- *Other*, regrouping park, agricultural, empty space, water, and golf field.

Finally, data about population has been divided by 1,000 to be of the same order of magnitude of the other covariates.

Table 1 Total number of calls and empirical mean divided by the type of zone

Type of zone	Number of zones	Total number of calls	Mean number of calls
Residential	266	95,394	358.62
Commercial	14	4,325	308.93
Office	7	3,361	480.14
Industrial	19	4,359	229.42
Institutional	46	18,004	391.39
Park	30	8,003	266.77
Street	184	67,738	368.14
Agricultural	4	506	126.50
Empty space	19	4,190	220.53
Water	2	405	202.50
Golf field	4	615	153.75

Table 2 Empirical mean and standard deviation of the number of calls divided by time slot

Time slot	Mean number of observations	Standard deviation of the number of observations
1	0.0606	0.2513
2	0.0535	0.2416
3	0.0409	0.2040
4	0.0529	0.2334
5	0.0908	0.3098
6	0.1039	0.3328
7	0.1010	0.3272
8	0.0991	0.3244
9	0.0937	0.3143
10	0.0915	0.3089
11	0.0893	0.3067
12	0.0776	0.2851

5.1 *Descriptive Statistics*

The dataset consists of 2,606,100 observations for $N^i_{z,t}$ ($I = 365$ days, $Z = 595$ zones and $T = 12$ slots) together with the related covariates.

Tables 1 and 2 report the main information about the data. Moreover, Fig. 1 shows a map of the territory together with the number of calls.

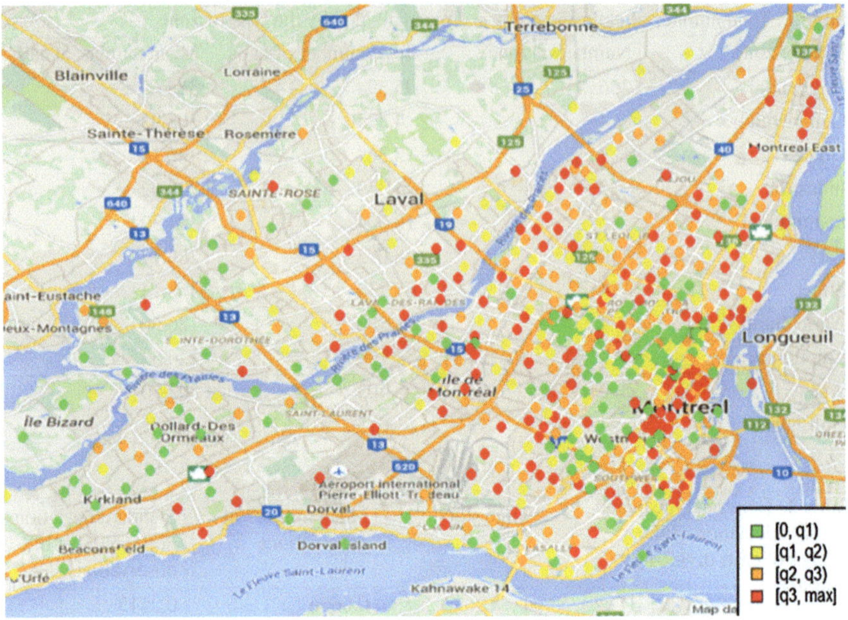

Fig. 1 Map of the city of Montréal together with the total number of calls. The number of calls for each zone is represented by a point in the center of the zone. *Green points* correspond to lower numbers of EMS calls, while *red points* represent higher numbers of EMS calls, according to the division in quartiles reported in the legend

5.2 *Posterior Analysis*

5.2.1 Convergence Analysis

The model is implemented in STAN (http://mc-stan.org/), which uses the Hamiltonian Monte Carlo algorithm to reduce the correlation and obtain faster convergence of the chains. Hence, 5,000 MCMC iterations have been run, with a burn-in of 1,000 iterations and a final sample size of 4,000.

Traceplots, autocorrelations and the Gelman–Rubin convergence statistics ($\hat{R}$) have been considered to verify that convergence is achieved. Moreover, we have estimated the Monte Carlo Standard Error (MCSE) with the MC error, the Naive SE and the Batch SE. See [8, 9] for further information.

Results show that $\hat{R}$ is equal to 1 and that the MCSE is always less than the 5% of the standard deviation for all parameters. Moreover, nice traceplots and autocorrelations are obtained, showing that the convergence of the chain is satisfactory.

5.2.2 Credible Intervals of Model Parameters

Inference for each model parameter is reported in terms of the posterior 95% credible interval (CI).

CIs of the fixed-effects parameters are reported in Table 3. The population parameter β_1 yields a positive effect, thus increasing number of calls, while the area parameter β_2 gives a negative effect. This is in agreement with the considered data, in which zones with large areas have small population densities; thus, the higher the population density of a zone is, the higher the number of calls is. Vector $\boldsymbol{\beta}_3$ gives the effect of the zone; results show that workplace zones and streets have more EMS calls, followed by Residential Zones. Finally, CI of parameter β_4 suggests that a lower number of calls is to be expected during holidays.

Posterior CIs for the random-effects vector γ_t are reported in Fig. 2. They suggest a clear distinction of the time slots: a higher number of calls arrive during the day (slots $t = 5, \ldots, 11$), while a lower number of calls arrive during night hours.

Table 3 95% CIs for the fixed-effects parameters

Parameter	Covariate	2.5%	50%	97.5%
β_1	Population	0.087	0.090	0.093
β_2	Area	−0.049	−0.047	−0.044
β_{31}	Residential	0.277	0.297	0.316
β_{32}	Workplace	0.347	0.369	0.389
β_{33}	Street	0.332	0.352	0.371
β_4	Holiday	−0.067	−0.055	−0.044

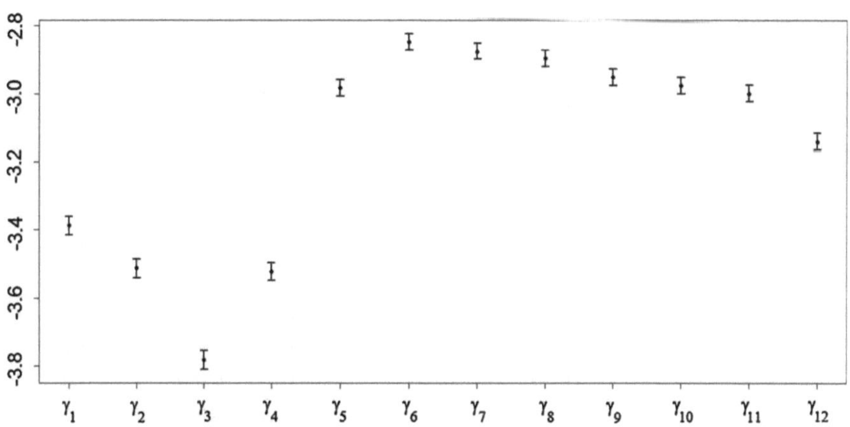

Fig. 2 95% CIs for the random-effects vector γ_t

5.2.3 Cross-Validation Prediction

A cross-validation approach is adopted to validate the model, by partitioning the complete dataset. The first 90% of the days (with $i = 1, \ldots, I - \tilde{I}$) is used as *training set* to determine the posterior density, while the remaining 10% (with $i = I - \tilde{I} + 1, \ldots, I$) is used as *testing set*. The predictive distributions of each $N^i_{z,t}$ (with $i = I - \tilde{I} + 1, \ldots, I$) are computed, and the predictions are checked with the corresponding observed data.

The accuracy of the predictions is evaluated in terms of the global Mean Absolute Error (MAE), defined as:

$$MAE = \frac{1}{\tilde{I} Z T} \sum_{i=I-\tilde{I}+1}^{I} \sum_{z=1}^{Z} \sum_{t=1}^{T} \left| N^{i\ obs}_{z,t} - \hat{N}^i_{z,t} \right|$$

where the product $\tilde{I}\, Z\, T$ is the numerousness of the sample in the testing set, and $N^{i\ obs}_{z,t}$ and $\hat{N}^i_{z,t}$ represent the observed number of calls and the number predicted by the model (median of the predictive distribution) at day i, zone z and slot t, respectively. The obtained value is 0.078, which is two orders of magnitude lower than the unit, showing a good fit of the model.

We have also detailed the MAE for each combination of type of zone k and time slot t. Results in Table 4 show quite similar values, whose maximum is 0.111, confirming a good fit of the model that does not significantly deteriorate for any pair k and t.

Table 4 MAE for each combination of type of zone k and time slot t

Type of zone k / Time slot t	0	1	2	3	All
1	0.035	0.061	0.060	0.062	0.059
2	0.036	0.048	0.046	0.051	0.048
3	0.025	0.043	0.036	0.043	0.040
4	0.033	0.060	0.057	0.056	0.056
5	0.066	0.103	0.100	0.096	0.097
6	0.063	0.111	0.111	0.104	0.104
7	0.054	0.105	0.104	0.102	0.099
8	0.052	0.102	0.103	0.104	0.098
9	0.061	0.093	0.093	0.097	0.091
10	0.057	0.093	0.085	0.092	0.088
11	0.063	0.087	0.083	0.089	0.084
12	0.047	0.077	0.074	0.078	0.074
All	0.049	0.082	0.079	0.081	**0.078**

Table 5 Comparison of the MAE between the proposed Bayesian model and the mean frequentist approach, grouped by type of zone z, time slot t, and holiday h

		Bayesian model	Frequentist mean
Type of zone	0	0.049	0.098
	1	0.082	0.152
	2	0.079	0.148
	3	0.081	0.153
Time slot	1	0.059	0.112
	2	0.048	0.096
	3	0.040	0.078
	4	0.056	0.103
	5	0.097	0.171
	6	0.104	0.188
	7	0.099	0.181
	8	0.098	0.178
	9	0.091	0.169
	10	0.088	0.164
	11	0.084	0.159
	12	0.074	0.141
Holiday	0	0.078	0.147
	1	0.077	0.142

5.2.4 Comparison with the Mean Estimate

In this Section we compare the outcomes of the proposed model with those of a very simple frequentist approach, in which the predictions are simply given by the historical means. This approach gives as a predictor the mean number of calls for the specific combination of type of zone z, time slot t and holiday parameter h. MAE values are computed considering the same training and testing sets as in Sect. 5.2.3.

The global MAE of the frequentist approach is equal to 0.145, while the values grouped by z, t and h are reported in Table 5. Results show that the MAE under the frequentist approach is higher, being about the double than the MAE under the proposed Bayesian approach. This further confirms the good fit of the proposed model to the data.

6 Conclusions

This paper presents a first attempt to deal with stochasticity in the EMS calls by using the Bayesian approach. A generalized linear mixed model has been proposed, with the aim of identifying relevant effects that influence the calls and giving predictions of future EMS calls.

Results from the Montréal case suggest that population, area and type of zone have a strong impact. Moreover, as expected, the time slot has a relevant effect, showing lower predicted values of number of calls during the night. Finally, the model shows good performance when used to make predictions, and documented by the low MAE values under cross-validation.

Moreover, the model is general, and can be easily applied to describe EMS demand in other cities. On the contrary, as for rural situations, we expect that some modifications are necessary to include the presence of rare events in an environment characterized by a usually low demand. Another extension will be to consider the area of each zone as an offset/exposure term of the Poisson regression.

References

1. Argiento, R., Guglielmi, A., Lanzarone, E., Nawajah, I.: A Bayesian framework for describing and predicting the stochastic demand of home care patients. Flex. Serv. Manuf. J. **28**, 254–279 (2016)
2. Argiento, R., Guglielmi, A., Lanzarone, E., Nawajah, I.: Bayesian joint modeling of the health profile and demand of home care patients. IMA J. Manag. Math. (2016). doi:10.1093/imaman/dpw001
3. Bélanger, V., Ruiz, A., Soriano, P.: Recent advances in emergency medical services management. Working document, Faculty of Business Administration of Laval University (2015)
4. Bélanger, V., Kergosien, Y., Ruiz, A., Soriano, P.: An empirical comparison of relocation strategies in real-time ambulance fleet management. Comput. Ind. Eng. **94**, 216–229 (2016)
5. Cadigan, R.T., Bugarin, C.E.: Predicting demand for emergency ambulance service. Ann. Emerg. Med. **18**, 618–621 (1989)
6. Channouf, N., L'Ecuyer, P., Ingolfsson, A., Avramidis, A.N.: The application of forecasting techniques to modeling emergency medical system calls in Calgary, Alberta. Health Care Manag. Sci. **10**, 25–45 (2007)
7. Crowe, J.: Une approche hybride pour la confection d'horaires des paramédics. Master thesis, HEC Montréal (2007)
8. Gelman, A., Rubin, D.B.: Inference from iterative simulation using multiple sequences (with discussion). Stat. Sci. **7**, 457–511 (1992)
9. Gelman, A., Carlin, J.B., Stern, H.S., Rubin, D.B.: Bayesian Data Analysis, 2nd edn. Chapman and Hall/CRC, Boca Raton (2003)
10. Kamenetsky, R., Shuman, L., Wolfe, H.: Estimating need and demand for prehospital care. Oper. Res. **30**, 1148–1167 (1982)
11. Kergosien, Y., Bélanger, V., Soriano, P., Gendreau, M., Ruiz, A.: A generic and flexible simulation-based analysis tool for EMS management. Int. J. Prod. Res. **53**, 7299–7316 (2015)
12. McConnell, C.E., Wilson, R.W.: The demand for prehospital emergency services in an aging society. Soc. Sci. Med. **46**, 1027–1031 (1998)

Flexible Parallel Split-Merge MCMC for the HDP

Debora Parisi and Stefania Perego

Abstract The Hierarchical Dirichlet Process Mixture model is useful to deal with topic modeling. Several sampling schemes have been developed to implement the model. We implemented the previously existing Sub-Cluster HDP Sampler introduced, since it can be parallelized. Our contribution consists in making the code as flexible as possible, in order to allow for an extension to several applications. We tested our code on synthetic and real datasets for topic modeling with categorical data.

Keywords Big data · Topic modeling · Hierarchical Dirichlet process mixture · Markov chain Monte Carlo · Split-merge

1 Introduction

In our work we deal with topic modeling, a particular example of models for clustering grouped data. The aim of topic modeling is to discover the latent structure of topics in a collection of documents, called corpus. A topic is seen as a probability distribution over words of some known vocabulary and documents are generally modeled under an exchangeability assumption (the "bag of words" assumption), in which the order of words in a document is ignored. Moreover, topics are shared between documents. For instance, if we have a document about Milan's fashion week and another about Milan's elections, the shared topic might be "Milan". Our contribution was in making the code as flexible as possible, in order to use the Sub-Cluster HDP Sampler algorithm of [1] not only for the topic modeling but also for other applications. Sections 2 and 3 give a brief description of the model and the algorithm. In Sect. 4 we describe the implementation of the code and the posterior analysis, while in Sect. 5 we report Bayesian inferences for simulated and real datasets.

D. Parisi (✉) · S. Perego
Department of Mathematics, Politecnico di Milano, Via Bonardi 9, 20133 Milan, Italy
e-mail: debora.parisi@mail.polimi.it

S. Perego
e-mail: stefania2.perego@mail.polimi.it

R. Argiento et al. (eds.), *Bayesian Statistics in Action*, Springer Proceedings in Mathematics & Statistics 194, DOI 10.1007/978-3-319-54084-9_20

2 Hierarchical Dirichlet Process Mixture

The Hierarchical Dirichlet Process Mixture (HDPM) model [4] can be described as follows. Data are organized into groups. A mixture model is associated to each group. In particular, we have several clustering problems, one for each group. We want the mixture components to be shared between groups, so these clustering problems must be linked together. The HDPM model posits that:

$$x_{ji}|\phi_{ji} \overset{indep}{\sim} f(x_{ji};\phi_{ji}) \tag{1}$$

$$\phi_{ji}|G_j \overset{indep}{\sim} G_j \tag{2}$$

$$G_j|\alpha, G_0 \overset{iid}{\sim} DP(\alpha, G_0) \tag{3}$$

$$G_0|\gamma, H \sim DP(\gamma, H) \tag{4}$$

where $j = 1, \ldots, D$ and $i = 1, \ldots, N_j$; D is the number of groups and N_j is the number of elements in group j. In the context of topic modeling, we have a vocabulary of W distinct words appearing in the corpus. To each word we associate a numeric identifier, so x_{ji} is the label for word i in document j. Here $f(x_{ji};\phi_{ji})$ is the distribution of a categorical variable, being ϕ_{ji} the parameter specific to the mixture component from which x_{ji} is generated. We choose a symmetric Dirichlet distribution with parameter λ for the base measure H: in this way, the inference is simplified, because the Dirichlet distribution is conjugate to the likelihood for categorical data.

A realization from a Dirichlet process prior is a discrete measure (with probability one). Therefore, we can use the stick-breaking representation of [3] for the global measure G_0:

$$G_0 = \sum_{k=1}^{+\infty} \beta_k \delta_{\theta_k} \tag{5}$$

where $\theta_k \overset{iid}{\sim} H$ and the probabilities β_k associated to the atoms are constructed as follows:

$$\beta_k^{'} \sim Beta(1, \gamma) \qquad \beta_k = \beta_k^{'} \prod_{l=1}^{k-1}(1 - \beta_l^{'}). \tag{6}$$

From (5) it can be seen that G_0 is a discrete measure and it has support at points $\{\theta_k\}_{k=1}^{+\infty}$. Given that each G_j is distributed according to a Dirichlet process and that G_0 is the base measure, each G_j is necessarily supported at the points $\{\theta_k\}_{k=1}^{+\infty}$. Thus we can rewrite as:

$$G_j = \sum_{k=1}^{+\infty} \pi_{jk} \delta_{\theta_k} \tag{7}$$

and for the probabilities π_{jk} we have:

$$\pi'_{jk} \sim Beta\left(\alpha\beta_k, \alpha\left(1 - \sum_{h=1}^{k} \beta_h\right)\right) \quad \pi_{jk} = \pi'_{jk} \prod_{h=1}^{k-1} (1 - \pi'_{jh}). \tag{8}$$

In (5), θ_k are the distinct values of the ϕ_{ji}'s in (2). The fact that, for each j, G_j has the same support as G_0 implies the sharing of clusters between groups.

The *Chinese Restaurant Franchise* metaphor describes how a draw from the HDP prior is sampled. We have a franchise of Chinese restaurants with a shared menu. When a client enters a restaurant, she sits at an occupied table with a probability proportional to the number of clients already seated at that table, or she can sit at an empty table. The first client who sits at a new table chooses the dish, which will be shared by the following clients that will sit at that table. She can choose a previously ordered dish, with a probability proportional to the number of tables at which that dish is served, or she can choose a new dish from the menu. It is necessary to keep counts of tables and customers, thus m_{jk} is the number of tables in restaurant j serving dish k and $n_{j.k}$ is the number of customers in restaurant j eating dish k. Partial counts are represented with dots, so for instance $m_{.k} = \sum_{j=1}^{D} m_{jk}$.

The Sub-Cluster HDP Sampler uses the *Direct Assignment* sampling scheme, which is simplified w.r.t. the *Chinese Restaurant Franchise*, because it directly associates customers to dishes. For the application of topic modeling, restaurants correspond to documents and dishes to topics, whereas the client choosing a table (and thus a particular dish) corresponds to the association of a word to a topic. The reader is referred to [4] for an exhaustive description of the *Chinese Restaurant Franchise* metaphor and the *Direct Assignment* sampling scheme. We introduce the notation that will be used in the following sections: β are the corpus-level topic proportions, π_j are the topic proportions specific to document j, θ_k is the latent parameter for topic k, z_{ji} is an indicator variable that associates word x_{ji} to a topic.

3 The Sub-cluster HDP Sampler Algorithm

To implement the model described in the previous section we choose the algorithm introduced in [1], which can be parallelized. The algorithm combines Gibbs sampler and Metropolis Hastings steps.

The Gibbs sampler updates the quantities β, π, θ, m and z for the non-empty clusters, using their full conditional distribution (see [1]). During these steps, the number of clusters is fixed.

In the Metropolis–Hastings steps new clusters are proposed. In particular, in a merge move two clusters are choosen to create a new cluster, while in a split move one cluster is choosen to be divided into two new clusters. In order to propose splits and merges it is necessary to introduce sub-clusters: each cluster k is characterized by the left and the right sub-clusters, kl and kr respectively. Each sub-cluster is described with the following variables: $\bar{\beta}_k = \{\bar{\beta}_{kl}, \bar{\beta}_{kr}\}$, $\bar{\pi}_{jk} = \{\bar{\pi}_{jkl}, \bar{\pi}_{jkr}\}$, $\bar{\theta}_k = \{\bar{\theta}_{kl}, \bar{\theta}_{kr}\}$, $\bar{m}_{jk} = \{\bar{m}_{jkl}, \bar{m}_{jkr}\}$ and $\bar{z}_{ji} \in \{l, r\}$.

The steps of the algorithm are: 1. initialize β and z randomly; 2. sample π, $\bar{\pi}$, θ and $\bar{\theta}$; 3. sample labels z and $\bar{z}$ for all data; 4. propose $\lfloor \frac{K}{2} \rfloor$ local merges and K local splits; 5. propose a global merge and a global split; 6. sample m and $\bar{m}$; 7. sample β and $\bar{\beta}$; 8. repeat from step 2 until the stop criterion is satisfied (in our case, number of iterations choosen by the user). Steps 4 and 5 refer to Metropolis Hastings moves; there are three versions of M–H, which differ in how labels $\hat{z}$ for the new clusters are proposed: one for local merge, one for local split and one for global moves. For the specific sampling equations and steps of the M–H algorithm we refer to [1].

4 Implementation of the Algorithm

We implemented model (1)–(4) according to the Sub-Cluster HDP Sampler described in the previous section. We introduced parallelization with OpenMP to decrease the execution time of simulations, especially when they involve large datasets. Moreover, we developed a posterior analysis suitable for the topic modeling. Our code is written in C++. Model (1)–(4) is flexible, because it is possible to choose the distribution for the base measure H and the likelihood $f(x_{ji}; \phi_{ji})$, according to the specific clustering problem. Our main effort was to keep this flexibility in the code: for this purpose, we separated the implementation of the algorithm from everything that depends on the model and we used template programming and inheritance techniques; this seems to make the implementation of the code more difficult, but a future extension to a different application is straightforward.

The *HDP_MCMC* class manages the steps of the algorithm and can be used for any model. *Document* is a class which manages the data in each group; it is responsible for sampling data-labels, as well as document specific topics' weights and tables. The *Model* class has the task of updating latent parameters, in other words it represents the choice of the prior; it is also in charge of managing all the clusters inferred during the algorithm's execution. Finally, the *PosteriorAnalisys* class manages the posterior analysis of results.

For all these classes, except for *HDP_MCMC*, we have decided to provide a generic interface for all possible models, then each generic class must be specialized for the chosen model. We implemented specialized class for our application of topic modeling, where data are categorical and H is the Dirichlet distribution. In extending the code to other models, the programmer must pay attention to the format of the quantities needed to update latent parameters and to the computation of likelihood and marginal distributions.

4.1 Parallelization

Bayesian algorithms are known to be heavy from a computational point of view. As the dataset size grows, the need of parallelizing the code becomes clear. We made extensive use of *for loop* throughout the code, both in Gibbs sampler and in Metropolis Hastings steps, so the choice of OpenMP is natural in order not to modify the code excessively with respect to the serial version. The user can choose the number of threads and this remains fixed during the execution.

The most expensive steps turned out to be the M–H moves and the update of labels z, especially when all data are involved; the execution time of these steps decreases as the number of threads increases, so the parallelization is useful. As the number of threads grows, other steps seem to increase the execution time, but this is negligible if compared to the most expensive steps and, in fact, the total execution time of the algorithm decreases. Moreover, if we increase the dataset dimension, the parallelization becomes convenient also for the less expensive steps. In the table below, for each dataset, we show the percentage reduction in the execution time when using T threads with respect to the serial execution; a negative percentage indicates a reduction, while a positive one represents an increase (Table 1).

As we can see, as the magnitude of the dataset grows, the reduction in the execution time becomes more and more important. However, it is difficult to understand which is the optimal number of threads. Indeed, if we fix the size but we change dataset, the posterior distribution of the number of cluster changes, so the computation might require more or less time.

It is necessary to pay attention to the correct generation of random numbers in parallel. As noted in [5], setting a different seed for each thread does not guarantee that the streams of random numbers are independent and an unlucky choice of seeds can lead to bad behaviour of the algorithm. The solution is to use a *parallel random numbers generator*. We chose the one of Matthew Bognar (http://www.stat.uiowa.edu/~mbognar/omprng), which is designed specifically for OpenMP and is simply to use. The generator is a C++ class that we slightly modified for our scopes.

Table 1 Percentage change in the mean execution time of one iteration

Size	1→2 (%)	1→4 (%)	1→6 (%)	1→8 (%)	1→10 (%)	1→12 (%)	1→14 (%)	1→16 (%)
1000	−9.23	−30.95	−45.05	−5.33	−17.27	+48.06	+102.52	+118.32
5000	−34.17	−55.78	−64.78	−70.04	−69.78	−64.31	−67.06	−65.39
10000	−21.32	−61.63	−71.74	−74.17	−75.72	−74.47	−74.92	−81.25
50000	−41.55	−68.14	−75.55	−80.68	−78.53	−81.01	−82.58	−79.66

4.2 Interfacing with R

We need to interface C++ to R in order to perform posterior inferences from the MCMC output. To this purpose we chose *RInside* and *Rcpp* packages. *RInside* opens a protected R session inside a C++ application, while *Rcpp* converts C++ containers into R data structures and viceversa. First we have analysed MCMC chains through standard techniques, using R package *coda*. We found the "best" posterior partition of the data according to the least square clustering criteria, as in [2]; we computed the number of groups in this partition and the number of items in each group.

At each iteration, the distribution of words in the inferred topics and the number of topics itself change. It is not possible to identify univocally a topic during the execution, because it might disappear at one iteration or it might not be characterized by exactly the same words with respect to a different iteration. Our aim is to find topics at different iterations which might be very similar, so we would like to assign the same label to these topics. We developed a method for tracking inferred topics in the last one hundred iterations. At each iteration we have a batch of inferred topics. First, we assign label zero to each topic in each batch. Then, we pick a topic in a batch, we call it *reference topic* and we compare it with all the topics in each successive batch. In each batch it can be found at most one topic similar to the *reference* one. To assert similarity between two topics we use the following criteria. We compute the *cosine similarity* between the vectors representing the topics, a standard technique in text analysis; this quantity must be close to one. The two topics must have the greatest number of words in common, but the sum of the weights associated to these words must be greater than a threshold of 0.7. Finally, the label assigned to a topic can be changed if it has more words in common with the *reference topic* than with the topics with its same label. After all the comparisons, some topics may be left with label zero: these topics appeared only once in the last hundred iterations. We also know the global weights β associated to each inferred topic, so we can follow their trend during the last iterations to see the dominant ones. We can represent the distribution of words in a topic through a wordcloud (see Fig. 2): the dimension of the word is proportional to its weight in the topic. The user can visualize the dominant topics when they have higher β. Finally, we use the LPML index to identify the best priors for γ and α hyperparameters and we choose the model with higher LPML.

5 Simulations

The user of our code is free of choosing fixed or random hyperparameters α and γ; in the latter case we put gamma priors on α and γ. We tested the algorithm using a synthetic dataset with 500 data. Model (1)–(4) tends to overestimate the real number of clusters, but we can fix α and γ in such a way that the number of clusters is underestimated a priori. The behaviour of the algorithm improves by using random α and γ: regardless of prior choice in terms of mean and variance of the hyperparameters, the algorithm identifies the real number of clusters.

When using likelihood for categorical data and a Dirichlet distribution H, the hyperparameter λ has to be fixed. In order to understand the algorithm's sensitivity with respect to this quantity, we run a simulation using $\lambda > 1$ and another using $\lambda < 0.5$, where the true value of λ is 0.5; for α and γ we assume gamma priors centered on their real values. When $\lambda > 1$, the algorithm underestimates the number of clusters, while with $\lambda < 0.5$ the number of clusters is overestimated. If the Dirichlet distribution is simmetric with $\lambda < 1$, the probability mass will be concentrated in few components of the vector, while the others are close to zero. This leads to distributions of words in the topics very different from each other, so it is easy to distinguish among all of them. If $\lambda > 1$, the mass will be distributed uniformly among all the components: this results in distributions of words which are very similar, so the algorithm finds it difficult to distinguish them.

As an application to a real case, we chose the KOS dataset from the UCI Machine Learning Repository. The data source is the Daily KOS (www.dailykos.com), an american web blog about politics. The corpus is then a collection of posts on this blog. We reduced a bit the dataset available on the Repository and we were left with 750 documents, 6366 distinct words, 96,942 total words. These posts were published in 2004 during the campaign for the presidency of the United States. The candidates were G.W. Bush for the Republican Party and J. Kerry for the Democratic Party. The campaign followed the September 11 attacks, so war and terrorism were some of the main topics discussed. We fitted model (1)–(4) to this dataset and we chose $\lambda = 0.5$, $\gamma \sim Gamma(0.25, 0.05)$, $\alpha \sim Gamma(2.0, 2.0)$.

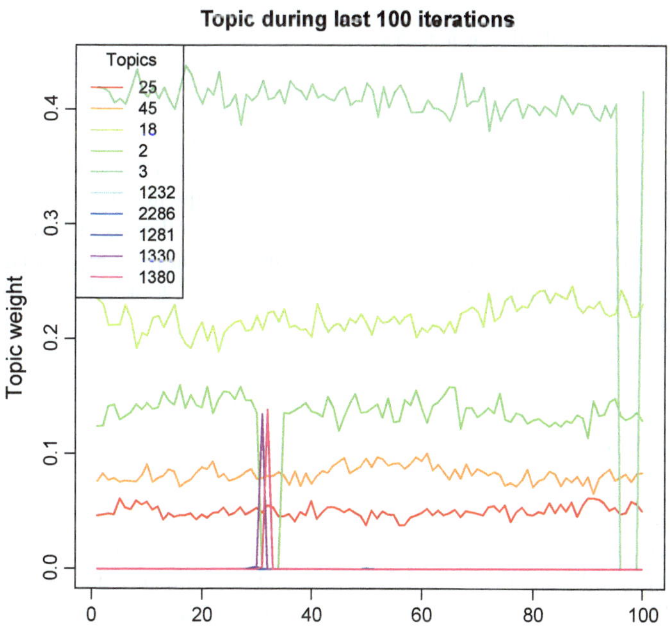

Fig. 1 Trend for the 10 most frequent topics

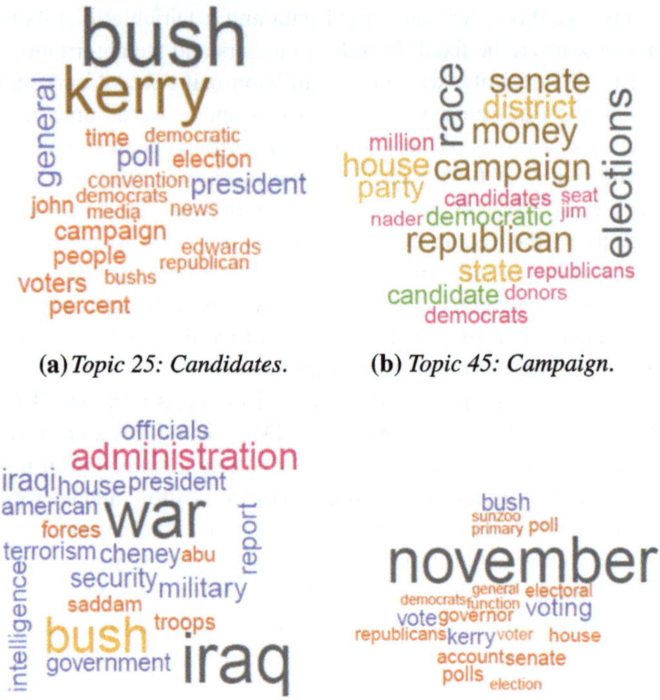

(a) *Topic 25: Candidates.* (b) *Topic 45: Campaign.*

(c) *Topic 18: War.* (d) *Topic 2: Elections.*

Fig. 2 Wordclouds for the dominant topics

We processed the MCMC output as we have described in the previous section. In particular, we applied our method to identify the inferred topics in the last iterations. We selected the ten most frequent topics, where the frequence means the number of times that a topic appears in the last iterations, then we followed their trend through the associated weights β. As Fig. 1 shows, there are 4–5 dominant topics in the corpus; Fig. 2 reports the wordclouds: as we can see, the inferred topics are consistent with our prior knowledge about the documents.

6 Conclusions

In this work we have focused on topic modeling, a particular example of clustering for grouped data. Following a BNP approach, we adopted the Hierarchical Dirichlet Process Mixture Model [4] to model the data and implemented the HDP Subcluster Sampler [1], which can be parallelized. We structured the code in such a way that the flexibility offered by the model is preserved, so it is possible to easily extend the code for other applications. We tested the parallelization with OpenMP on synthetic datasets with different dimension; the gain in terms of reduction of execution time increases as the dataset size increases, but it is difficult to identify the optimal number of threads. The model is sensitive with respect to the hyperparameters λ, γ and α. We run simulations with a synthetic dataset in order to understand the best choice for them. It turned out that putting a gamma prior on γ and α and setting $\lambda < 1$ leads to better results. We overcame the label switching problem through a method that identifies the topics inferred in the last iterations. We fitted the model and the algorithm to the KOS dataset.

Acknowledgements We would like to thank Professor Alessandra Guglielmi for helping us to find a suitable model and algorithm for our applications on topic modeling and for her helpful advice.

References

1. Chang, J., Fisher III, J.W.: Parallel sampling of HDPs using sub-clusters splits. Adv. Neural Inf. Process. Syst. **27**, 235–243 (2014)
2. Dahl, D.B.: Model-based clustering for expression data via a Dirichlet process mixture model. In: Do, K.-A., Müller, P., Vannucci, M. (eds.) Bayesian Inference for Gene Expression and Proteomics, pp. 201–218. Cambridge University Press, Cambridge (2006)
3. Sethuraman, J.: A constructive definition of Dirichlet priors. Stat. Sin. **4**, 639–650 (1994)
4. Teh, Y.W., Jordan, M.I., Beal, M.J., Blei, D.M.: Hierarchical Dirichlet processes. J. Am. Stat. Assoc. **101**, 1566–1581 (2006)
5. Wilkinson, D.: Getting started with parallel MCMC. https://darrenjw.wordpress.com/2010/12/14/getting-started-with-parallel-mcmc/

datasets with different dimensions, the gain in terms of reduction of execution time increases as the dataset size increases, but it is difficult to identify the optimal number of threads. [illegible]

[illegible]

References

Bayesian Inference for Continuous Time Animal Movement Based on Steps and Turns

Alison Parton, Paul G. Blackwell and Anna Skarin

Abstract Although animal locations gained via GPS, etc. are typically observed on a discrete time scale, movement models formulated in continuous time are preferable in order to avoid the struggles experienced in discrete time when faced with irregular observations or the prospect of comparing analyses on different time scales. A class of models able to emulate a range of movement ideas are defined by representing movement as a combination of stochastic processes describing both speed and bearing. A method for Bayesian inference for such models is described through the use of a Markov chain Monte Carlo approach. Such inference relies on an augmentation of the animal's locations in discrete time that have been observed with error, with a more detailed movement path gained via simulation techniques. Analysis of real data on an individual reindeer *Rangifer tarandus* illustrates the presented methods.

Keywords Movement modelling · Random walk · Rangifer tarandus · Data augmentation · GPS data

1 Introduction

Movement ecology is a fast growing area of research concerned with addressing questions of patterns in animal movements, their underlying mechanisms, driving causes and constraints [6]. Animal movement data gained via GPS, etc. are commonly given as 2-dimensional locations at a sequence of discrete—but not necessarily regular—points in time. A widespread group of models for analysing such data are based on

A. Parton (✉) · P.G. Blackwell
School of Mathematics and Statistics, University of Sheffield, Sheffield, UK
e-mail: aparton2@shef.ac.uk

P.G. Blackwell
e-mail: p.blackwell@shef.ac.uk

A. Skarin
Department of Animal Nutrition and Management, Swedish University of Agricultural Sciences, Uppsala, Sweden
e-mail: anna.skarin@slu.se

R. Argiento et al. (eds.), *Bayesian Statistics in Action*, Springer Proceedings in Mathematics & Statistics 194, DOI 10.1007/978-3-319-54084-9_21

parameterising movement by turning angles and step lengths (see e.g. [4, 5, 7, 8]). This movement is formulated in discrete time, with each subsequent location defined by a 'turn' and 'step' following some circular and positive distribution, respectively.

Discrete time 'step and turn' models are intuitive to ecologists and statistical inference given observed data can be carried out using a range of existing methods and software. The reliance on a discrete time scale, however, poses a number of issues. The chosen scale must be ecologically relevant to the movement decisions of the animal, but is more often dictated by the sampling rate of the received data. These models are not invariant under a change of time scale, leading to no guarantee of a coherent extension to a different time scale, or how to interpret such an extension. Irregular or missing data can therefore be difficult to model, and there is often no way to compare multiple analyses defined on differing scales.

Movement models that are formulated in continuous time avoid the discrete time difficulties; the true underlying mechanism of continuous movement is maintained, no user-defined time frame is needed and flexibility is introduced by time scale invariance. The following introduces a continuous time approach to modelling that preserves the familiar description of movement based on 'steps' and 'turns'.

2 The Continuous Time Model

At any time $t \geq 0$, let the animal have a bearing $\theta(t)$ and a speed $\psi(t)$ that evolve according to the stochastic differential equations

$$\begin{aligned} \mathrm{d}\theta(t) &= F_1\left(t, \theta(t)\right)\mathrm{d}t + F_2\left(t, \theta(t)\right)\mathrm{d}W_1(t), \\ \mathrm{d}\psi(t) &= F_3\left(t, \psi(t)\right)\mathrm{d}t + F_4\left(t, \psi(t)\right)\mathrm{d}W_2(t), \end{aligned} \tag{1}$$

where $W_i(t),\ i \in \{1, 2\}$ is Brownian motion and $F_i(t, \cdot),\ i \in \{1, \ldots, 4\}$ are known functions.

Many discrete time 'step and turn' models make the assumption that animals move with persistence, using a correlated random walk to reflect this. Such an assumption can be made within this continuous time framework by assuming $\theta(t)$ follows Brownian motion with volatility σ_B^2 by taking $F_1(t, \theta(t)) = 0$ and $F_2(t, \theta(t)) = \sigma_B$. Note that although the direction the animal is facing is constrained to be within $[-\pi, \pi]$, $\theta(t)$ itself is not constrained in this way. Although discussed no further here, a range of other movement modes could be modelled under this framework, including directional bias and attraction to centres.

A 1-dimensional Ornstein–Uhlenbeck (OU) process is assumed for $\psi(t)$ with parameters $\{\mu, \lambda, \sigma_S^2\}$, reflecting the idea that the animal's speed is stochastic but correlated over time, with some long-term average speed. This is achieved by taking $F_3(t, \psi(t)) = \lambda(\mu - \psi(t))$ and $F_4(t, \psi(t)) = \sigma_S$. This choice is similar to [3], in which movement is modelled by a 2-dimensional OU velocity process. In the classic examples of discrete time models, the 'step' process is assumed to be independent over disjoint time intervals. Although discussed no further here, this form of movement can easily be emulated by basing $\psi(t)$ on a distance process that follows Brownian motion with drift—where the drift describes the average speed of the animal.

The continuous time movement model can be simulated by taking an Euler approximation over the small increment δt. Given $\theta(t)$ and $\psi(t)$ at time $t \geq 0$,

$$\theta\left(t+\delta t\right)|\theta(t) \sim \mathrm{N}\left(\theta(t), \sigma_B^2 \delta t\right),$$
$$\psi\left(t+\delta t\right)|\psi(t) \sim \mathrm{N}\left(\mu + \mathrm{e}^{-\lambda \delta t}\left(\psi(t)-\mu\right), \frac{\sigma_S^2}{2\lambda}\left(1-\mathrm{e}^{-2\lambda\delta t}\right)\right). \tag{2}$$

The familiar notion of a 'turn' is then given by $\theta(t+\delta t) - \theta(t)$ and a 'step' by $\nu(t) = \psi(t)\delta t$.

3 Inference for the Continuous Time Model

An animal's location $(\mathbf{X}, \mathbf{Y})$ at a series of discrete times $\mathbf{t}$ has been observed with error. Throughout the following, observation errors are assumed to be independent and identically distributed in both space and time, following a circular bivariate Normal distribution with variance σ_E^2. The aim of the following is to describe a method for statistical inference on the movement and error parameters $\mathbf{\Phi} = \left\{\sigma_B^2, \mu, \lambda, \sigma_S^2, \sigma_E^2\right\}$, given $(\mathbf{X}, \mathbf{Y})$.

It is not possible to evaluate the likelihood of $(\mathbf{X}, \mathbf{Y})$, given $\mathbf{\Phi}$. The approach for inference described is to therefore augment $(\mathbf{X}, \mathbf{Y})$ with a 'refined path' defined by $(\boldsymbol{\theta}, \boldsymbol{\nu})$. This refined path is given as a set of bearings, $\boldsymbol{\theta}$, and steps, $\boldsymbol{\nu}$, on some δt time scale—assuming throughout that δt is small enough that such a refined path can be taken as an approximation to the continuous time model of Eq. 1. A representation of the relationship between $(\mathbf{X}, \mathbf{Y})$ and $(\boldsymbol{\theta}, \boldsymbol{\nu})$ is given in Fig. 1. The joint likelihood of $(\mathbf{X}, \mathbf{Y})$ and $(\boldsymbol{\theta}, \boldsymbol{\nu})$ can be evaluated, given by

$$\mathscr{L}\left(\mathbf{X}, \mathbf{Y}, \boldsymbol{\theta}, \boldsymbol{\nu} \mid \mathbf{\Phi}\right) = \mathscr{L}\left(\boldsymbol{\theta}, \boldsymbol{\nu} \mid \mathbf{\Phi}\right) \mathscr{L}\left(\mathbf{X}, \mathbf{Y} \mid \boldsymbol{\theta}, \boldsymbol{\nu}, \mathbf{\Phi}\right). \tag{3}$$

The first term on the r.h.s. of Eq. 3 is the likelihood of the refined path, given by

$$\mathscr{L}\left(\boldsymbol{\theta}, \boldsymbol{\nu} \mid \mathbf{\Phi}\right) = \pi_\theta\left(\theta_1 \mid \mathbf{\Phi}\right) \pi_\nu\left(\nu_1 \mid \mathbf{\Phi}\right) \prod_{i=2} \pi_\theta\left(\theta_i \mid \theta_{i-1}, \mathbf{\Phi}\right) \pi_\nu\left(\nu_i \mid \nu_{i-1}, \mathbf{\Phi}\right), \tag{4}$$

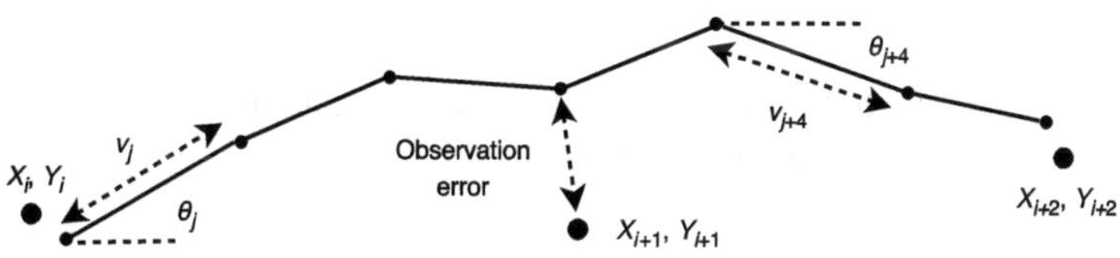

Fig. 1 Representation of the refined movement path $(\boldsymbol{\theta}, \boldsymbol{\nu})$ and observed locations $(\mathbf{X}, \mathbf{Y})$

where

$$\theta_1 \mid \mathbf{\Phi} \sim \mathrm{U}(-\pi, \pi), \tag{5}$$

$$\nu_1 \mid \mathbf{\Phi} \sim \mathrm{N}\left(\delta t \mu, \frac{\delta t^2 \sigma_S^2}{\lambda}\right), \tag{6}$$

and $\pi_\theta(\theta_i \mid \theta_{i-1}, \mathbf{\Phi}), \pi_\nu(\nu_i \mid \nu_{i-1}, \mathbf{\Phi})$ are given by Eq. 2 for $i \geq 2$. The second term on the r.h.s. of Eq. 3 is the likelihood of the observation error when $(\boldsymbol{\theta}, \boldsymbol{\nu})$ is the 'true' path.

The refined path is unknown and is simulated using a Metropolis-within-Gibbs sampler to carry out full inference on the movement parameters. This sampler alternately updates $\mathbf{\Phi}$ and $(\boldsymbol{\theta}, \boldsymbol{\nu})$, both also conditioned on $(\mathbf{X}, \mathbf{Y})$. The respective full conditional distributions cannot be directly sampled from, and so the Metropolis-Hastings (MH) algorithm is used within each of the two steps.

3.1 Approach for Sampling the Movement Parameters

The full conditional distribution of the movement parameters $\mathbf{\Phi}$, given the refined path $(\boldsymbol{\theta}, \boldsymbol{\nu})$ and the observed positions $(\mathbf{X}, \mathbf{Y})$ is given by

$$\mathscr{L}(\mathbf{\Phi} \mid \boldsymbol{\theta}, \boldsymbol{\nu}, \mathbf{X}, \mathbf{Y}) \propto \pi_\Phi(\mathbf{\Phi}) \mathscr{L}(\mathbf{X}, \mathbf{Y}, \boldsymbol{\theta}, \boldsymbol{\nu} \mid \mathbf{\Phi}), \tag{7}$$

where $\mathscr{L}(\mathbf{X}, \mathbf{Y}, \boldsymbol{\theta}, \boldsymbol{\nu} \mid \mathbf{\Phi})$ is given in Eq. 3 and $\pi_\Phi(\mathbf{\Phi})$ is the prior distribution of the movement parameters. The movement parameters are proposed within the MH algorithm simultaneously using independent Normal random walks (truncated below at zero and centred on the current realisation). Acceptance is then based on the standard MH ratio using Eq. 7.

3.2 Approach for Sampling the Refined Path

The augmentation of refined movement paths is complicated by observed locations. Forward simulation based only on movement parameters will be unlikely to agree well with observations, proving infeasible within a MH step. The following describes a simulation method that, in part, takes the observations into account.

The full conditional distribution of the refined path $(\boldsymbol{\theta}, \boldsymbol{\nu})$, given the movement parameters $\mathbf{\Phi}$ and the observed positions $(\mathbf{X}, \mathbf{Y})$ can be expressed as

$$\begin{aligned} \mathscr{L}(\boldsymbol{\theta}, \boldsymbol{\nu} \mid \mathbf{\Phi}, \mathbf{X}, \mathbf{Y}) &= \mathscr{L}(\boldsymbol{\theta} \mid \mathbf{\Phi}, \mathbf{X}, \mathbf{Y}) \mathscr{L}(\boldsymbol{\nu} \mid \boldsymbol{\theta}, \mathbf{\Phi}, \mathbf{X}, \mathbf{Y}) \\ &\propto \mathscr{L}(\boldsymbol{\theta} \mid \mathbf{\Phi}) \mathscr{L}(\mathbf{X}, \mathbf{Y} \mid \boldsymbol{\theta}, \mathbf{\Phi}) \mathscr{L}(\boldsymbol{\nu} \mid \boldsymbol{\theta}, \mathbf{\Phi}, \mathbf{X}, \mathbf{Y}), \end{aligned} \tag{8}$$

where $\mathscr{L}(\boldsymbol{\theta} \mid \boldsymbol{\Phi})$ is given by the product of $\pi_\theta(\cdot)$ in Eq. 2. Each observed location (X_i, Y_i) can be expressed as

$$X_i = X_0 + \sum_j \nu_j \cos(\theta_j) + \varepsilon_{i,x},$$
$$Y_i = Y_0 + \sum_j \nu_j \sin(\theta_j) + \varepsilon_{i,y}, \qquad (9)$$

which, given $\boldsymbol{\theta}$, are linear combinations of the Normally distributed $\boldsymbol{\nu}, \boldsymbol{\varepsilon}$, and so $(\mathbf{X}, \mathbf{Y} \mid \boldsymbol{\theta}, \boldsymbol{\Phi})$ is Normally distributed with known mean and variance. The final term in Eq. 8 is obtained by taking the Normally distributed $(\boldsymbol{\nu} \mid \boldsymbol{\theta}, \boldsymbol{\Phi})$, with likelihood given by $\pi_\nu(\cdot)$, and conditioning this on $(\mathbf{X}, \mathbf{Y} \mid \boldsymbol{\theta}, \boldsymbol{\Phi})$. The mean and variance of $(\boldsymbol{\nu} \mid \boldsymbol{\theta}, \boldsymbol{\Phi}, \mathbf{X}, \mathbf{Y})$ are therefore given by the standard results for multivariate conditioned Normal distributions.

Within the MH algorithm, a refined path proposal is made by first proposing bearings with density proportional to $\mathscr{L}(\boldsymbol{\theta} \mid \boldsymbol{\Phi})$. Conditional on both these bearings and observed locations, steps are proposed with density proportional to $\mathscr{L}(\boldsymbol{\nu} \mid \boldsymbol{\theta}, \boldsymbol{\Phi}, \mathbf{X}, \mathbf{Y})$. Acceptance of a simulated refined path is then based only on $\mathscr{L}(\mathbf{X}, \mathbf{Y} \mid \boldsymbol{\theta}, \boldsymbol{\Phi})$, by Eq. 8 and the standard MH acceptance ratio.

Proposing an entire refined path in this way is likely to yield a very low acceptance rate due to the high dimensionality. In reality, only sections of the refined path are updated at a time. This is carried out as above, but with additional conditioning upon the fixed bearings, steps and locations at the endpoints of the chosen section—i.e. $\pi_\theta(\cdot)$ is given by a Brownian bridge and $\pi_\nu(\cdot)$ is given by an OU bridge. The additional condition that the chosen section will need to meet its fixed end locations leads to the step proposal distribution being singular, and so realisations are proposed using singular value decomposition.

4 Reindeer Movement Example

The method described above for statistical inference is demonstrated using observations of *Rangifer tarandus* movement. A subset of 50 observations of the reindeer 'b53.10' walking in the Malå reindeer herding community in northern Sweden was used, taken at mostly 2 min intervals and shown as the points in Fig. 2, with a refined path defined on a time scale of 0.5 min. The inference method described above was carried out with flat priors for all parameters apart from a dependence between the speed parameters to reduce the possibility of negative speeds. The refined path was sampled in short sections of between 5–12 points chosen randomly from the entire refined path, with 50 updates to the path for every parameter update in the Gibbs sampler.

A burn-in time of 10^5 iterations was discarded and the sampler run for a further 10^5 iterations, thinned by a factor of 10^2. Posterior 90% credible intervals for

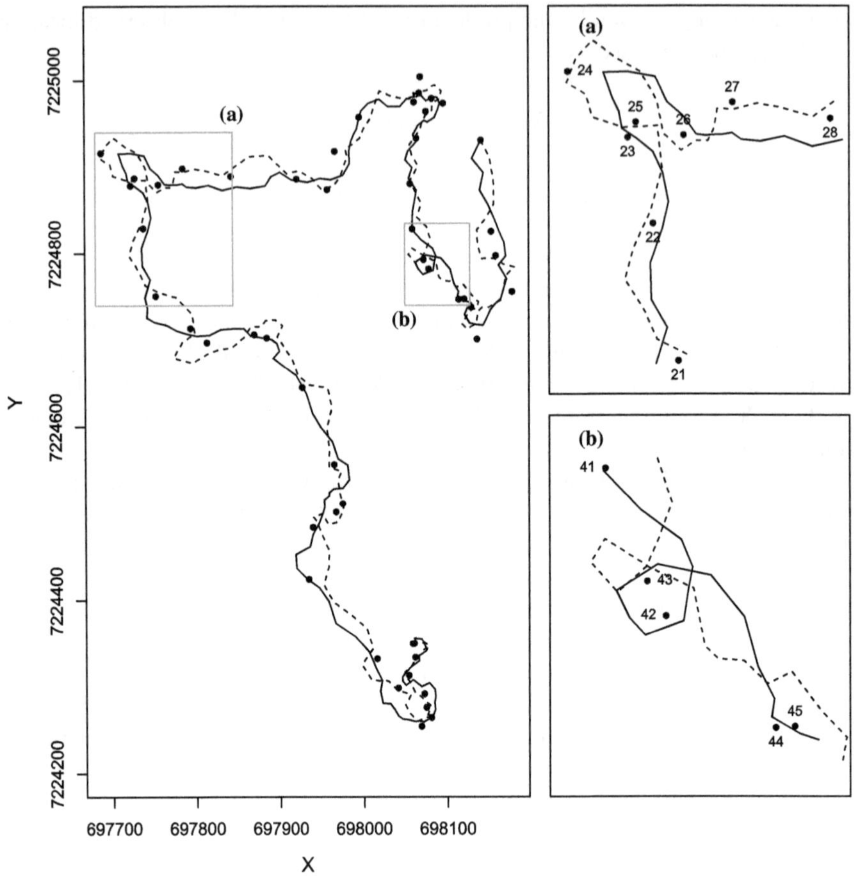

Fig. 2 Observations of reindeer 'b53.10' (*points*) and examples of two sampled paths (*block* and *dashed lines*). Sub-figures **a** and **b** show zoomed in sections of the path, indicated by the *grey boxes*, with numbering showing the temporal ordering of observations

the remaining 10^3 samples of $\mathbf{\Phi}$ are given as σ_B^2 : (0.670, 1.53), μ : (24.2, 29.3), λ : (0.465, 0.668), σ_S^2 : (116.4, 135.4), σ_E^2 : (80.4, 100.9). The posterior credible interval for σ_E^2 agrees well with current levels of observation error, expected to be up to 20 m.

Examples of two sampled paths from throughout the run are shown in Fig. 2. The marked difference in the reconstruction between some pairs of observations exhibited by the example sampled paths suggests that the linear interpolation employed by discrete time methods could be ignoring important characteristics of movement. Furthermore, in sub-plots (a) and (b) there are a number of 'sharp' turns between observations 23–25 and 42–43 that have been 'smoothed out' in the example path reconstructions. In a discrete time analysis this would amount to multiple turns of approximately π radians, leading to large estimates of the turning volatility.

5 Conclusion

We have introduced a framework for modelling animal movement in continuous time based on the popular movement metrics of step lengths and turning angles. A method for statistical inference via the augmentation of a refined path that is assumed to approximate the continuous time path has been described and demonstrated on a subset of reindeer location observations.

Parameter estimates for the proposed movement model give insight into the characteristics of an animal's movement in a form that is immediately interpretable, such as the mean speed at which the animal travels. These judgements are useful in addressing ecological questions relating to space/resource use, such as the size of an animal's 'home range'. The augmentation method employed further supports accessible inference by supplying reconstructions of the movement path at a finer time scale than the observations. Therefore, the space use of the animal at the local scale can immediately be estimated and this enables its combination with environmental covariates, such as land cover data, whose resolution is fast increasing. The interpretation of the estimated parameters is also furthered by the ability to visualise the actual movement paths they describe.

The method here assumes a simplistic model for observation error, being Normally distributed and independent through time. A common feature of telemetry data is autocorrelation in observation error, and so in further applications more realistic models for observation error will be sought that account for this feature.

In all of the work presented here, movement has been assumed to follow a single behavioural mode, which is unrealistic in practice for animal tracks covering an extended period of time. Behavioural switching for this model in continuous time is currently being implemented based on the works of [1, 2], allowing switching between a finite number of 'behavioural states' that represent quantitative or qualitative differences in movement.

References

1. Blackwell, P.G., Niu, M., Lambert, C., LaPoint, S.D.: Exact Bayesian inference for animal movement in continuous time. Methods Ecol. Evol. (2015). doi:10.1111/2041-210X.12460
2. Harris, K.J., Blackwell, P.G.: Flexible continuous-time modelling for heterogeneous animal movement. Ecol. Model. **255**, 29–37 (2013). doi:10.1016/j.ecolmodel.2013.01.020
3. Johnson, D.S., London, J.M., Lea, M.A., Durban, J.W.: Continuous-time correlated random walk model for animal telemetry data. Ecology **89**(5), 1208–15 (2008). doi:10.1890/07-1032.1
4. Langrock, R., King, R., Matthiopoulos, J., Thomas, L., Fortin, D., Morales, J.M.: Flexible and practical modeling of animal telemetry data: hidden Markov models and extensions. Ecology **93**(11), 2336–2342 (2012). doi:10.1890/11-2241.1
5. McClintock, B.T., King, R., Thomas, L., Matthiopoulos, J., McConnell, B.J., Morales, J.M.: A general discrete-time modeling framework for animal movement using multistate random walks. Ecol. Monogr. **82**(3), 335–349 (2012). doi:10.1890/11-0326.1
6. Minerva Center for Movement Ecology. Available via DIALOG. http://move-ecol-minerva.huji.ac.il/. Accessed 29 Jan 2016

7. Morales, J.M., Haydon, D.T., Frair, J., Holsinger, K.E., Fryxell, J.M.: Extracting more out of relocation data: building movement models as mixtures of random walks. Ecology **85**(9), 2436–2445 (2004). doi:10.1890/03-0269
8. Patterson, T.A., Thomas, L., Wilcox, C., Ovaskainen, O., Matthiopoulos, J.: State-space models of individual animal movement. Trends Ecol. Evol. Trends Ecol. Evol. **23**(2), 87–94 (2008). doi:10.1016/j.tree.2007.10.009

Explaining the Lethality of Boko Haram's Terrorist Attacks in Nigeria, 2009–2014: A Hierarchical Bayesian Approach

André Python, Janine Illian, Charlotte Jones-Todd and Marta Blangiardo

Abstract Since 2009, Nigeria has been the scene of numerous deadly terrorist attacks perpetrated by the terrorist group Boko Haram. In response to this threat, stakeholders in the fight against terrorism have deployed various counterterrorism policies, the costs of which could be reduced through efficient preventive measures. Statistical models able to integrate complex spatial dependence structures have not yet been applied, despite their potential for providing guidance to assess characteristics of terrorist attacks. In an effort to address this shortcoming, we use a flexible approach that represents a Gaussian Markov random field through stochastic partial differential equation and model the fine-scale spatial patterns of the lethality of terrorism perpetrated by Boko Haram in Nigeria from 2009 to 2014. Our results suggest that the lethality of terrorist attacks is contagious in space and attacks are more likely to be lethal at higher altitudes and far from large cities.

Keywords Bayesian hierarchical model · Boko Haram · GMRF · Terrorism · SPDE

1 Introduction

In 2014, Nigeria experienced the world's largest increase in deaths due to terrorism, mostly due to the intensification of Boko Haram's presence [11]. In the same year, Boko Haram became the most lethal of the world's extremist organisations before the Islamic State (ISIS), to whom it pledged allegiance in March 2015 [11, 21]. Boko Haram, also known as "Jama'atu Ahlis Sunna Lidda'Awati Wal-Jihad" (People Committed to the Propagation of the Prophet's Teachings and Jihad), had conducted mainly non-violent actions until it clashed with Nigerian authorities in July 2009, which resulted in more than 800 deaths [40].

A. Python (✉) · J. Illian · C. Jones-Todd
University of St-Andrews, St-Andrews, UK
e-mail: ap215@st-andrews.ac.uk

M. Blangiardo
Imperial College London, London, UK

R. Argiento et al. (eds.), *Bayesian Statistics in Action*, Springer Proceedings in Mathematics & Statistics 194, DOI 10.1007/978-3-319-54084-9_22

Research in terrorism confirmed that terrorist groups do not act randomly [20, 26, 34] and terrorist targets are usually not randomly distributed in space but tend to form "hotspots" [4, 38]. However, most studies have analysed terrorism at country-level [16, 23, 33], which does not provide an understanding of spatial mechanisms at sub-national level [7]. Moreover, when sub-national analyses were carried out, it was typically in a descriptive perspective [5, 15, 25]. In an effort to address these shortcomings, we apply a spatial Bayesian model based on a stochastic partial differentiation equation (SPDE) approach implemented through computationally efficient integrated nested Laplace approximation (INLA) techniques [28, 35].

This research, which integrates spatially explicit covariates and data on terrorist events, allows to capture local-scale spatial patterns of the lethality of terrorism and hence, to better understand the mechanisms behind the patterns. Ultimately, this study may provide complementary tools to enhance the efficacy of preventive counterterrorism policies. First, the subnational nature of this work provides policy makers with key information about the spatial heterogeneity of the lethality of terrorism observed locally within Nigeria.[1] Second, since terrorism activity could be accurately measured in any location before and after counterterrorism missions, this study could be used as a tool to evaluate the impact of counterterrorism policies [32].

2 Data

We put particular emphasis on the *spatial accuracy* of the data since this study investigates terrorism, which is a phenomenon observed at subnational level (terrorist attacks occur in specific locations and are geolocalised at city-level). We use the *Global Terrorism Database* (GTD), which is currently the only available public terrorism database that includes a variable assigning the spatial accuracy of each individual observation [18]. Spatial accuracy is represented by an ordinal variable called *specificity*, with five possible levels (from low to high spatial accuracy) (for further information, see GTD codebook: https://www.start.umd.edu/gtd/downloads/Codebook.pdf).

The selected model (described in Sect. 3) includes five covariates (Fig. 1).[2] First, we assess the role of economy on terrorism, which has been under debate [1, 16, 23, 33]. We use NOAA satellite lights at night (Version 4 DMSP-OLS), which are high-resolution data (30 arc-second grid ≈1 km at the equator). They

[1] Note that in line with Kegley [22] and Sáanchez-Cuenca and De la Calle [36], we do not distinguish "domestic" from "transnational" terrorism. Since the 1970s, almost all terrorist acts have generated impacts across national borders through the international diffusion of media or external funding from foreign countries for example. Boko Haram is no exception and its influence extends beyond Nigerian national borders. Therefore, this study encompasses both types of terrorism without any distinction.

[2] Note that from a total of six potential covariates, the most parsimonious selected model uses five covariates. More detail about the procedure used to select the variables is provided in Sect. 3.

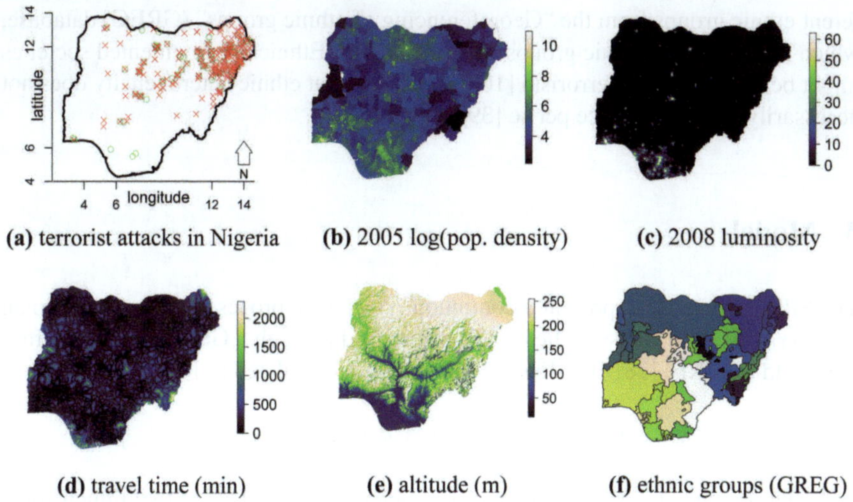

(a) terrorist attacks in Nigeria (b) 2005 log(pop. density) (c) 2008 luminosity

(d) travel time (min) (e) altitude (m) (f) ethnic groups (GREG)

Fig. 1 Nigeria (*polygon*), non-lethal (*green circle*) and lethal (*red cross*) terrorist attacks perpetrated by Boko Haram in Nigeria from 2009 to 2014 (**a**). The covariates are: Gridded Population of the World (GPW) - v4 (**b**), 2008 NOAA satellite lights at night - v4 DMSP-OLS (**c**), Travel time to major cities (**d**), NOAA Global Relief Model - ETOPO1 (**e**), and the Georeferencing of ethnic groups (GREG) (**f**)

have been used as a proxy for socio-economic development measures, including per capita GDP estimation [12, 19, 41].[3]

Second, we assess the role of demography. Cities provide anonymity, mobility, audiences and a larger recruitment pool in comparison to rural areas [10, p. 115], [27, p. 41]. Densely populated areas are usually more prone to terrorism, since they tend to provide a higher density of symbolic, human, and public targets [9, 37, 42]. We use the Gridded Population of the World (GPW) (v4), which provides population density at high-resolution (30 arc-second grid) [8].[4] Third, we assess the role of communication infrastructure based on Travel Time to Major Cities [29], which provides the travel time from each terrorist event to the nearest large city (more than 50,000 inhabitants) at a high spatial resolution (30 arc-second grid).

Fourth, we assess the role of altitude, which is expected to have a positive association with terrorism since insurgent and terrorist groups have tactical advantages to carry out attacks in less accessible areas, such as mountainous regions [1, 14, 30]. Data is obtained from NOAA Global Relief Model (ETOPO1) (1 arc-minute grid) [2]. Fifth, we assess the role of ethnic diversity by computing the number of dif-

[3]Note that to avoid *endogeneity* (simultaneity), we use values of luminosity observed a year before the study period (i.e. year 2008), calibrated according to the procedure described in Elvidge et al. [13, Chap. 6]. Moreover, we use the *Stable Lights* version of the data, which filters background noise and identifies zero-cloud free observations [31].

[4]Note that to avoid *endogeneity* (simultaneity), we use the first available year of population density prior the study period (i.e. year 2005).

ferent ethnic groups from the "Georeferencing of ethnic groups" (GREG) database, which counts 1, 276 ethnic groups worldwide [44]. Ethnically fragmented societies might be more prone to terrorism [16, 24] despite that ethnic heterogeneity does not necessarily lead to violence per se [39, p. 68].

3 Model

The SPDE is used to represent a continuous stochastic process in two-dimensional space ($\boldsymbol{s} \in \mathscr{D}$) as a Gaussian field (GF) discretised through a Gaussian Markov random field (GMRF) [28]. The linear SPDE can be formulated as [6, Chap. 6]:

$$(\kappa^2 - \Delta)^{\alpha/2}(\tau\xi(\boldsymbol{s})) = \mathscr{W}(\boldsymbol{s}), \quad \boldsymbol{s} \in \mathscr{D}\,, \tag{1}$$

with the Laplacian Δ, smoothness parameter $\alpha = \lambda + 1$ (for two-dimensional processes), scale parameter $\kappa > 0$, precision parameter τ, domain $\mathscr{D}$, and Gaussian spatial white noise $\mathscr{W}(\boldsymbol{s})$. The stationary solution of Eq. 1 is the GF ($\xi(\boldsymbol{s})$) with Matérn covariance function (Fig. 2, *solid line*):

$$Cov(\xi(\boldsymbol{s}_i), \xi(\boldsymbol{s}_j)) = \sigma^2_{\xi_i} \frac{1}{\Gamma(\lambda)2^{\lambda-1}} \left(\kappa \left\| \boldsymbol{s}_i - \boldsymbol{s}_j \right\|\right)^{\lambda} K_{\lambda}\left(\kappa \left\| \boldsymbol{s}_i - \boldsymbol{s}_j \right\|\right), \tag{2}$$

where $\left\| \boldsymbol{s}_i - \boldsymbol{s}_j \right\|$ denotes the Euclidean distance between two locations, $\sigma^2_{\xi_i}$ the marginal variance and K_λ the modified Bessel function of the second kind and order $\lambda > 0$. The distance from which the spatial correlation becomes negligible (for $\lambda > 0.5$) is given by the range r (Fig. 2, *dotted vertical line*), which can be empirically derived from the scale parameter $r = \sqrt{8\lambda}/\kappa$ to be estimated. The GF defined through Eqs. 1 and 2 appears in the so-called *latent field* as given in Eq. 3b. Hence, the GF ($\xi(\boldsymbol{s})$) is approximated as a GMRF through a finite element method using basis functions defined on a Constrained Refined Delaunay Triangulation (mesh) over Nigeria (Fig. 3a), which includes $n = 141$ vertices [28].

We extract 1,147 terrorist events perpetrated by Boko Haram in Nigeria in the years 2009–2014. This number includes 895 lethal and 252 non-lethal observations y_i, which we assume to follow a Bernoulli distribution with probability of observing a lethal terrorist attack π_i and a non-lethal terrorist attack $1 - \pi_i$ (Eq. 3a). Hence, we use a three-stage Bayesian hierarchical modelling framework [3]:

$$y_i|\eta_i, \theta \sim \text{Bernoulli}(\pi_i), \quad i = 1, \ldots, 1147. \tag{3a}$$

$$\eta_i|\theta = \beta_0 + \sum_{k=1}^{n_\beta} \boldsymbol{\beta}_k z_{k,i} + \xi_i, \quad i = 1, \ldots, 1147. \tag{3b}$$

$$\theta = p(\theta). \tag{3c}$$

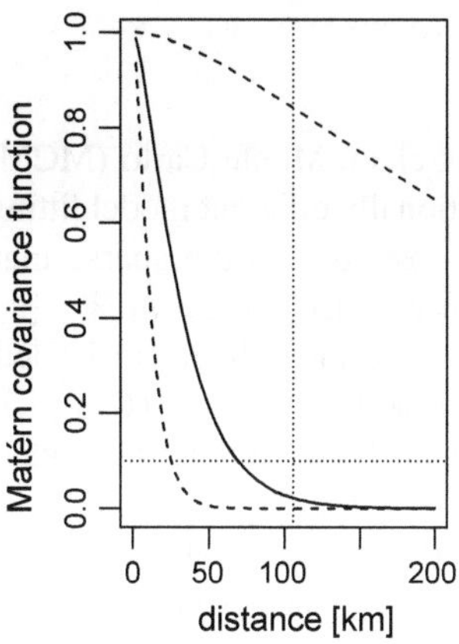

Fig. 2 Matérn correlation function (*solid line*) with parameters $\lambda = 1$, posterior mean $\kappa \approx 3.9$ and its 95% credible intervals (*dashed lines*), posterior mean range $r \approx 105$ km (*dotted vertical line*) and correlation ≈ 0.1 (*dotted horizontal line*)

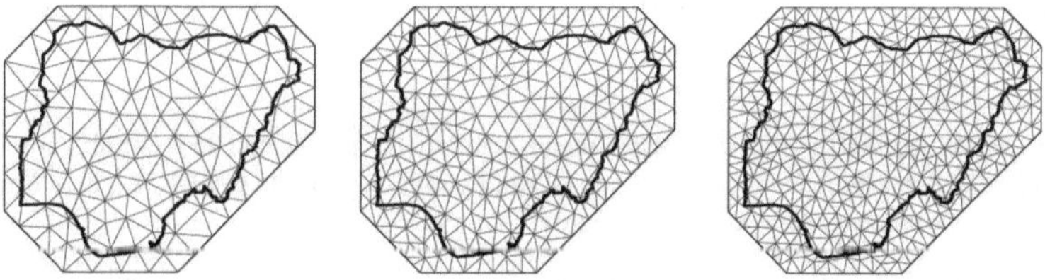

(**a**) *original mesh* (**b**) *mesh 2* (**c**) *mesh 3*

Fig. 3 *Original mesh* ($n = 141$) (**a**) which covers our study area, Nigeria (*thick line*). For robustness tests, we used higher resolution meshes, including *mesh 2*, with the number of vertices $n_2 = 283$ (**b**) and *mesh 3* with $n_3 = 425$ (**c**)

Equation 3b describes the linear predictor η_i =logit(π_i), which represents the *latent field*, including: the intercept β_0, n_β coefficients of the covariates $z_{k,t}$, and the GMRF field ξ_i with Matérn covariance function (Eq. 2). We apply a *backward selection* procedure to select the covariates, using the Watanabe–Akaike information criterion (WAIC) [43] as selection criterion, as suggested by Gelman et al. [17]. Hence, we select the most parsimonious model which includes: satellite night light (β_{lum}), the natural logarithm of altitude (β_{logalt}), the number of different ethnic groups (β_{greg}), and travel time to the nearest large city (β_{tt}). The hyperparameters $\theta = \{\kappa, \sigma_\xi^2\}$ are assigned a joint prior distribution (Eq. 3c) through $\log(\tau)$ and $\log(\kappa)$, which are assumed to follow a joint Normal distribution.

4 Results

As an alternative to Markov chain Monte Carlo (MCMC) methods, we use INLA as an accurate and computationally efficient model fitting method, which drastically reduces the computational costs due to the sparse matrix of GMRF ($\mathcal{O}(n^{3/2})$ in contrast to $\mathcal{O}(n^3)$ for a dense matrix of GF in $\mathbb{R}^2$) [6, 35]. Our results (Table 1) suggest that terrorist attacks are more likely to be lethal at higher altitudes and far from large cities (95% credible intervals ($CI_{95\%}$ β_{logalt}, $\beta_{tt} > 0$). We did not find evidence of a relationship between satellite night light, ethnic diversity, and the lethality of terrorism ($0 \in CI_{95\%}$ β_{lum}, β_{greg}). In addition, Fig. 4 provides the posterior probability of a lethal attack π_s at each mesh vertex and in interpolated values in all locations $s \in \mathcal{D}$. The uncertainty in the spatial prediction is provided through the posterior mean variance of the GMRF $\sigma^2_{\xi_s}$ (Fig. 4b). Moreover, our model suggests that lethality is contagious in space ($CI_{95\%}$ $\kappa \approx [0.5 - 14]$) with a highly uncertain range ($CI_{95\%}$ $r \approx [6 - 258]$ km, posterior mean $\approx$105 km).

As robustness test, we change the prior distribution of κ and τ through modifications of the prior distribution of the variance of the GMRF σ^2_ξ and range r for a

Table 1 SPDE Bernoulli model: posterior mean, standard deviation, and 95% credible intervals of the standardised coefficients of the covariate (β) and parameters of the GMRF (ξ), including scale parameter κ, variance σ^2_ξ, and empirical range r

Covariate coefficients (β)	Post. mean	Post. std.	Post. 95% CI
$\beta_0(intercept)$	1.24	0.19	(0.87; 1.62)
β_{lum}	0.004	0.121	(−0.23; 0.24)
β_{logalt}	0.22	0.07	(0.076; 0.37)
β_{tt}	0.31	0.12	(0.08; 0.55)
β_{greg}	−0.20	0.13	(−0.46; 0.05)
GMRF (ξ) parameters			
κ	5.18		(0.49; 14.2)
σ^2_ξ	0.90		(0.04; 2.84)
r (km)	105		(6.04; 258)

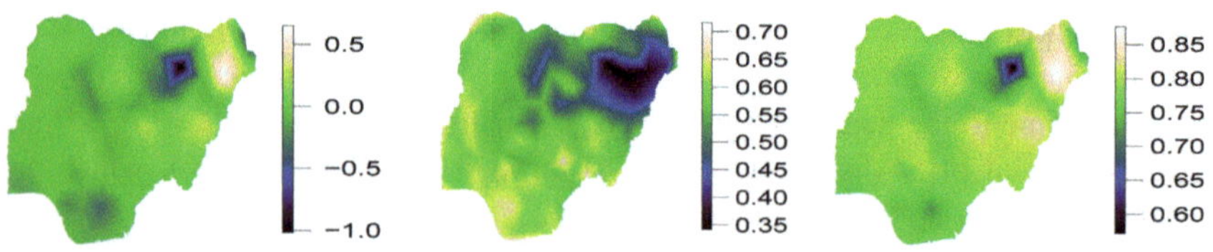

(a) posterior mean ξ_s **(b)** posterior variance $\sigma^2_{\xi_s}$ **(c)** posterior probability π_s

Fig. 4 GMRF posterior mean ξ_s (**a**), variance $\sigma^2_{\xi_s}$ (**b**), and probability of lethal terrorist attack π_s (**c**) interpolated in the spatial domain $\mathcal{D}$, which covers Nigeria better interpretation of the results. We use a relative low standard deviation value ($\sigma_\xi = 0.007$) along with range $r_1 = 50$ km and $r_2 = 500$ in the first and second robustness model respectively. The results are not affected by this change in prior (not reported). As an alternative to the original mesh ($n = 141$) (Fig. 3a), we use *mesh 2* ($n_2 = 283$) (Fig. 3b) and *mesh 3* ($n_3 = 425$) (Fig. 3c). The alternative meshes (*mesh 2* and *mesh 3*) count roughly two and three more vertices and increase the computation time required to fit the model by approximately 24% and 33%, respectively. However, this does not affect the results (not reported).

Undoubtedly, this study could benefit from further improvement in using e.g. elicited priors based on the opinion from experts in terrorism studies and the use of additional covariates. Moreover, more accurate data along with access to more computational power would allow investigating the phenomenon in finer spatial and/or temporal resolution, and/or through the use of non-stationary spatial models (i.e. models in which spatial correlation varies in space). Despite its shortcomings, the findings of this research work contribute to a better understanding of the spatial pattern of the lethality of terrorist attacks perpetrated by Boko Haram in Nigeria from 2009 to 2014.

References

1. Abadie, A.: Poverty, political freedom and the roots of terrorism. Am. Econ. Rev. **96**, 50–56 (2006)
2. Amante, C., Eakins, B.: ETOPO1 1 Arc-Minute Global Relief Model. Procedures, Data Sources and Analysis. NOAA Technical Memorandum NESDIS NGDC-24. National Geophysical Data Center, NOAA (2009). http://www.ngdc.noaa.gov/mgg/global/global.html. Accessed 7 Dec 2014
3. Banerjee, S., Carlin, B.P., Gelfand, A.E.: Hierarchical Modeling and Analysis for Spatial Data. CRC Press, Boca Raton (2014)
4. Behlendorf, B., LaFree, G., Legault, R.: Microcycles of violence: evidence from terrorist attacks by ETA and the FMLN. J. Quant. Criminol. **28**, 49–75 (2012)
5. Berrebi, C., Lakdawalla, D.: How does terrorism risk vary across space and time? An analysis based on the Isreali experience. Def. Peace Econ. **18**, 113–131 (2007)
6. Blangiardo, M., Cameletti, M.: Spatial and Spatio-Temporal Bayesian Models with R-INLA. Wiley, New York (2015)
7. Braithwaite, A., Li, Q.: Transnational terrorism hot spots: identification and impact evaluation. Confl. Manag. Peace Sci. **24**, 281–296 (2007)
8. CIESIN: Gridded Population of the World, Version 4 (GPWv4): Population Density Grid. Data from Center for International Earth Science Information Network (CIESIN). Columbia University and Centro Internacional de Agricultura Tropical (2005). http://dx.doi.org/10.7927/H4XK8CG2. Accessed 12 July 2016
9. Coaffee, J.: Protecting vulnerable cities: the UK's resilience response to defending everyday urban infrastructure. Int. Aff. **86**(4), 939–954 (2010)
10. Crenshaw, M.: The Causes of Terrorism. St. Martin's Press Inc., New York (1990)
11. Cummings, R.: World Deadliest? Boko Haram and the Challenge of Quantifying Violence (2015). https://theglobalobservatory.org/2015/11/boko-haram-worlds-deadliest-terrorists-isis-paris/. Accessed 18 Feb 2015

12. Elvidge, C.D., Safran, J., Tuttle, B., Sutton, P., Cinzano, P., Pettit, D., Arvesen, J., Small, C.: Potential for global mapping of development via a nightsat mission. GeoJournal **69**(1–2), 45–53 (2007)
13. Elvidge, C.D., Hsu, F.C., Baugh, K.E., Ghosh, T.: National Trends in Satellite Observed Lighting: 1992–2012. CRC Press, Boca Raton (2013)
14. Fearon, J.D., Laitin, D.D.: Ethnicity, insurgency, and civil war. Am. Polit. Sci. Rev. **97**(1), 75–90 (2003). http://www.jstor.org/stable/3118222
15. Gao, P., Guo, D., Liao, K., Webb, J.J., Cutter, S.L.: Early detection of terrorism outbreaks using prospective space-time scan statistics. Prof. Geogr. **65**, 676–691 (2013)
16. Gassebner, M., Luechinger, S.: Lock, stock, and barrel: a comprehensive assessment of the determinants of terror. Public Choice **149**, 235–261 (2011)
17. Gelman, A., Hwang, J., Vehtari, A.: Understanding predictive information criteria for Bayesian models. Stat. Comput. **24**(6), 997–1016 (2014)
18. GTD: Global Terrorism Database (GTD) Codebook: Inclusion Criteria and Variables (2014). http://www.start.umd.edu/gtd/downloads/Codebook.pdf. Accessed 9 Dec 2014
19. Henderson, J.V., Storeygard, A., Weil, D.N.: Measuring economic growth from outer space. no. w15199. In: National Bureau of Economic Research (2009)
20. Hoffman, B.: Inside Terrorism, rev. and expanded edn. Columbia University Press, New York (2006)
21. Institute for Economics and Peace: Global terrorism index 2015 (2015)
22. Kegley, C.W.: The Characteristics, Causes, and Controls of International Terrorism: An Introduction. St. Martin's Press Inc., New York (1990)
23. Krueger, A.B., Laitin, D.D.: Kto Kogo?: A Cross-Country Study of the Origins and Targets of Terrorism, pp. 148–173. Cambridge University Press, Cambridge (2008)
24. Kurrild-Klitgaard, P., Justesen, M., Klemmensen, R.: The political economy of freedom, democracy and transnational terrorism. Public Choice **128**, 289–315 (2006)
25. LaFree, G., Morris, N.A., Dugan, L.: Cross-national patterns of terrorism comparing trajectories for total, attributed and fatal attacks, 1970–2006. Br. J. Criminol. **50**(4), 622–649 (2010)
26. LaFree, G., Dugan, L., Xie, M., Singh, P.: Spatial and temporal patterns of terrorist attacks by ETA 1970 to 2007. J. Quant. Criminol. **28**, 7–29 (2012)
27. Laqueur, W.: The New Terrorism: Fanaticism and the Arms of Mass Destruction, 1st edn. Oxford University Press, New York (1999)
28. Lindgren, F., Rue, H., Lindström, J.: An explicit link between Gaussian fields and Gaussian Markov random fields: the stochastic partial differential equation approach. J. R. Stat. Soc. Ser. B (Statistical Methodology) **73**(4), 423–498 (2011)
29. Nelson, A.: Estimated travel time to the nearest city of 50,000 or more people in year 2000 (2008). http://bioval.jrc.ec.europa.eu/products/gam/. Accessed 10 June 2014
30. Nemeth, S.C., Mauslein, J.A., Stapley, C.: The primacy of the local: identifying terrorist hot spots using Geographic Information Systems. J. Polit. **76**, 304–317 (2014). doi:10.1017/S0022381613001333
31. NOAA: Version 4 DMSP-OLS Nighttime Lights Time Series. National Oceanic and Atmospheric Administration, National Geophysical Data Center (2014). http://ngdc.noaa.gov/eog/dmsp/downloadV4composites.html
32. Perl, A.: Combating terrorism: The challenge of measuring effectiveness. Technical report, Technical Re-port RL33160, Congressional Research Services (2007). http://fpc.state.gov/documents/organization/57513.pdf
33. Piazza, J.: Rooted in poverty? Terrorism, poor economic development, and social cleavages. Polit. Violence **18**, 159–177 (2006)
34. Richardson, L.: What Terrorists Want, 1st edn. John Murray, London (2006)
35. Rue, H., Martino, S., Chopin, N.: Approximate Bayesian inference for latent Gaussian models by using integrated nested Laplace approximations. J. R. Stat. Soc. Ser. B (Statistical Methodology) **71**(2), 319–392 (2009)
36. Sánchez-Cuenca, I., De la Calle, L.: Domestic terrorism: the hidden side of political violence. Ann. Rev. Polit. Sci. **12**, 31–49 (2009)

37. Savitch, H.V., Ardashev, G.: Does terror have an urban future? Urban Stud. **38**, 2515–2533 (2001)
38. Siebeneck, L.K., Medina, R.M., Yamada, I., Hepner, G.F.: Spatial and temporal analyses of terrorist incidents in Iraq, 2004–2006. Stud. Confl. Terror. **32**, 591–610 (2009)
39. Silberfein, M.: Insurrections. In: Cutter, Sl., Richardson, D.B., Wilbanks, T.J. (eds.) The Geographical Dimension of Terrorism, pp. 67–73. Routledge, New York (2003)
40. START: FTO Designation: Boko Haram and Ansaru (2013). http://www.start.umd.edu/sites/default/files/files/publications/br/STARTBackgroundReport_BokoHaram_Nov2013.pdf. Accessed 18 Feb 2015
41. Sutton, P.C., Elvidge, C.D., Ghosh, T.: Estimation of gross domestic product at sub-national scales using nighttime satellite imagery. Int. J. Ecol. Econ. Stat. **8**(S07), 5–21 (2007)
42. Swanstrom, T.: Are fear and urbanism at war? Urban Aff. Rev. **38**(1), 135–140 (2002)
43. Watanabe, S.: Asymptotic equivalence of Bayes cross validation and widely applicable information criterion in singular learning theory. J. Mach. Learn. Res. **11**, 3571–3594 (2010)
44. Weidmann, N.B., Rød, J.K., Cederman, L.E.: Representing ethnic groups in space: a new dataset. J. Peace Res. **47**(4), 1–9 (2010)

Optimizing Movement of Cooperating Pedestrians by Exploiting Floor-Field Model and Markov Decision Process

Vladimíra Sečkárová and Pavel Hrabák

Abstract Optimizing movement of pedestrians is a topic of great importance, calling for modeling crowds. In this contribution we address the problem of evacuation, where pedestrians choose their actions in order to leave the endangered area. To address such decision making process we exploit the well-known floor-field model with modeling based on Markov decision processes (MDP). In addition, we also allow the pedestrians to cooperate and exchange their information (probability distribution) about the state of the surrounding environment. This information in form of probability distributions is then combined in the Kullback–Leibler sense. We show in the simulation study how the use of MDP and information sharing positively influences the amount of inhaled CO and the evacuation time.

Keywords Optimization of cooperating pedestrians · Floor-field model · Markov decision process · Combination of transition probabilities

1 Introduction

The crowd modeling (pedestrian modeling) is of great importance and arises in many situations such as transportation, strike action and other areas of artificial intelligence and multi-agent systems. Generally used models can yield undesirable behavior of the crowd, e.g., no movement when the path is clear and collisions among pedestrians, in situations requiring somewhat smooth and organized movement, such as evacuation. We address the former problem by allowing more uncertainty in decision making for the next move. To solve the latter problem we allow pedestrians to cooperate so

V. Sečkárová (✉) · P. Hrabák
Institute of Information Theory and Automation of the CAS,
Pod Vodárenskou věží 4, 18200 Prague 8, Czech Republic
e-mail: seckarov@utia.cas.cz

P. Hrabák
Faculty of Civil Engineering, Brno University of Technology,
Veveří 95, 60200 Brno, Czech Republic
e-mail: pavel.hrabak@vutbr.cz

R. Argiento et al. (eds.), *Bayesian Statistics in Action*, Springer Proceedings in Mathematics & Statistics 194, DOI 10.1007/978-3-319-54084-9_23

that they can exploit information about decision making of other pedestrians. The comparison of regular pedestrian model and more sophisticated decision making technique with cooperation is of interest in this contribution.

There are two distinct approaches to modeling the pedestrians. First focuses on individuals and works well for small crowds, second takes the whole group as one for large crowds [5]. We focus on the former way of modeling in case of evacuation. If the details about the environment are not available, the specification of the whole design of evacuation is needed, e.g., arrangement of exits, fire-reducing equipment and evacuation scenario [15]. Here, we model the environment by the two-dimensional (predetermined) grid and refer to pedestrians as the agents on the grid. We assume pedestrians would like to leave the endangered area and reach the exit as soon as possible. To describe the decision making process of the agents on the grid we consider two types of models: *the floor-field* (FF) model and theory of *Markov decision processes* (MDP). The simplicity of FF model (static, dynamic) in applying to evacuation problem in, e.g., [3, 7], is appealing. However, the evacuation time of agents (the number of time instants leading to agent-free state of the grid) can be high, since we predetermine how attracted they are towards the exit. To overcome this, we focus on MDP, often exploited in area of autonomous robot systems for prediction of pedestrians trajectories: 'softened' version of MDP accounting for decision uncertainty [16], jump MDP for long-time predictions [6]. MDP includes optimization step, which significantly decreases the evacuation time and still offers a computationally efficient way how to model movement of agents.

If beside the evacuation time another variable such as upper bound on the inhaled CO is present, we should prevent occurrence of collision between agents. In case of collision, the agent moves to the recently occupied position and has to step back to the original position. Despite it does not change the position on the grid, it inhales more CO than in case without a move. We address this issue by focusing on the interaction among agents. Although some interaction is included in case of MDP agents by assigning subjective transition probabilities to states of the grid, it deserves more attention. In particular, we suggest to combine agents' opinions about the situation on the grid, i.e., their transition probabilities, according to the combination of probability distributions described in [12]. This combination exploits the theory of Bayesian decision making, but because the likelihood is not provided we use information theory, i.e., minimum cross-entropy (Kullback–Leibler divergence) principle instead of the Bayes rule. It is useful especially in cases when we would like to combine opinions/preferences and we have little or no prior knowledge about global model of the studied problem.

This contribution consists of two main parts: a theoretical overview and a comparison of suggested approaches in a simulation study for scenarios with FF, MDP and without/with combining from perspective of evacuation time and inhaled CO.

2 Theoretical Background

Let the environment be described by a finite grid $\mathscr{X}$ with a finite number of positions $x \in \mathscr{X}$, see Fig. 1 on the left. Let $\mathscr{S}$ denote the set of all states of the grid, where the term *state* refers to a specific allocation of the agents on the grid. Actions, $a \in \mathscr{A}$, allowed on the grid, are depicted in Fig. 1 on the right.

Let us have K agents ($K \in \{1, 2, \ldots\}$), each of them would like to choose an action $a \in \mathscr{A}$ and then move to a future position based on the current state s of the grid, $s \in \mathscr{S}$. In case $K > 1$ we also assume that agents are aware of the allocation of other agents within the grid state s and they are

- interested in the whole future state $\tilde{s}$ (depending on scenario), $\tilde{s} \in \mathscr{S}$,
- willing to cooperate and exchange their information with *neighbors*. A neighbor is the agent at one-move distance.

As said, we consider modeling of pedestrians movement with FF model and MDP together with combination of probabilities based on Kullback–Leibler divergence with basic information given below. Such conception of modeling pedestrian evacuation is based on the cellular automata models of pedestrian flow reviewed, e.g., in [11]. In such models, the cell of the grid is usually considered to cover an area of 40 cm times 40 cm, therefore the cells can be occupied by at most one agent.

2.1 Floor-Field Scenario for Pedestrian Movement

For agents following the floor-field (FF) model for pedestrian movement we assume that the position of the exit $e \in \mathscr{X}$ is known. The choice of the destination position is for the agent following FF model based on probability of a future position $p_{\tilde{x}}$. This probability includes information on the distance towards the exit:

$$p_{\tilde{x}} \propto \exp^{-k_S S^e_{\tilde{x}}}, \tag{1}$$

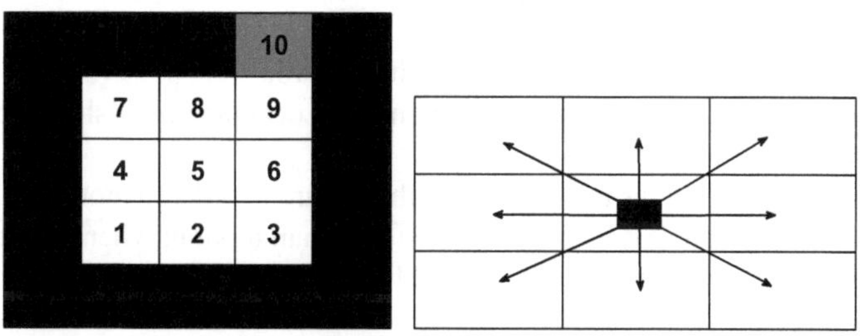

Fig. 1 *Left* Example of environment for pedestrians – a grid with 10 positions. Position 10 represents the exit. *Right* 9 possible movement (actions) of the agent on the grid – 8 directions represented by arrows + 1 "direction" with *no movement*

where k_S is the coupling constant and $S^e_{\tilde{x}}$ is the distance between exit and $\tilde{x}$. For higher value of the coupling constant the agent is more attracted towards the exit. More details on floor-field model in discrete time and space can be found in [8].

2.2 Scenario Exploiting Markov Decision Process

To choose the new destination position for agents following Markov decision processes (MDP) scenario we assume existence of

- a reward $r(s, a)$, $s \in \mathcal{S}, a \in \mathcal{A}$,
- a transition probability function $p(\tilde{s}|s, a)$, $\tilde{s} \in \mathcal{S}$, which is composed of local hopping probabilities (1).

We search for the action a maximizing total expected reward consisting of immediate reward and expected terminal reward in one-period MDP

$$a^* = \arg\max_{a \in \mathcal{A}} \left\{ r(s, a) + \sum_{\tilde{s}} p(\tilde{s}|s, a) v(\tilde{s}) \right\} \tag{2}$$

where $v(\tilde{s})$ is a specified function (terminal reward). More details on finite-horizon MDP can be found in [9], their application to evacuation modeling is given in [4].

2.3 Description of Reward

Let us suppose the situation of people trying to escape from some premises due to fire emergency.

The agents are motivated to leave the premises as soon as possible (i.e., every second spend in the system decreases their comfortability) keeping the amount of inhaled CO minimal (i.e., every action related to increased necessity of breathing decreases the comfortability as well). Thus,

1. every attempt to move should decrease the reward,
2. collision (i.e., unsuccessful attempt to enter an occupied cell) should cost more than motion to empty cell,
3. standing in current cell should decrease the reward in order to motivate the agents to leave as fast as possible. It should cost less than motion; when waiting for the target cell to be emptied, its cost should be lower than the costs for a collision.

The reward that we exploit later on reflects

- the possibility of collisions: agent inhales more by jumping to the occupied position (see Fig. 2),

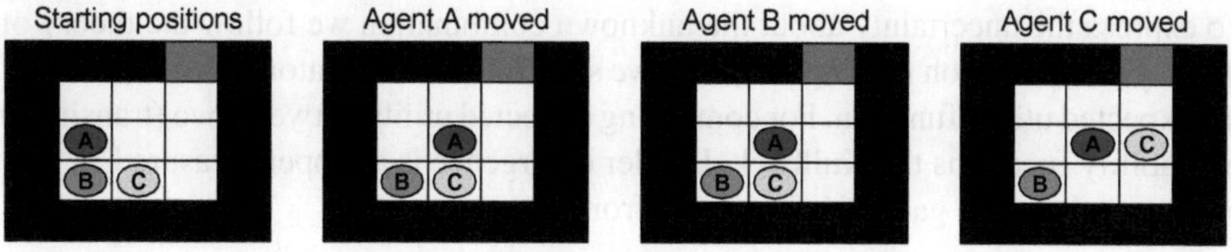

Fig. 2 Example of collision. In the beginning agents A,B,C choose their action. Agent A moves first, his targeted position was vacant. Second, Agent B moves to targeted position and steps back, because this position is occupied. Third, Agent C moves to targeted position, which is vacant

- no activity: if the decision of the agent for the next time step is to stay in the current position, agent should still inhale some amount of CO, but lower than in the case of movement,

and has the following form (using negative sign since maximization of reward is related to minimization of inhaled CO):

$$r(.,.) = \begin{cases} -1 & \text{jumps to vacant position,} \\ -2 & \text{jumps to occupied position, has to jump back,} \\ -1/2 & \text{avoids collision - stays in current position,} \\ -10 & \text{hits the wall.} \end{cases} \quad (3)$$

For the agent with FF model the attraction towards the exit is predetermined by the coupling constant and the distance from the exit, both independent of the current state of the grid (position of other agents). For MDP we assume that the situation is reflected in the transition probability function, i.e., low probability for staying in the current position if neighboring positions are empty.

2.4 Combining Transition Probabilities

To minimize the number of collisions between agents on the grid and thus to minimize the evacuation time and inhaled carbon oxide we assume agents are willing to cooperate in scenario introduced in [12]. There, the so-called sources shared the probability functions in order to obtain *a compromise* - a weighted linear combination of their probability functions. With the agents on the grid representing the sources, we would like to combine their transition probabilities

$$p_j = p_j(\tilde{s}|s, a), \quad s, \tilde{s} \in \mathcal{S}, a \in \mathcal{A}, \quad (4)$$

to obtain new transition probabilities including the information from other agents. We now briefly summarize the basic steps leading to the combination.

Assume that there exists an unknown transition probability function $q = q(\tilde{s}|s, a)$ representing the unknown compromise of $p_1, \ldots, p_K$ in MDP scenario on the grid.

To express the uncertainty about the unknown combination we follow the theory of the Bayesian decision making [10], i.e., we search for its estimator $\hat{q}$ as minimizer of the expected utility function. For computing expected utility between two (transition) probability functions the Kullback–Leibler divergence is a proper measure [2], as it expresses the mean gain of information from estimator $\check{q}$ to q

$$\hat{q} \in \arg\min_{\check{q}} \mathrm{E}_{\pi(q|p_1,\dots,p_K)} \mathrm{KLD}(q||\check{q}) = \arg\min_{\check{q}} \mathrm{E}_{\pi(q|p_1,\dots,p_K)} \sum_{\tilde{s}} q_{\tilde{s}} \ln \frac{q_{\tilde{s}}}{\check{q}_{\tilde{s}}}. \quad (5)$$

$\pi(q|p_1, \dots, p_K)$ is a probability density function (pdf) over set of all possible q. Minimizer of (5) is the conditional expected value of q with respect to $\pi(q|p_1, \dots, p_K)$

$$\hat{q} = \mathrm{E}_{\pi(q|p_1,\dots,p_K)}[q|p_1, \dots, p_K]. \quad (6)$$

The non-existence of the likelihood prevents us from a direct use of the Bayes theorem in the search for $\pi(q|p_1, \dots, p_K)$. In such case, the minimum cross-entropy (Kullback–Leibler divergence) [14] is axiomatically recommended

$$\arg\min_{\pi(q|p_1,\dots,p_K)} \mathrm{KLD}(\pi(q|p_1, \dots, p_K)||\pi_0(q)), \quad (7)$$

where $\pi_0(q) = \pi_0(q|p_1, \dots, p_K)$ denotes the prior guess on $\pi(q|p_1, \dots, p_K)$. Restrictions on the solution are formulated as the following $K - 1$ equations

$$\mathrm{E}_{\pi(q|p_1,\dots,p_K)}[\mathrm{KLD}(p_K||q)|p_1, \dots, p_K] = \mathrm{E}_{\pi(q|p_1,\dots,p_K)}[\mathrm{KLD}(p_j||q)|p_1, \dots, p_K], \quad (8)$$

$j = 1, \dots, K - 1$, expressing that expected information gain when transitioning from q to particular p_j is equal across the group of agents on the grid, shortly referred to as agent's *selfishness*, cf. [2].

The formula for conditional pdf resulting from (7) is

$$\pi(q|p_1, \dots, p_K) \propto \pi_0(q|p_1, \dots, p_K) \prod_{\tilde{s}} q_{\tilde{s}}^{\sum_{j=1}^{K-1} \lambda_j (p_{j,\tilde{s}} - p_{K,\tilde{s}})}, \quad (9)$$

where λ_j are the Lagrange multipliers. For the Dirichlet distribution $Dir(\nu_{0,\tilde{s}}, \tilde{s} \in \mathscr{S})$ as the prior distribution in (9) we obtain $\pi(q|p_1, \dots, p_K)$ as the pdf of the Dirichlet distribution $Dir(\nu_{\tilde{s}}, \tilde{s} \in \mathscr{S})$ with parameters

$$\nu_{\tilde{s}} = \nu_{0,\tilde{s}} + \sum_{j=1}^{K-1} \lambda_j (p_{j,\tilde{s}} - p_{K,\tilde{s}}) = \nu_{0,\tilde{s}} + \sum_{j=1}^{K-1} \lambda_j p_{j,\tilde{s}} - p_{K,\tilde{s}} \left(\sum_{j=1}^{K-1} \lambda_j \right). \quad (10)$$

Then, according to (10) the combination (6) is

$$\hat{q}_{\tilde{s}} = \frac{\nu_{0,\tilde{s}}}{\sum_{\tilde{s}} \nu_{0,\tilde{s}}} + \sum_{j=1}^{K-1} \lambda_j \frac{(p_{j,\tilde{s}} - p_{K,\tilde{s}})}{\sum_{\tilde{s}} \nu_{0,\tilde{s}}}. \tag{11}$$

When no specific prior information is available, we use the arithmetic mean of $p_1, \ldots, p_K$

$$\nu_{0,\tilde{s}} = \sum_{j=1}^{K} \frac{p_{j,\tilde{s}}}{K}, \tag{12}$$

as prior guess on parameters of the Dirichlet distribution. The sum of parameters $\nu_{0,\tilde{s}}$ and $\nu_{\tilde{s}}$ is according to (10) equal; combination (11) is thus viewed as reallocation of the original (prior) guess. In this contribution we assume that $\sum_{\tilde{s}} \nu_{0,\tilde{s}} = 1$. Recent development [13] showed, that choice $\sum_{\tilde{s}} \nu_{0,\tilde{s}} = K$ is more suitable from theoretical and practical point of view.

3 Simulation Experiment

In this section we show the results based on previously described theory on a simple example. Let us have 3 agents on the grid shown in Fig. 1 on the left following scenario:

- FF: with the coupling constant k_S having values in {3, 8},
- MDP: with predefined transition probabilities and reward,
- starting positions: 4, 2, 1 for agents A, B, C,
- order of action execution is random,
- the data are averaged over 40 of simulations for each setting.

First, we inspect the non-cooperative scenario (without combining probabilities). While action of an FF agent is chosen (randomly) according to hopping probability, decision making of an MDP agent is based on deterministic transition probabilities. Thus, collisions should occur and the agents should inhale more CO.

Second, we incorporate information about the current state of the grid by letting the MDP agent(s) to combine its transition probability, i.e., its local hopping probability with those from FF agent(s). The resulting amount of CO and evacuation time should decrease.

3.1 Scenario - Without Combining

First, we consider all agents having FF model with

$kS_1 = 3, \qquad kS_2 = 3, \qquad kS_3 = 8.$

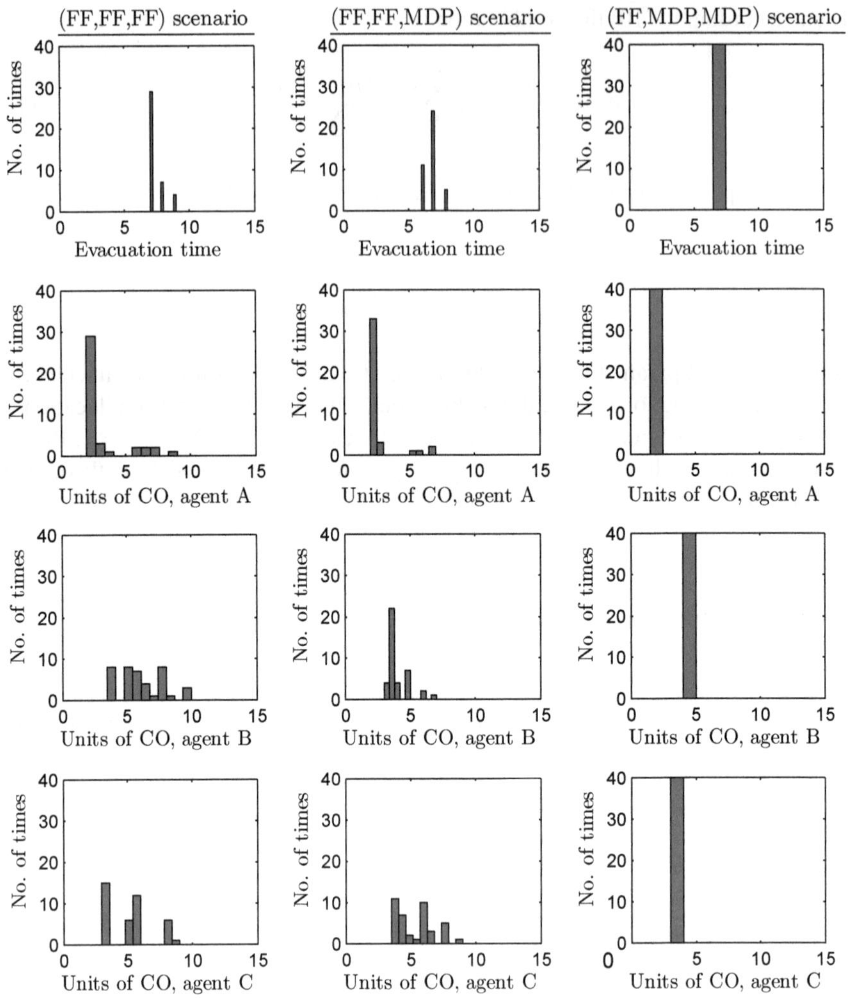

Fig. 3 Histograms: *Left column* Case (FF,FF,FF) with $kS_1 = 3, kS_2 = 3, kS_3 = 8$. *Middle column* Case (FF,FF,MDP) with $kS_1 = 3$, $kS_2 = 3$ and MDP. *Right column* Case (FF,MDP,MDP) with $kS_1 = 8$

How well the agents choose their actions to proceed to 10th position on the grid (the exit) and the inhaled CO are shown in Fig. 3 on the left.

Next, Fig. 3 in the middle shows histograms of evacuation times and inhaled CO for agents (FF,FF,MDP), FF agents with

$$kS_1 = 3, \qquad kS_2 = 3.$$

Finally, Fig. 3 on the right shows histograms of evacuation times and inhaled CO for agents (FF,MDP,MDP), FF agent with $kS_1 = 8$.

We see that the evacuation time (number of time instants leading to agent-free grid) improved with incorporation of MDP agent(s). Table 1 gives the average amount

Table 1 Average amount of CO for agents A, B, C and different scenarios

Scenarios	Agent A	Agent B	Agent C	Together
Without combining				
(FF,FF,FF)	2.98	6.09	5.16	14.23
(FF,FF,MDP)	2.5	3.98	5.41	11.89
(FF,MDP,MDP)	2	4.5	3.5	10
With combining				
(FF,FF,FF)	–	–	–	–
(FF,FF,MDP)	2.37	4.14	5.19	11.70
(FF,MDP,MDP)	2	4.5	3.5	10

of inhaled CO with decrease when including one MDP agent and with significant decrease when including two MDP agents.

3.2 *Scenario with Combining*

We used the same scenarios for the case of combining agents' probabilities (note that only MDP agents can exploit this). The incorporation of combination had positive influence on the amount of inhaled CO in case of one MDP agent (see Table 1). The results in case of two MDP agents coincide with results without combining since the dimension of the grid is low. In this case, the agents were able to reach the exit in the shortest possible way.

3.3 *Performing Actions - Fixed Order of Agents*

In the above part the order in which the agents perform their actions is random. If we fix the order in case of one MDP agent, e.g., we let the 'reasonable' MDP agent to go first (agent B second, agent A third), we can achieve even better results for the amount of inhaled CO:

With combining	Agent no.1	Agent no.2	Agent no.3	Together
(FF,FF,MDP)	2	3.5	6	11.5

4 Conclusions and Future Work

In this work we focused on improvement of pedestrian movement based on the Markov decision processes in case of evacuation. We considered also commonly used floor-field model and in a simple simulation study we showed how both modeling approaches performed in terms of the evacuation time and the amount of inhaled CO. The incorporation of MDP agents immediately decreased the amount of inhaled CO.

We also suggested that, because of possible collisions, agents should cooperate, exchange their transition probabilities and combine them. This approach positively influenced the results and yielded lower values of inhaled CO than in case without combining.

The authors are now motivated to study more complex situations for larger grid containing more cells and more agents. In this contribution we assumed that the agent was choosing his decision according to the overall state of the system, i.e., the state of every cell in the grid. Increasing the number of considered cells therefore significantly increases the cardinality of the state space. Thus, for larger grid, agents can possibly provide information only on a subset of the possible states of the grid, e.g., a sub-grid containing cell with agent and neighboring cells. Agents' sub-grids then overlap partially or do not overlap, which yields incompatibility of transition probabilities that we would like to combine. In such case, the outlined combining approach is expected to demonstrate its full strength. It can be then enhanced by employing partially observable MDP [1].

Acknowledgements This research has been supported by the grants GAČR 13-13502S and GAČR 16-09848S.

References

1. Bandyopadhyay, T., Won, K.S., Frazzoli, E., Hsu, D., Lee, W.S., Rus, D.: Intention-aware motion planning. In: Frazzoli, E., Lozano-Perez, T., Roy, N., Rus, D. (eds.) Algorithmic Foundations of Robotics X: Proceedings of the Tenth Workshop on the Algorithmic Foundations of Robotics, pp. 475–491. Springer, Berlin (2013)
2. Bernardo, J.M.: Expected information as expected utility. Ann. Stat. **7**, 686–690 (1979)
3. Guo, R., Huang, H.: A modified floor field cellular automata model for pedestrian evacuation simulation. J. Phys. A Math. Theor. **41**(38), 385104 (2008)
4. Hrabák, P., Ticháček, O., Sečkárová, V.: Estimation of discretized motion of pedestrians by the decision-making model. In: Knoop, V. L., Daamen, W. (eds.) Traffic and Granular Flow '15, pp. 313–320. Springer International Publishing (2016). http://arxiv.org/abs/1602.09107
5. Hughes, R.L.: A continuum theory for the flow of pedestrians. Transp. Res. Part B Methodol. **36**(6), 507–535 (2002)
6. Karasev, V., Ayvaci, A., Heisele, B., Soatto, S.: Intent-aware long-term prediction of pedestrian motion. In: 2016 IEEE International Conference on Robotics and Automation (ICRA), pp. 2543–2549, May 2016
7. Kirchner, A., Schadschneider, A.: Simulation of evacuation processes using a bionics-inspired cellular automaton model for pedestrian dynamics. Physica A Stat. Mech. Appl. **312**(12), 260–276 (2002)
8. Kretz, T.: Pedestrian Traffic, Simulation and Experiments. Ph.D. thesis, Universität Duisburg-Essen, Germany (2007)
9. Puterman, M.L.: Markov Decision Processes: Discrete Stochastic Dynamic Programming, 1st edn. Wiley, New York (1994)
10. Savage, L.J.: The Foundations of Statistics. Dover Books on Mathematics Series. Dover Publications, New York (1972)

11. Schadschneider, A., Chowdhury, D., Nishinari, K.: Stochastic Transport in Complex Systems: From Molecules to Vehicles. Elsevier Science B. V., Amsterdam (2010)
12. Sečkárová, V.: Dynamic parameter estimation based on minimum cross-entropy method for combining information sources. Pliska Stud. Math. **24**, 181–188 (2015)
13. Sečkárová, V.: Performance of Kullback–Leibler based expert opinion pooling for unlikely events (to be published)
14. Shore, J.E., Johnson, R.W.: Axiomatic derivation of the principle of maximum entropy and the principle of minimum cross-entropy. IEEE Trans. Inf. Theory **26**, 26–37 (1980)
15. Tanaka, T.: A study for performance based design of means of escape in fire. In: Cox, G., Langford, B. (eds.) Proceedings of the 3rd International Symposium on Fire Safety Science, pp. 729–738. Elsevier (1991)
16. Ziebart, B.D., Ratliff, N., Gallagher, G., Mertz, C., Peterson, K., Bagnell, J.A., Hebert, M., Dey, A.K., Srinivasa, S.: Planning-based prediction for pedestrians. In: Proceedings of the 2009 IEEE/RSJ International Conference on Intelligent Robots and Systems, IROS'09, pp. 3931–3936. IEEE Press, Piscataway (2009)

MIX
Papier aus verantwortungsvollen Quellen
Paper from responsible sources
FSC® C105338

If you have any concerns about our products,
you can contact us on
ProductSafety@springernature.com

In case Publisher is established outside the EU,
the EU authorized representative is:
Springer Nature Customer Service Center GmbH
Europaplatz 3, 69115 Heidelberg, Germany

Printed by Libri Plureos GmbH
in Hamburg, Germany